Equine Viruses

Equine Viruses

Special Issue Editor
Romain Paillot

MDPI • Basel • Beijing • Wuhan • Barcelona • Belgrade • Manchester • Tokyo • Cluj • Tianjin

Special Issue Editor
Romain Paillot
LABÉO-UniCaen
France

Editorial Office
MDPI
St. Alban-Anlage 66
4052 Basel, Switzerland

This is a reprint of articles from the Special Issue published online in the open access journal *Viruses* (ISSN 1999-4915) (available at: https://www.mdpi.com/journal/viruses/special_issues/equine_viruses).

For citation purposes, cite each article independently as indicated on the article page online and as indicated below:

LastName, A.A.; LastName, B.B.; LastName, C.C. Article Title. *Journal Name* **Year**, *Article Number*, Page Range.

ISBN 978-3-03928-320-0 (Pbk)
ISBN 978-3-03928-321-7 (PDF)

Cover image courtesy of G. Sutton & C. Thieulent, PhD students (LABÉO, France). It shows EHV-1 induced cytopathic effect on RK13 cell line culture.

Contents

About the Special Issue Editor

Romain Paillot, dEPHE, PhD, HDR. Dr. Paillot's main research looks at equine immunity and equine respiratory pathogens (EIV, EHV, and streptococci). His work aims to improve our understanding of immunity induced by pathogens during infection and by vaccination. His research also seeks to elucidate specific mechanisms used by pathogens to evade or deregulate the equine immune response (primarily streptococcal superantigens). Dr. Paillot is strongly involved with veterinary vaccine manufacturers in the evaluation of equine influenza vaccines.

Dr. Paillot's research career began in 1995 in the field of transplantation immunology. He studied the immunosuppressive drug methotrexate and described a new mechanism of action. From 1998 to 2004, he worked in the discovery research department for a veterinary vaccine manufacturer to develop new immunological assays in several veterinary species.

In 2004, he moved to the Animal Health Trust (AHT, United Kingdom) to start a PhD on cell-mediated immunity induced by EIV and EHV-1. He held the position of Immunology Team Leader in the department of infectious diseases at the AHT from 2007 to 2012. His work focused on immunity and protection induced by several modern equine vaccines, and he was strongly involved with all European veterinary vaccine manufacturers for the evaluation, characterization, and/or registration of equine influenza vaccines and other equine pathogens. In 2013, he developed a collaborative research program between the U2RM unit of the University of Caen Basse-Normandie, the Frank Duncombe Laboratory, and the AHT in order to investigate poor response in the field to equine influenza vaccination (Chair of Excellence, Equine Immunology). In 2016, he was awarded a French Habilitation to Direct Research (HDR, Higher Doctorate D.Sc/Habilitation equivalent) from Normandy University. Since February 2018, he has been in charge of the Equine Health Department for LABÉO (Research Division, formerly Frank Duncombe Laboratory) and the Director of CENTAURE (GIS), a cluster for equine research in Normandy.

Editorial

Special Issue "Equine Viruses": Old "Friends" and New Foes?

Romain Paillot [1,2]

1   LABÉO, Equine Health Department, Research Division, 14280 Saint-Contest, France;
    romain.paillot@laboratoire-labeo.fr; Tel.: +33-2-3147-1926
2   NORMANDIE UNIV, UNICAEN, BIOTARGEN, 14000 Caen, France

Received: 27 January 2020; Accepted: 27 January 2020; Published: 29 January 2020

The Food and Agriculture Organization of the United Nations has recently estimated that the world equid population exceeds 110 million (FAOSTAT 2017). Working equids (horses, ponies, donkeys, and mules) remain essential to ensure the livelihood of poor communities around the world. In many developed countries, the equine industry has a significant economical weight, with around 7 million horses in Europe alone. The close relationship between humans and equids, and the fact that the athlete horse is the terrestrial mammal that travels the most worldwide after humans, are important elements to consider in the transmission of pathogens and diseases, amongst equids and to other species. The potential effect of climate change on vector ecology and vector-borne diseases is also of concern for both human and animal health.

With this Special Issue, which assembles a collection of communications, research articles, and reviews, we intend to explore our understanding of a panel of equine viruses, looking at their pathogenicity, their importance in terms of welfare and potential association with diseases, their economic importance and impact on performance, and how their identification can be helped by new technologies and methods. Beyond their potential risk to other species, including humans, equine viruses may also represent an interesting model for reproducing virus infection in the host species.

Dennis et al. [1] contributed a review on African Horse Sickness (AHS). This disease, caused by the *orbivirus* AHS virus (AHSV), induces a very high mortality rate that can exceed 95% in its most severe form. This disease mostly occurs in southern African countries, but its transmission by *Culicoides* biting midges is of great concern in the current context of global warming and its consequence on displacement of vector populations. In the absence of treatment, prevention is essential, and Dennis et al. also provide a comprehensive review of the different vaccine strategies and technologies available and in current development against AHSV. While live attenuated and inactivated AHSV vaccines have played a role to reduce the impact and occurrence of AHS in affected areas, the number of AHSV serotypes in circulation and their lack of DIVA markers (differentiating infected from vaccinated animals) is a drawback that leads to the development of a new generation of vaccines, such as poxvirus-vectored or reverse genetics vaccines. Lecollinet et al. [2] reviewed major viruses inducing encephalitis in equids and their growing importance as a threat to the European horse population. Amongst them, equid herpesviruses (EHVs) are some of the most frequently isolated equine viruses worldwide. The equine herpesvirus type 1 (EHV-1) is of particular interest to the equine industry because of the different forms of disease it can induce, from a mild respiratory infection to abortion, neonatal death, and myeloencephalopathy (EHM) [3]. A communication from Preziuso et al. [4] and an article from Sutton et al. [5] specifically focused on EHV-1 strain characterization in order to better understand EHV circulation in Italy and France. Different approaches were compared, from the single-nucleotide point (SNP) mutation in ORF30 (historically associated with abortive or neuro-pathogenic strains), to other ORF gene sequences and the newly described multilocus strain typing methods (MLST; [6]). The MLST method is an interesting new approach for EHV-1 and a potential epidemiological tool that could provide an alternative until the development of more accessible EHV whole-genome sequencing methods. EHV-1 strain characterization by Sutton et al. allowed to conclude that the surge of EHV-1 outbreaks reported in France in 2018 was not linked to the introduction and/or circulation of a new EHV-1 strain in the French horse population. The origin of this crisis could be linked to a shortage of EHV vaccine and a subsequent reduced rate of EHV vaccination in the preceding years [7]. Lecollinet et al. also reviewed less frequently isolated encephalitis viruses, which may be zoonotic, such as rabies virus, borna disease virus, and West Nile virus (WNV). In the case of WNV, both horses and humans are highly susceptible to viral infection through infected mosquitoes. Unprecedented circulations of WNV have been observed in several European countries in the last decade, with a potential role of climatic and environmental conditions. Both species are considered as dead-end hosts. However, the horse could be used as a sentinel species to monitor and control vector-borne virus activity. Other enzootic flaviviruses were also reviewed, such as Tick-Borne Encephalitis virus (TBEV) and Louping Ill virus (LIV). In Europe, vaccination is only available against some of these pathogens (i.e., EHV-1 and 4, Rabies virus, and WNV), which highlights the importance of surveillance. Taking into account that most of these viruses will induce similar clinical signs of disease, the development of discriminative diagnostic tools is also of increasing importance. Finally, the review presents some other vector-borne (mosquitoes or midges) equine encephalitis viruses, not currently circulating in Europe, from the *Flaviridae* family (i.e., the Japanese encephalitis virus (JEV), Saint Louis encephalitis virus (SLEV), and Murray Valley encephalitis virus (MVEV)) or the alphaviruses from the *Togaviridae* family (i.e., eastern, western, and Venezuelan equine encephalitis viruses; EEEV, WEEV, and VEEV, respectively). In relation to VEEV, Rusnak et al. [8] presented systematic approaches for strain selection and propagation of virus and challenge material for the development and approval of a VEEV vaccine under the FDA Animal Rule and the different animal models available (rodents and non-human primates).

Altan et al. [9] have used metagenomics to identify viruses in horses with neurological and respiratory diseases. The equine hepacivirus (EqHV) was detected in the plasma from several neurological cases. This virus, which was first reported in horses in 2012 [10], was further investigated by Badenhorst et al. [11], with a specific focus on its circulation in Austria and the potential role of mosquitoes in its transmission. The prevalence of EqHV in the Austrian horse population studies reached 45% (based on serological evidence), with around 4% of samples positive for EqHV RNA. No EqHV RNA was found in mosquitoes collected across Austria, raising questions about its methods of transmission. Some aspects of this particular question of EqHV transmission were treated by Pronost et al. [12], who presented evidence to support a potential in utero transmission of EqHV from the mare to the foal, based on three positive clinical cases amongst 394 dead foals screened for the presence of EqHV RNA (prevalence of 0.76%). Altan et al. [9] also detected two copiparvoviruses, the equine parvovirus-hepatitis (EqPV-H) and a new one named *Eqcopivirus* by the authors, with no specific and/or statistical association with disease. Equine parvovirus-hepatitis was also the subject of the article from Meister et al. [13], which reported an EqPV-H infection occurrence in a quarter of the actively breeding Thoroughbred horse population from northern and western Germany. EqPV-H prevalence reached 7% and 35% (EqPV-H DNA positive detection and seroconversion, respectively). This study concerned mostly Thoroughbred brood mares, which represented 97% of the analyzed cohort. Concerning Thoroughbred stallions, Li et al. [14] identified a new equine papillomavirus (EcPV9) in the semen from an Australian Thoroughbred stallion suffering from a genital wart. The clinical significance of this new equine papillomavirus remained to be determined and will require further investigation. A similar question was raised by Nemoto et al. [15], who reported the first detection of equine coronavirus (ECoV) in Irish equids suffering from diarrhea. At five occasions, ECoV RNA was detected in feces from more than 400 equids with enteric diseases. However, the association with disease remains to be substantiated. While ECoV prevalence in Irish equids was 1.2% when measured by rRT-PCR in feces samples, evidence of ECoV infection was significantly higher when measured by serology in 984 serum samples from Dutch horses, 100 serums from Icelandic horses, and 27 paired serum samples from an ECoV outbreak in the USA. Zhao et al.

[16] developed and validated an S1-protein-based ELISA for this purpose. Seroprevalence ranged from 26% in young horses to nearly 83% in adults. The authors highlighted the potential use of this ELISA as a diagnostic test to confirm ECoV outbreaks, as a complement to feces samples analysis by qRT-PCR. The study from Back et al. [17] shed some light on the potential role of equine rhinitis A virus (ERAV) infection in poor performance. This longitudinal study, which involved 30 Thoroughbred racehorses, significantly associated seroconversion to ERAV and subsequent failure to attend races. However, similarly to EqPV-H and ECoV infections previously reported in this Special Issue, a direct association of ERAV infection with clinical signs of disease could not be confirmed in this study.

Finally, Fatima et al. [18] investigated the antiviral activity of the equine interferon-mediated host factors myxovirus (Mx) protein (eqMx1) against a range of influenza A viruses (IAVs). The authors highlight the potential protective role of eqMx1, which primarily targets the virus nucleoprotein (NP), against the transmission of new IAVs in horses (i.e., eqMx1 could only inhibit the polymerase activity of IAVs of avian and human origin but remained inactive against the equine IAVs tested). Introduction of a new IAV in the equine population is considered a rare event. In 1989, an equine influenza epizootic was reported in the Jilin and Heilongjiang provinces of northeastern China, with up to 20% mortality, which is quite high when compared with conventional equine influenza outbreaks. The IAV strain representative of this outbreak (i.e., A/equine/Jilin/1/1989) was closely related to an avian H3N8 IAV [19]. The authors show that the IAV strain A/equine/Jilin/1/1989 bears two adaptive NP mutations that confer resistance to eqMx1. To date, equine influenza virus remains one of the most important respiratory pathogens of horses worldwide, with a potential damaging impact on the equine industry, as clearly illustrated in 2007 in Australia and in 2019 in Europe [20,21].

We hope this Special Issue helps to highlight the diversity of equine viruses and their importance, in terms of welfare and/or economic impact, to equids and humans.

Acknowledgments: I wish to express my sincere thanks to all authors for their contribution to the Special Issue "Equine Viruses". I am also pleased to acknowledge all the Editorial Office team from *Viruses* for their great help and support and all reviewers involved in the peer-review process.

Conflicts of Interest: The author declares no conflict of interest.

References

1. Dennis, S.J.; Meyers, A.E.; Hitzeroth, II; Rybicki, E.P. African Horse Sickness: A Review of Current Understanding and Vaccine Development. *Viruses* **2019**, *11*, 174.
2. Lecollinet, S.; Pronost, S.; Coulpier, M.; Beck, C.; Gonzalez, G.; Leblond, A.; Tritz, P. Viral Equine Encephalitis, a Growing Threat to the Horse Population in Europe? *Viruses* **2019**, *12*, 23.
3. Oladunni, F.S.; Horohov, D.W.; Chambers, T.M. EHV-1: A Constant Threat to the Horse Industry. *Front. Microbiol* **2019**, *10*, 2668.
4. Preziuso, S.; Sgorbini, M.; Marmorini, P.; Cuteri, V. Equid alphaherpesvirus 1 from Italian Horses: Evaluation of the Variability of the ORF30, ORF33, ORF34 and ORF68 Genes. *Viruses* **2019**, *11*, 851.
5. Sutton, G.; Garvey, M.; Cullinane, A.; Jourdan, M.; Fortier, C.; Moreau, P.; Foursin, M.; Gryspeerdt, A.; Maisonnier, V.; Marcillaud-Pitel, C., et al. Molecular Surveillance of EHV-1 Strains Circulating in France during and after the Major 2009 Outbreak in Normandy Involving Respiratory Infection, Neurological Disorder, and Abortion. *Viruses* **2019**, *11*, 916.
6. Garvey, M.; Lyons, R.; Hector, R.D.; Walsh, C.; Arkins, S.; Cullinane, A. Molecular Characterisation of Equine Herpesvirus 1 Isolates from Cases of Abortion, Respiratory and Neurological Disease in Ireland between 1990 and 2017. *Pathogens* **2019**, *8*, 7.
7. Paillot, R.; Marcillaud Pitel, C.; D'Ablon, X.; Pronost, S. Equine Vaccines: How, When and Why? Report of the Vaccinology Session, French Equine Veterinarians Association, 2016, Reims. *Vaccines* **2017**, *5*, 46.
8. Rusnak, J.M.; Glass, P.J.; Weaver, S.C.; Sabourin, C.L.; Glenn, A.M.; Klimstra, W.; Badorrek, C.S.; Nasar, F.; Ward, L.A. Approach to Strain Selection and the Propagation of Viral Stocks for Venezuelan Equine Encephalitis Virus Vaccine Efficacy Testing under the Animal Rule. *Viruses* **2019**, *11*, 807.

9. Altan, E.; Li, Y.; Sabino-Santos, G., Jr.; Sawaswong, V.; Barnum, S.; Pusterla, N.; Deng, X.; Delwart, E. Viruses in Horses with Neurologic and Respiratory Diseases. *Viruses* **2019**, *11*, 942.

10. Burbelo, P.D.; Dubovi, E.J.; Simmonds, P.; Medina, J.L.; Henriquez, J.A.; Mishra, N.; Wagner, J.; Tokarz, R.; Cullen, J.M.; Iadarola, M.J., et al. Serology-enabled discovery of genetically diverse hepaciviruses in a new host. *J. Virol.* **2012**, *86*, 6171–6178, doi:10.1128/JVI.00250-12.

11. Badenhorst, M.; de Heus, P.; Auer, A.; Rumenapf, T.; Tegtmeyer, B.; Kolodziejek, J.; Nowotny, N.; Steinmann, E.; Cavalleri, J.V. No Evidence of Mosquito Involvement in the Transmission of Equine Hepacivirus (Flaviviridae) in an Epidemiological Survey of Austrian Horses. *Viruses* **2019**, *11*, 1014.

12. Pronost, S.; Fortier, C.; Marcillaud-Pitel, C.; Tapprest, J.; Foursin, M.; Saunier, B.; Pitel, P.H.; Paillot, R.; Hue, E.S. Further Evidence for in Utero Transmission of Equine Hepacivirus to Foals. *Viruses* **2019**, *11*, 1124.

13. Meister, T.L.; Tegtmeyer, B.; Bruggemann, Y.; Sieme, H.; Feige, K.; Todt, D.; Stang, A.; Cavalleri, J.V.; Steinmann, E. Characterization of Equine Parvovirus in Thoroughbred Breeding Horses from Germany. *Viruses* **2019**, *11*, 965.

14. Li, C.X.; Chang, W.S.; Mitsakos, K.; Rodger, J.; Holmes, E.C.; Hudson, B.J. Identification of a Novel Equine Papillomavirus in Semen from a Thoroughbred Stallion with a Penile Lesion. *Viruses* **2019**, *11*, 713.

15. Nemoto, M.; Schofield, W.; Cullinane, A. The First Detection of Equine Coronavirus in Adult Horses and Foals in Ireland. *Viruses* **2019**, *11*, 946.

16. Zhao, S.; Smits, C.; Schuurman, N.; Barnum, S.; Pusterla, N.; Kuppeveld, F.V.; Bosch, B.J.; Maanen, K.V.; Egberink, H. Development and Validation of a S1 Protein-Based ELISA for the Specific Detection of Antibodies against Equine Coronavirus. *Viruses* **2019**, *11*, 1109.

17. Back, H.; Weld, J.; Walsh, C.; Cullinane, A. Equine Rhinitis A Virus Infection in Thoroughbred Racehorses-A Putative Role in Poor Performance? *Viruses* **2019**, *11*, 963.

18. Fatima, U.; Zhang, Z.; Zhang, H.; Wang, X.F.; Xu, L.; Chu, X.; Ji, S.; Wang, X. Equine Mx1 Restricts Influenza A Virus Replication by Targeting at Distinct Site of its Nucleoprotein. *Viruses* **2019**, *11*, 1114.

19. Guo, Y.; Wang, M.; Kawaoka, Y.; Gorman, O.; Ito, T.; Saito, T.; Webster, R.G. Characterization of a new avian-like influenza A virus from horses in China. *Virology* **1992**, *188*, 245–255.

20. Paillot, R.; El-Hage, C.M. The Use of a Recombinant Canarypox-Based Equine Influenza Vaccine during the 2007 Australian Outbreak: A Systematic Review and Summary. *Pathogens* **2016**, *5*, 42.

21. Fougerolle, S.; Fortier, C.; Legrand, L.; Jourdan, M.; Marcillaud-Pitel, C.; Pronost, S.; Paillot, R. Success and Limitation of Equine Influenza Vaccination: The First Incursion in a Decade of a Florida Clade 1 Equine Influenza Virus that Shakes Protection Despite High Vaccine Coverage. *Vaccines* **2019**, *7*, 174.

Review

African Horse Sickness: A Review of Current Understanding and Vaccine Development

Susan J Dennis [1,*], Ann E Meyers [1], Inga I Hitzeroth [1] and Edward P Rybicki [1,2]

[1] Biopharming Research Unit, Department of Molecular and Cell Biology, University of Cape Town, Rondebosch 7701, Cape Town, South Africa; ann.meyers@uct.ac.za (A.E.M.); inga.hitzeroth@uct.ac.za (I.I.H.); ed.rybicki@uct.ac.za (E.P.R.)

[2] Institute of Infectious Disease and Molecular Medicine, Faculty of Health Sciences, University of Cape Town, Observatory 7925, Cape Town, South Africa

* Correspondence: sue.dennis@uct.ac.za; Tel.: +27-21-650-5712

Received: 31 July 2019; Accepted: 4 September 2019; Published: 11 September 2019

Abstract: African horse sickness is a devastating disease that causes great suffering and many fatalities amongst horses in sub-Saharan Africa. It is caused by nine different serotypes of the orbivirus African horse sickness virus (AHSV) and it is spread by Culicoid midges. The disease has significant economic consequences for the equine industry both in southern Africa and increasingly further afield as the geographic distribution of the midge vector broadens with global warming and climate change. Live attenuated vaccines (LAV) have been used with relative success for many decades but carry the risk of reversion to virulence and/or genetic re-assortment between outbreak and vaccine strains. Furthermore, the vaccines lack DIVA capacity, the ability to distinguish between vaccine-induced immunity and that induced by natural infection. These concerns have motivated interest in the development of new, more favourable recombinant vaccines that utilize viral vectors or are based on reverse genetics or virus-like particle technologies. This review summarizes the current understanding of AHSV structure and the viral replication cycle and also evaluates existing and potential vaccine strategies that may be applied to prevent or control the disease.

Keywords: African horse sickness; virus structure; replication; vaccine strategies

---

1. Introduction

For several centuries, the devastating African horse sickness (AHS) has been a cruel scourge to horse owners in sub-Saharan Africa. The disease is infectious but non-contagious and causes high fatality rates in susceptible hosts. It is listed as a notifiable viral disease by the World Organization for Animal Health (OIE) because of its severity and the potential risk it poses for rapid global spread [1]. AHS remains the most economically significant equine disease worldwide.

The first known historical reference to AHS was recorded in an Arabian document entitled "Le Kitab El-Akoual El-Kafiah Wa El Chafiah", which apparently relates to an epidemic that occurred in the Yemen in 1327 [2]. However, the virus is believed to have originated in Africa, with the first record of the disease on the continent being made by Father Monclaro in his account of the journey of Francisco Barreto to East Africa in 1569 [1]. Unlike zebras, which are endemic to the region, horses are not native to southern Africa and reference to AHS in South Africa was first made about fifty years after the introduction of horses and donkeys to the Cape of Good Hope by the early Dutch Settlers in 1657. A major outbreak occurred in 1719 when almost 1700 animals were reported to have succumbed to the dreaded "perreziekte" or "pardeziekte" [2]. Prior to 1953, periodic outbreaks seemed to occur at roughly 20–30 year intervals, the most severe being the outbreak in South Africa in 1854–1855, which claimed the lives of nearly 70,000 horses, more than 40% of the entire horse population of the Cape at the time [3]. Indeed in South Africa, the economic impact of the disease has been such that it directly and significantly influenced the progress and development of the field of veterinary science itself [4].

AHS continues to occur regularly in southern African countries, but the virus has also occasionally escaped its geographical limitations and extended further afield to countries in North Africa, the Middle East, the Arabian peninsula, South-West Asia and the Mediterranean region (Figure 1) [5–7]. The severe epizootic in the Middle East and South West Asia between 1959 and 1963 was responsible for the deaths of over 300,000 equines and was finally only arrested as a result of a concerted vaccination campaign and widespread depletion of susceptible animals [7,8]. AHS-free countries with milder climate conditions are believed to be increasingly at risk for outbreaks of the disease due to the northward migration of the midge vector as a result of global warming and climate change [9–11]. Such an AHS outbreak in Europe would have significant economic and emotional consequences for horse owners on the continent, indicating the pressing need to develop new, safe, efficacious and cost-effective vaccines which would additionally allow differentiation between vaccinated and infected animals (DIVA). Such vaccines would not only address the concerns of the South African equestrian community but would also serve as acceptable prophylactic or rapid response vaccines in the European and other emerging outbreak contexts.

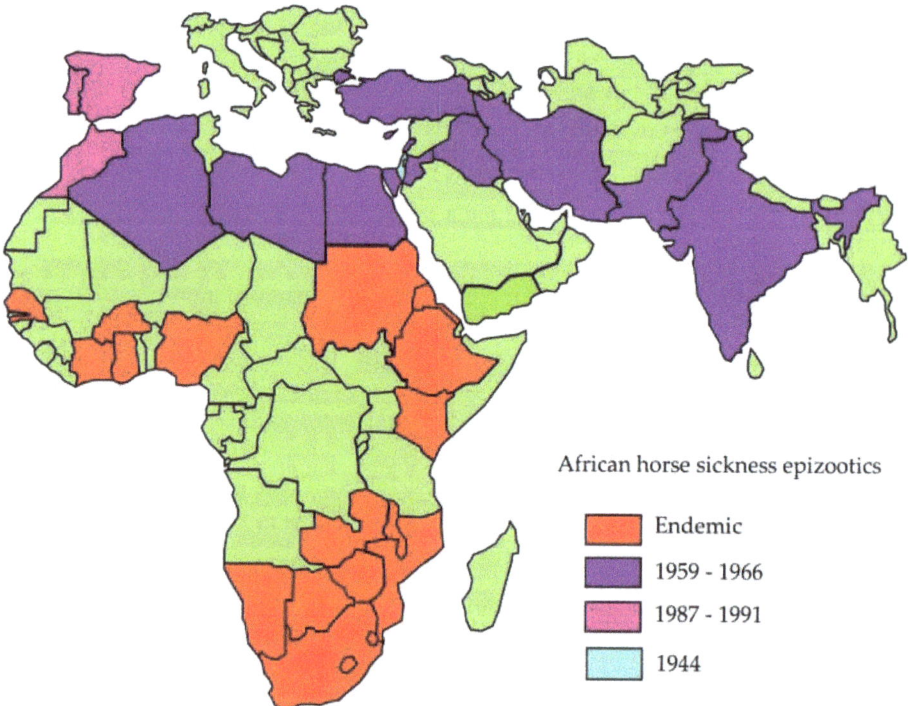

Figure 1. A map of African horse sickness outbreaks that have occurred worldwide during the last century.

2. African Horse Sickness Virus

The first sign that the causative agent of AHS may be a virus was provided by M'Fadyean [12], who demonstrated the filterability of the infectious organism by successfully transmitting the disease using a bacteria-free blood filtrate from an infected horse. This finding was later confirmed by Theiler and Nocard, who concluded that the disease was caused by a virus [2]. Further research done by Theiler led to the suggestion that more than one strain of the virus may exist, and that acquired immunity against one strain would not necessarily afford protection against a different heterologous strain [3].

It is now known that the disease is caused by nine distinct serotypes of African horse sickness virus (AHSV) [13,14], a virus with a multicomponent linear double stranded RNA (dsRNA) genome belonging to the *Orbivirus* genus of the family *Reoviridae* [1,15]. The idea that AHSV may possibly be transmitted by hematophagous arthropods of the *Culicoides* genus was first suggested by Pitchford and Theiler in 1903 [3]. Du Toit [16] subsequently demonstrated that mixed pools of wild-caught *Culicoides* species were infected with AHSV. This was confirmed by Mellor, et al. [17] and Boorman, et al. [18], who demonstrated the occurrence of AHSV replication within a *Culicoides* species after oral ingestion. Field and laboratory-based trials have implicated *C. imicola* and, to a lesser extent, *C. bolitinos* as the primary vectors of AHSV, although some evidence does exist for possible AHSV transmission by other arthropod vectors [19]. The ability of AHSV to propagate in both arthropod and mammalian cells is a notable feature shared with all orbiviruses, and one which distinguishes them from some other members of the family *Reoviridae* [20].

Zebra are generally resistant to AHS and have been identified as asymptomatic maintenance hosts of AHSV, while mules and donkeys are much less susceptible than horses to the disease [21]. Dogs are the only non-equine animal species that have been shown to contract AHS, with evidence suggesting that the route of infection may, although not exclusively, be via the ingestion of infected meat [22,23]. However, dogs do not appear to be important hosts for AHSV, most likely due to the fact that they are not preferential feeding targets of the midge vector.

The African horse sickness virion is a structurally complex and highly organized non-enveloped isometric particle with a diameter of ±80 nm [24,25]. Like the genus prototype bluetongue virus (BTV), with which it is morphologically almost identical, the non-enveloped virion is quasi-icosahedrally symmetrical and is composed of three concentric protein layers [26–28]. The innermost layer encloses the AHSV genome, which consists of 10 segments of linear dsRNA, encoding seven structural (four major and three minor) and five non-structural proteins [29]. Two of the major structural proteins, VP5 and VP2, make up the outer capsid layer, while the other two major structural proteins VP3 and VP7, and the three minor structural proteins, VP1, VP4 and VP6, make up the AHSV core particle (Figure 2).

In recent years, considerable advances have been made in determining the BTV atomic structure and mechanism of assembly as well as the functions of the individual protein components of the viral particles [20,30–33]. This has recently been clearly and concisely reviewed elsewhere [34]. Less extensive studies have been conducted on the molecular biology of AHSV, but the morphological and biochemical similarity between BTV and AHSV is indicative of a similar mode of replication and assembly for both viruses.

The inner core layer of AHSV is composed of 60 asymmetric dimers of protein VP3 (103 kD), the most conserved protein among the different serotypes [35]. Each dimer consists of two VP3 isoforms comparable to the A and B conformations of BTV VP3 [31] and the inner core is thus completed by the assembly of 12 decamers, each of which consists of five copies of the VP3 (A) molecule with five VP3 (B) molecules in between. This architecture is not in agreement with the hypothesis for icosahedral symmetry proposed by Caspar and Klug [36], but a model of geometrical quasi-equivalence has been proposed whereby the symmetry of the VP3 layer is T = 2 rather than T = 1 [31]. The thin VP3 shell thus defines the overall shape and size of the viral particle and provides a scaffold for the deposition of the outer core and capsid protein layers [37].

In both BTV and AHSV, the minor structural proteins VP1 (RNA-dependent RNA polymerase, 150kD), VP4 (capping enzyme, 78kD) and VP6 (helicase and ATPase, 36kD) form 12 flower-shaped transcription complexes (TC) attached to the VP3 layer directly under each of the fivefold vertices [31,33,38]. Internal concentric layers of RNA comprising the full dsRNA genome are situated around these transcription complexes in shallow grooves in the inner VP3 surface [37,39].

BTV contains pores in the VP3 layer at the fivefold axes which are lined with four arginine (Arg) residues that are conserved across all the serotypes, and which are believed to play a role in electrostatic steering of RNA entering or leaving the sub-core [31]. In the 3D image construction of AHSV, these pores were closed [38], but the same 4 Arg residues are strictly conserved across all 9 AHSV serotypes. It is possible that, as has been shown to be the case for BTV [40], these pores may be enlarged by activation of VP1 during transcription, permitting the exit of nascent mRNA.

The AHSV core is made rigid by the addition of 780 monomers of protein VP7 (38 kD) which is highly conserved across the serotypes and is the group-specific antigen currently used in AHSV ELISA-based diagnostic tests [41]. The atomic structure of BTV VP7 has been determined by x-ray crystallography [42], but only the upper domain of AHSV VP7 has been crystallized [43], as the protein was unfortunately cleaved in half during the crystallization process. The VP7 monomers contain a helical lower domain and an upper anti-parallel-ß-sandwich domain [44]: these trimerize in solution by twisting around each other so that the top domain of one monomer rests on the lower domain of the adjacent VP7 subunit.

These 260 VP7 trimers conform to the principle of quasi-equivalence as, although they are chemically identical, five different trimer types can be identified based on slightly different side chain arrangements. These are named P, Q, R, S and T, denoting their position with regard to the five-fold vertices. They crystallize perpendicularly onto the VP3 sub-core, making thirteen unique contacts so that each icosahedral subunit contains a P, Q, R and S trimer and one monomer of a shared T trimer located on the adjacent three-fold axis [31,45].

The trimers are robust building blocks, which also make extensive connections between each other [42]. They are arranged as either six-member rings, or five-member rings at the fivefold vertices, thus forming 132 channels over the core particle surface. Symmetries between the inner and outer core layers are best matched at the three-fold axes and, as the T trimers situated here also seem to be the most tightly attached, it has been suggested that these are probably the first set of trimers to attach to the VP3 layer while the P trimers, which are more loosely attached, assemble last [46].

An interesting phenomenon which distinguishes AHSV VP7 from BTV VP7 is the fact that despite the 70% amino acid sequence homology [47], BTV VP7 is soluble while AHSV VP7 is not. AHSV VP7 is, in fact, highly hydrophobic, and the VP7 trimers have been shown to aggregate into flat hexagonal crystals up to 250 um in length and up to 25 um wide, both when expressed in insect cells via baculovirus-mediated expression and in naturally infected mammalian cells [41,48].

The viral particle is completed by the addition of 120 globular trimers of VP5 (57 kD) and 60 triskelion-like VP2 (123 kD) spikes which together form the outer capsid layer. VP2 is the most variable protein among the serotypes and contains the antigenic determinants which elicit serotype-specific neutralizing antibodies [49]. The viral particles contain 180 monomers of VP2, each of which contains a hub, body, hairpin and tip domain, the latter containing the neutralizing antibody binding sites [30,38]. Cryo-EM and 3D reconstruction have revealed that the VP2 spikes are formed by trimerization of the hub domains of three VP2 monomers. The base of each VP2 spike interacts with a Q type VP7 trimer, while the body domains of the three VP2 monomers connect with a P, R and S - type VP7 trimer, respectively. Furthermore, for BTV, Zhang et al. identified a zinc finger motif and a putative sialic acid (SA) binding domain in the VP2 trimer hub [32].

The VP5 trimers exist as two quasi-equivalent conformers and are positioned between the propeller-like arms of the VP2 triskelions, bridging the 120 channels formed by the six-member VP7 trimer rings. The viral outer shell therefore covers nearly all of the inner shell, but significantly, leaves the 20 VP7 T trimer spikes on the icosahedral three-fold axes accessible to possible antibody binding [50].

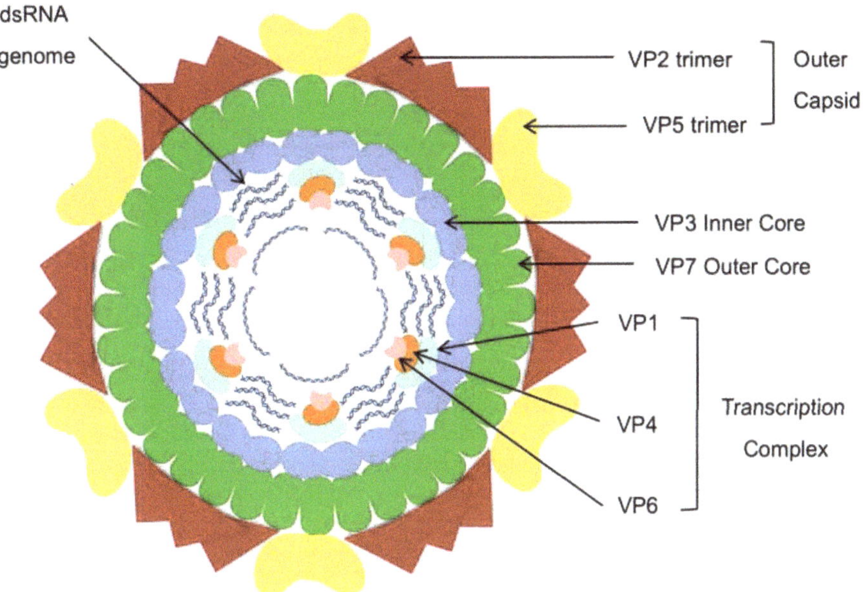

Figure 2. Schematic representation of the AHSV virion. The genome contains 10 segments of linear dsRNA coding for 12 proteins. The virion is non-enveloped with a triple capsid structure and is about 80 nm in diameter, enclosing the genome and transcription complexes. The inner core layer has T = 1 symmetry with each of the 60 units composed of a homodimer of VP3, while the outer core is composed of 260 trimers of VP7 and has T = 13 icosahedral symmetry. The outer capsid layer consists of 120 globular trimers of VP5 and 60 triskelion-shaped spikes of VP2. *Image created with Biorender.com.*

In addition to the seven structural proteins that make up the viral particles, five non-structural proteins, NS1, NS2, NS3, NS3a (lacking 13 N-terminal amino acids) and NS4, are also synthesized in infected cells. These are involved in virus replication, assembly and transport from infected cells [29]. The tubular structures commonly observed in the cytoplasm of infected cells are composed of NS1, which has been shown to play a role in the preferential up-regulation of viral protein synthesis [51]. The single-stranded RNA (ssRNA)-binding protein NS2 forms viral inclusion bodies (VIBs), which recruit viral ssRNA and form a scaffold for viral replication and core assembly [52,53]. Viral particle release is mediated by NS3/NS3a, the only AHSV glycosylated membrane protein. This, unlike BTV NS3/NS3a, which is highly conserved, is the second most variable protein across the different serotypes. Once the outer capsid proteins VP2 and VP5 have been acquired by the newly formed cores, NS3 facilitates egress of fully formed viral particles from infected cells [54,55]. The most recently identified non-structural protein, NS4, is thought to play a role in modulating host innate immunity by counteracting the interferon response in infected cells [56,57].

3. Viral Infection and Replication

AHSV host infection is believed to be initiated by the outer capsid protein VP2, the host cellular surface receptor and capsid protein VP5, which contains a characteristic coiled-coil motif typical of membrane fusion proteins [32,58]. Due to the location of the putative sialic acid (SA) binding domain in the VP2 trimer hub, Zhang et al. have postulated that the VP2 trimers may attach to the surface of cells through two different interactions [32]. First, the antigenic tip domains bind to certain surface cell receptors which have yet to be identified. There is some preliminary evidence implicating heparin sulphate as one possible candidate [59], but due to the high sequence variability in the AHSV outer protein VP2, it is possible that different receptors and entry mechanisms may be utilized by the different serotypes to enter distinct cell types [60]. These initial bonds are then stabilised by a second connection between the SA-binding domain and a surface glycoprotein such as glycophorin A [61,62], which is a heavily glycosylated sialoglycoprotein abundantly present on the surface of equine erythrocytes. The sensitivity of VP2 to serum proteases has been established [48,63], and it has been suggested [38] that a central domain at the top of the AHSV VP2 triskelion hub, which is absent in BTV VP2 trimers, could be the target site for such a horse serum protease. This domain is situated directly above the BTV putative SA-binding site and cleavage of AHSV VP2 in this region would thus increase accessibility to this potential binding site.

Due to the structural similarity between the two viruses, a model for the entry of AHSV into the host cell can be derived from the current understanding of BTV infection. The process is believed to be initiated by proteolytic cleavage of VP2 either in infected insect saliva or in the host serum, after which the virion attaches to the host cell membrane [64]. These initial studies suggested that viral cell entry was accomplished by clathrin-mediated endocytosis. However, more recently evidence has been presented which describes a macropinocytosis-like entry route dependant on actin and dynamin [60]. In both instances, the low pH (6.0–6.5) within the early endosome disturbs the interactions between VP2 and VP7, facilitating detachment of the VP2 trimers and disrupting the zinc finger motif situated at the interface between the VP2 hub and body domains, which is believed to play a role in controlling conformational changes [30]. Together with a further lowering of the pH (~ 5.0) in the late endosome, the removal of VP2 causes a re-folding of VP5, which in turn leads to the outward protrusion of barb-like structures from the particle surface with the VP5 protein trimers remaining tethered to the particle by their anchoring domains [32,65]. These barb-like fusion peptides insert themselves into the endosomal membrane, causing release of the viral core particle into the cytoplasm.

The removal of the outer capsid proteins, and the release of viral cores into the cellular environment containing the necessary host substrates and transcription factors causes the core to become transcriptionally active. Each of the gene segments is then simultaneously and repeatedly transcribed by VP1 to produce ssRNAs [30,66], which are modified by the capping and methylation activity of enzyme VP4 within the core before being released into the cytoplasm [67]. The viral dsRNA is thus kept within the core particle protected from detection by components of the host cell innate immunity. The nascent ssRNAs act as mRNAs for the synthesis of viral proteins using the host cell machinery and later, in the newly formed cores, as templates for dsRNA gene synthesis. The VIBs act as the sites of viral assembly and protein NS2 plays a role in directly and specifically sequestering the 10 ssRNAs, together with the three enzymatic proteins VP1, VP4 and VP6 and inner core protein VP3, for encapsidation and the formation of new sub-core particles [53,68].

The deposition of VP7 trimers serves to stabilize the particles, and phosphorylation of NS2 then regulates their exit from the VIBs in order to acquire the two outer capsid proteins VP5 and VP2 for the formation of mature progeny virions [69,70]. Although the release of these virions from infected mammalian cells is predominantly affected by cell lysis accompanied by significant cytopathic effects, the viruses have also been shown to use a budding mechanism for viral egress earlier on in the infection cycle. The latter is mediated by utilising the host exocytosis pathway and the membrane destabilising action of non-structural glycoprotein NS3, which functions as a viroporin and also interacts with calpactin to function as a bridging molecule between the new virions and the host cell export machinery [71,72]. In the insect vector, where the establishment of a persistent viral infection is important, there is no observable cytopathic effect and viral release is mediated exclusively via vesicle formation at the cytoplasmic membrane [53] (Figure 3). It is interesting to note that the more variable AHSV NS3 causes a much greater cytopathic effect (CPE) earlier on in the infection cycle than BTV NS3, possibly indicating that AHSV NS3 is expressed at a higher level or is more toxic than BTV NS3 [73].

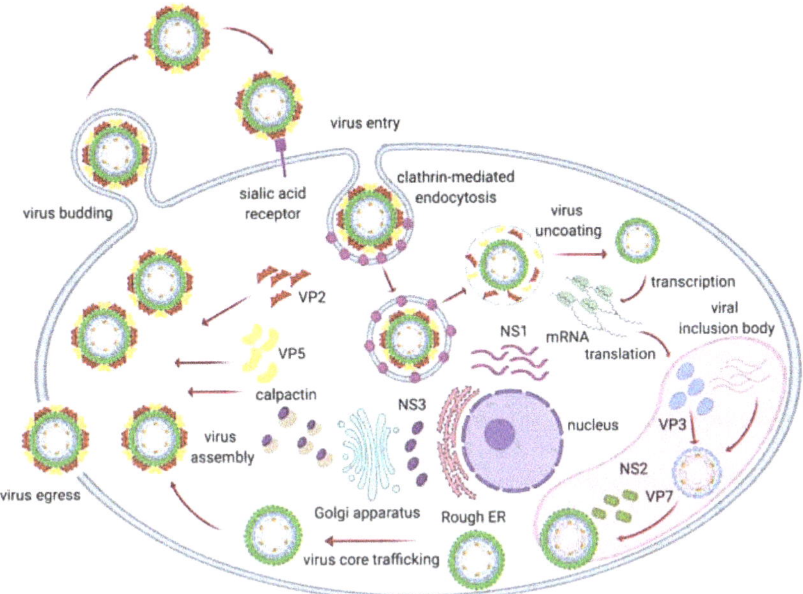

Figure 3. Diagrammatic representation of the replication cycle of BTV/AHSV. The virus enters the cell by the attachment of VP2 to sialic acid receptors and either clathrin-mediated endocytosis or macropinocytosis. The acidic pH in the endosome causes the loss of VP2 and mediates VP5 membrane permeabilization, which results in uncoating of the virion and release of the transcriptionally active core particle into the host cell cytoplasm. Transcription and translation of viral proteins occurs, utilizing the host cell machinery and the VIBs act as sites of assembly for the progeny virions. Assembled core particles are then trafficked from the VIB on exocytotic vesicles by NS3 interaction with calpactin. The outer capsid proteins VP5 and VP2 are acquired during this process to produce mature virions. Particles are released from the cell via budding mediated by NS3 or via host cell lysis. *Image adapted from* [53] *and created with* Biorender.com.

Although the primary route of BTV and AHSV host infection is believed to be initiated by the outer capsid proteins, there is evidence to suggest that BTV core-like particles (CLPs), i.e., particles that have lost the outer capsid proteins, are also able to infect both insect and, to a lesser extent, mammalian cells [74]. Interestingly, in this regard, the upper domains of both BTV and AHSV VP7 trimers have characteristic Arg-Gly-Asp (RGD) motifs, albeit in slightly different locations [43]. RGD domains in biological systems are associated with integrin-ligand recognition and fusion of molecules to cell membranes [75]. The fact that there are holes in the surface of the outer capsid layer of both BTV and AHSV particles makes it tempting to speculate that these RGD sites on VP7 may play a role in the ability of viral CLPs to infect cells.

4. African Horse Sickness Disease

Three distinct forms as well as a mixed form of AHS have been described. However, it is possible that these represent points on a continuum of virulence as some disease outbreaks are characterized by only one form of the disease, while during other outbreaks, multiple forms occur [76]. The most severe form of AHS, with mortality rates exceeding 95%, is the pulmonary form or "dunkop" (thin head). This is an acute febrile disease accompanied by mild depression, sweating, spasmodic coughing, anorexia and respiratory distress, with a possible frothy nasal discharge in the terminal stages [1,3]. The cardiac form or "dikkop" (thick head), with a mortality of about 50%, is characterized by fever, swelling of the head, neck and supraorbital fossae and sometimes, petechial hemorrhages in the eyes.

The mildest form of AHS is generally not fatal and is accompanied by a low-grade fever, often more pronounced in the afternoon, anorexia, depression and congestion of the mucous membranes. The most common form, with a 70% mortality rate, is a mix of the pulmonary and cardiac forms.

The bite of an infected *Culicoides* midge signals the potential initiation of infection in a susceptible host, after which initial AHSV replication occurs in the regional lymph nodes. This primary viraemia is responsible for disseminating the virus to all parts of the body [1]. Viral particles are known to associate with erythrocytes and monocytes and are transported in the bloodstream to the endothelial cells of the lungs, spleen and other lymphoid tissue, which are the main sites of secondary replication [76,77]. Although the level of replication is relatively low in these organs, the virus causes severe injury to the endothelial cells and the symptoms of oedema and pleural effusion, which characterize the severe form of AHS, are believed to be the result of increased vascular permeability and the impairment of circulatory and respiratory systems.

The primary factors which influence the severity and duration of the disease in horses are related to the virulence of the virus and the immune status and susceptibility of the animal. Host genetics must play a role, as is evidenced by the susceptibility of both horses and zebra to AHSV, yet only horses contract AHS disease. An animal which has recovered from a prior infection is fully protected by re-infection with the same serotype and may well only contract a fever or the cardiac form of AHS upon exposure to a heterotypic viral strain. However, Erasmus [78] noted that excessive vaccine administration may lead to an immunological unresponsiveness or even hypersensitivity.

Virulence is related to tissue tropism, and it would seem that the virus itself is the primary determinant of the resultant form of disease [79]. AHSV field isolates are believed to be composed of a mixed virus population with regard to the host tissue target, and the virulence of the particular isolate may thus be determined by the percentage of particles affecting vital organs, such as the lungs, for example. In this sense, Erasmus [79] suggests that viral attenuation may be the result of specific selection of viruses which do not have affinity for the tissues of vital organs.

Alexander [80] was the first to demonstrate the humoral nature of the AHSV immune response when he utilized a mouse neutralization test to indicate a strong association between the production of neutralizing antibodies and protective immunity. The antigenic determinants responsible for this neutralising antibody response are located on capsid protein VP2: several studies have suggested putative sites for these epitopes, mainly in the 5′ terminal half of the protein [38,49,81–84]. However, the probable contribution of cell-mediated immunity cannot be ignored—at least three CD8 T-cell epitopes have been identified on VP2 or NS1 [85] and others have reported the stimulation of IFN-y responses in vaccinated animals [86,87]. The degree to which the presence of neutralizing antibodies may be regarded as an adequate correlate of protection therefore remains to be established.

African horse sickness is on the OIE list of notifiable viral diseases, which means that it is compulsory for member states like South Africa to inform the organization of any change of disease status. In South Africa, the Western Cape Province has historically been relatively free of AHS; in order to maintain the country's horse export status, an AHS controlled area—to and from which the movement of all horses is strictly monitored—was established in 1997 [88]. The area consists of an AHS-free zone, which is the small Cape Metropolis where no cases of AHS have ever been recorded, a surrounding surveillance zone and beyond that, a zone of protection. Whenever a new outbreak occurs in the surveillance zone, horse exports to the EU are suspended for a minimum of 2 years.

According to the EU surveillance requirements, every month, at least 60 identified unvaccinated horses distributed throughout the free and surveillance zones are serologically tested for AHSV [89]. Local law requires all suspected cases of AHS to be reported to the state veterinary authority, and all equine deaths due to AHS undergo official equine necroscopy examination. As a further measure of control, it is compulsory to obtain permission prior to vaccinating horses with the AHS polyvalent live attenuated vaccine (LAV) produced by Onderstepoort Biological Products (OBP, Pretoria) in the free and surveillance zones.

The clinical presentation of AHS is often sufficient to make a tentative diagnosis of the disease; however, particularly in the early stages, differentiation from other equine diseases, such as equine encephalosis (EEV), equine viral arteritis (EVA) and West Nile virus (WNV), may not be possible. Traditional methods of virus isolation and serotyping by virus neutralization assay can be used to make a definitive diagnosis, but these tests rely on the availability of appropriate reference strains and antisera and can take weeks before a result can be obtained. They are therefore unsuitable for early detection of AHS disease as often the animals will die before a detectable humoral response is raised [90]. The currently used indirect ELISA test is based on detection of the group-specific VP7 antigen in the serum sample [91]. However, although accurate, it is neither possible to determine the virus serotype nor to differentiate between vaccinated and infected animals using this method.

Over the last two decades, several real-time reverse transcription polymerase chain reaction (RT-PCR) assays have been developed and made available to the scientific and veterinary communities [92]. Some tests were group-specific [93–96], based on amplifying AHSV VP7 or one of the non-structural proteins and two of these are recognized as official screening tests by the OIE [97]. The type-specific (TS) tests generally require separate PCR assays to diagnose each of the nine serotypes [90,98,99], but the most recently described test was designed such that three serotypes can be detected in the same assay [92]. These triplex AHSV TS RT-quantitative PCR assays can be directly applied to nucleic acid extracted from blood samples from AHSV-infected horses, meaning that samples can be extracted and evaluated within 4 h of their arrival at the laboratory. The AHSV serotype in a field outbreak can thus be rapidly determined, making this test most useful in directing the timeous implementation of appropriate vaccination and control strategies.

5. Prevention and Control

There is no cure for AHS and no specific treatment aside from rest and good animal husbandry. Various interventions, such as non-steroidal anti-inflammatory drugs for alleviating pain and reducing fever, antimicrobials to fight secondary bacterial infection or corticosteroids to help stabilize cell membranes and preserve vascular membrane integrity, have been employed, but all these treatments are supportive rather than curative (African Horse Sickness Trust). Anecdotal reference to homeopathic remedies has also been made, but there is no scientific evidence to prove the efficacy of such treatments in AHS cases. The implementation of certain husbandry modifications, such as stabling animals before dark in vector-proof housing, using insect repellants and encouraging natural vector predators like fish, frogs and bats, may assist in prevention [1,100]; however, ultimately, vaccination of animals remains the most successful method of prevention and control.

5.1. Live Attenuated Vaccines

Alexander [101] was the first to demonstrate that a mouse-adapted strain of AHSV could be propagated in chicken embryos, and that serial passage in embryonated hens' eggs caused attenuation of the virus without loss of immunogenicity. His studies also supported the existence of multiple strains of the virus, as he found a large variation (26%–81%) in the number of horses that were protected after being challenged a second time with a different virus isolate. The successful propagation of AHSV in mammalian tissue culture by Erasmus [102], aided by the discovery that viral plaque size could be used as a genetic marker to identify avirulent clones of AHSV from a mixed population [78], further advanced the development of live-attenuated vaccine (LAV) strains. The polyvalent LAVs that have been successfully used to vaccinate horses over the past six decades are based on these viral formulations. However, although the frequency and severity of outbreaks has declined significantly since these vaccines have come into use, many horse deaths due to AHS still occur in South Africa every year.

The currently used LAV is supplied in two polyvalent vials containing 3 (serotypes 1, 3 and 4) and 4 (serotypes 2, 6, 7 and 8) AHSV serotypes each. Neither AHSV 5 nor AHSV 9 are included in the vaccine [103]: serotype 5 was originally included but was withdrawn in 1990 following reports of residual virulence, believed to be the result of re-assortment between serotypes 4 and 5 in the vaccine formulation [104]. Serotype 9 has never been included due to its low incidence in southern Africa and, because cross protection between serotypes 1 & 2, 3 & 7, 5 & 8 and 6 & 9 has been documented [3,78,104], protection against AHSV 9 is expected to be provided by AHSV 6. There is no cross-protection between AHSV 4 and any of the other serotypes. Of concern, however, is the fact that in 2006 both AHSV 5 and 9 dominated outbreaks in South Africa, particularly in the Western Cape Province: this raises questions about the competency of the LAV to provide sufficient protection against these two AHSV serotypes.

Although there is little argument that the LAV is still the best option in the fight against AHS, its use has raised concerns with regard to other important issues. The serotype-specific immune response within horse populations, as well as between the different serotypes appears to be quite variable, and it may take as many as 8 vaccination courses over 6 years before an animal is fully protected against all nine AHSV serotypes [104–106]. Furthermore, gene segment re-assortment between outbreak and vaccine strains may lead to the establishment of new genetic variants or reversion to virulence of attenuated vaccine strains [1]. Indeed, a study comparing the whole genome sequences from AHSV isolates responsible for outbreaks between 2004 and 2014 in the controlled area of the Western Cape with LAV and AHSV reference strains demonstrated conclusive evidence of re-assortment between and reversion to virulence of viruses within the LAV itself [88]. The outcome of this study highlights the importance of employing judicious LAV vaccination strategies and genetic screening of circulating field strains during AHS outbreaks.

Another shortcoming of the LAV is the inability to serologically differentiate vaccine-induced immunity from that induced by natural infection, i.e., the absence of what is known as DIVA capacity. This differentiation is important both for early detection of disease and sero-surveillance and can also limit unnecessary culling of animals in an outbreak situation. A further major issue regarding routine vaccination with the LAV is the fact that it is not licensed for use outside of the African sub-continent, which has a hugely negative impact on the international equine trade and export industry.

5.2. Inactivated Vaccines

Inactivated or "killed" vaccines have been prepared by treating mammalian cell-cultured AHSV with formaldehyde or ß-propiolactone [107] or with bromoethylenimine [108]. The latter acts on nucleic acid but not protein, ensuring that the immunogenic properties of the vaccine are not compromised. An inactivated AHSV 9 vaccine tested in guinea pigs and horses [109] elicited a comparable neutralizing antibody response in both animal species, confirming the usefulness of the guinea pig as a small animal model to test the efficacy of potential vaccine candidates [110]. The vaccine proved to be safe and all horses survived a challenge with the same virus used to generate the killed vaccine.

A formalin-inactivated AHSV vaccine was commercially produced and used during the 1987–1991 AHS outbreak in Spain, Portugal and Morocco, but although it proved to be efficacious at the time, this vaccine is no longer available [1]. The main drawbacks with regard to inactivated vaccines are firstly that they are expensive to produce, requiring large-scale isolation of infectious virus, which poses a significant bio-containment risk; secondly, repeated inoculations may be required to ensure long-lasting protective immunity. Furthermore, although the risk of gene segment re-assortment and reversion to virulence are mitigated with this type of vaccine, differentiation between vaccinated and infected animals is not possible.

5.3. Recombinant Vaccines

Due to raised international awareness and local dissatisfaction with the current vaccine, in recent years, AHSV research has focussed on the development of recombinant vaccines. These have largely been based on producing the antigenic AHSV proteins involved in eliciting a protective immune response, particularly the outer capsid proteins VP2 and VP5, and investigating the best ways in which to present them to the host's immune system.

5.3.1. DNA Vaccines

Besides the AHSV proteins themselves, there has been some investigation of the efficacy of using naked AHSV VP2 DNA as a vaccine candidate [111]. Although a VP2-specific humoral and cellular immune response following inoculation of a single horse was observed, and the same horse survived an AHS outbreak during the following rainy season, the neutralizing antibody titre reported was sub-optimal and no experimental challenge ensued. Furthermore, vaccination of hens with cloned VP2 cDNA stimulated the production of egg yolk IgY antibodies with a serum neutralization titre 80-fold less than that obtained following vaccination with purified AHSV. The potential for producing a suitable AHSV DNA vaccine thus seems limited.

5.3.2. Subunit Vaccines

Recombinant AHSV VP2 produced via the baculovirus expression system has been used either singly or in combination with VP5 and/or VP7 as a subunit vaccine, and was shown to induce protective immunity against the virus [112–114]. However, recombinant soluble antigens are generally poorly immunogenic and the aggregation of baculovirus-expressed VP2 purified from insect cell lysates together with the requirement for repeated boost inoculations [115], and the use of potent adjuvants to enhance immunogenicity [116] have limited the usefulness and application of this type of vaccine. Furthermore, although subunit vaccines are advantageous in that they permit differentiation between AHSV-vaccinated and infected animals, baculovirus expression requires growth under sterile conditions and is uneconomical for an animal vaccine due to the high cost of media required to culture insect cells.

5.3.3. Poxvirus-Vectored Vaccines

Poxvirus vectored vaccines are recombinant poxvirus strains which have been genetically modified to contain a copy of the gene of interest within the viral genome. The vaccine is delivered directly to the cells where the viral protein is expressed and presented to the host immune system for the stimulation of both humoral and cellular immunity.

Vaccination of horses with an adjuvant-formulated canarypox vaccine (ALVAC-AHSV) expressing both AHSV 4 VP2 and VP5 was shown to elicit serotype-specific neutralizing antibodies against the virus [117]. Upon challenge, horses which received a sufficiently high dose of vaccine developed sterilizing immunity against AHSV 4 with serum neutralization titres ranging from 10 to 80 (expressed as the reciprocal of the highest dilution that provided >50% cell protection). However, stimulation of neutralizing antibodies has yet to be established as a definite correlate of protection: this was emphasized by the survival of a seronegative horse after challenge with virulent AHSV. Furthermore, in a follow-up study using the same vaccine, gamma interferon-producing cells were detected after stimulation with both VP2 and VP5, indicating that cell-mediated immune responses most likely also played a role in protecting against the viral challenge [86].

Since 2009, several studies on the development of an alternative AHSV candidate vaccine based on a modified vaccinia Ankara (MVA) virus have been published [85,118–123]. The MVA strain was originally produced by extensive passage of the chorioallantosis vaccinia virus Ankara (CVA) in chicken embryo fibroblast cells (CEF). This resulted in the loss of 12% of the viral genome, including genes that interfere with the host immune response and caused inability to replicate in most mammalian cells [124]. However, replication is only blocked after DNA synthesis, so gene expression still occurs, with resulting expression of recombinant antigens inside infected cells. Recombinant MVA vaccines are therefore non-replicative live viral vectors that can induce both a humoral response as well as stimulate T-cell immunity by facilitating intracellular presentation of the antigen of interest via MHC molecules on the cell surface. Interestingly, MVA vaccines have been shown to be most effective in prime-boost regimens, i.e., when administered as a heterologous boost following a strong prime vaccination [125]. Boosting of an existing T-cell response to a recombinant antigen causes an amplification of the initial response and reduces the response to antigens on the viral vector itself, thus avoiding the issue of pre-existing immunity to the viral vector and allowing re-use of the vaccine.

However, homologous prime-boost vaccination with MVA expressing AHSV VP2 (MVA-VP2) has also been shown to elicit neutralizing antibodies which provided complete protection in both mice and horses [118–121]. The type 1 interferon receptor knockout (IFNAR -/-) mice used in these experiments were chosen as small animal challenge models as AHSV infection causes similar clinical pathologies and mortality rates to those observed in horses. In these studies, protection was provided by vaccination with MVA vaccines expressing only outer capsid protein VP2, indicating that VP5 is not essential for, but may improve vaccine efficacy.

The role of cell-mediated immunity appears to be less important than antibody responses, as the transfer of splenocytes from MVA-VP2-vaccinated mice to unvaccinated mouse recipients did not cause a statistically significant reduction in viraemia [123,126]. Furthermore, passive immunization with vaccinated donor sera protected recipient mice from infection, demonstrating the primary protective role of the neutralizing antibody response. However, cell-mediated immunity is likely to play an additional protective role to some extent, as mice immunized with either MVA-VP2 or MVA-NS1 have been shown to develop gamma-interferon-producing cells when stimulated with peptide sequences on VP2 and NS1 [85].

In a study to investigate a polyvalent AHSV vaccination approach, horses were immunized and boosted four weeks later with either MVA-VP2(4) or MVA-VP2(9) or both, simultaneously or sequentially [119]. Simultaneous vaccination with recombinant MVA-VP2 of both serotypes, induced a statistically significant virus neutralizing antibody (VNAb) response against AHSV 4 and AHSV 9 as well as a cross-protective response to AHSV 6 in the horses which received MVA-VP2(9). Furthermore, four months later, when the VNAb titres had decreased dramatically, vaccination with MVA-VP2(5) representing a third AHSV serotype, not only elicited VNAb against AHSV 5, but also induced an anamnestic response towards AHSV 4, 6 and 9 as well as the cross-reactive AHSV 8. These results demonstrate the suitability of MVA-VP2 to be used as a polyvalent vaccine mixture providing protection against more than one AHSV serotype. The results also suggest the possibility that other sub-dominant cross-reactive epitopes may exist between AHSV serotypes 5, 6, 8 and 9. This study further confirms that any possible pre-existing immunity to the viral vector does not impact negatively the usefulness of the MVA-VP2 vaccines.

The immune responses induced by four different AHSV 4 MVA-VP2 vaccines, namely live MVA-VP2, heat, inactivated MVA-VP2, UV, inactivated MVA-VP2 and sucrose gradient-purified MVA-VP2, proved that both pre-formed VP2 in the MVA vaccine and transient expression of VP2 in the vaccinated host's cells contribute to inducing a protective immune response [122]. The inactivated MVA-VP2 vaccines, containing only pre-formed VP2, induced lower VNAb titres than the live MVA-VP2, yet the gradient purified MVA-VP2, containing no pre-formed VP2, also induced a weaker immune response than that induced by live MVA-VP2. In fact, sterilizing immunity was only induced by the live MVA-VP2 vaccine. It is likely that the transient intracellular expression of conformationally intact VP2 in infected cells activates T-cells, lending credence to the possible supportive role of cellular-mediated immunity in the protective immune response.

5.3.4. Reverse Genetics Vaccines

Over the last decade, reverse genetics systems have been used to generate novel live virus-based BTV and AHSV vaccine candidates, engineered according to a rational design rather than by random serial passage attenuation [127–138]. These live vaccine strains depend on the availability of cloned cDNA copies of the viral genes and are produced in mammalian cell lines via a double transfection strategy. Firstly, a primary viral replication complex is pre-expressed by transfection with expression plasmids encoding the five viral proteins, VP1, VP3, VP4, VP6 and NS2. A second transfection with ten exact copy capped T7 viral RNA transcripts, which serve as effective substitutes for authentic core-derived viral transcripts, then triggers full replication and enables virus rescue. Different AHSV serotypes can be rescued by using the same primary transcription complex, and then exchanging the T7 RNA transcripts of one or more capsid proteins [130]. More importantly for vaccine purposes, genes encoding these proteins can be incorporated into a common viral genome which has been precisely engineered to contain one or more defective genes.

Two main vaccine platforms to produce defective virus strains have been developed using this technology. Entry Competent Replication Abortive (ECRA) vaccine strains, previously also referred to as Disabled Infectious Single Cycle (DISC) vaccines, lack a functional VP6 gene and are therefore unable to complete even a single replication cycle in infected cells [127,131]. The defective vaccine is rescued and propagated in a complementary cell line expressing VP6 *in trans* and viral antigens capable of eliciting the expected antibody response are expressed in normal cells, but no active infection ensues. In contrast, Disabled Infectious Single Animal (DISA) vaccine strains lack a functional gene for expression of non-essential non-structural protein NS3/NS3a [129,132,135]. The absence of these proteins prevents viral egress, thus inhibiting viraemia and allowing only local replication in infected cells, with no propagation in nor transmission by midges. Both these vaccine candidates fulfil the criteria for DIVA compliance, as antibodies to the missing viral protein in each case would be absent in vaccinated animals but present in animals which have been infected.

The main goal in new AHSV vaccine development is to provide protection against all nine serotypes of the virus. Initially, an attempt was made to develop a set of defective AHSV virus strains, each consisting of a common core coated with a different serotype-specific outer capsid protein VP2 [129]. However, the exchange of only a single protein resulted in unequal and significantly lower re-assortant viral titres compared to the parental virus strain. To produce suitably-replicating defective vaccine strains for each serotype, it appears necessary to exchange between two (VP2 and VP5) and five (VP2, VP3, VP5, VP7 and NS3) proteins on the common backbone, the number depending on the desired viral serotype [127]. The safety and immunogenicity of both monospecific (AHSV 4) and multivalent cocktail (AHSV 1/4/6/8) ECRA vaccines was tested in ponies; these were protected against virulent challenge with AHSV 4 [131]. Pre-challenge serum neutralization titres were in the range of 8–64 (expressed as the reciprocal of the highest dilution that provided >50% cell protection), below those generally obtained following vaccination with the AHS LAV [106], but nevertheless demonstrating the potential efficacy of a reverse genetics vaccine candidate to protect against the disease. Although the technology looks promising, further research is necessary to determine the minimum dose requirement and longevity of the immune response. Furthermore, the associated cost and upscaling requirements may deter successful commercialization of these potential vaccine candidates.

5.3.5. Virus-Like Particle Vaccines

Virus-like particles (VLPs) are safe, non-replicating protein complexes which mimic the structure of intact virions. They possess self-adjuvanting properties and have the advantage of being highly immunogenic compared to subunit vaccines, as epitopes are displayed in ordered repetitive arrays on the particle surface [139]. The size of VLPs ensures appropriate drainage into the lymph nodes and is also optimal for uptake by antigen-presenting cells and MHC cross presentation [140,141]. This efficient trafficking of VLPs and their interaction with the host immune cells induces both innate and adaptive humoral and cellular immune responses, making them particularly attractive vaccine candidates [142,143]. Furthermore, such vaccines present no risk of reversion to virulence nor of dsRNA segment re-assortment with wild virus or live vaccine strains, because they do not contain viral RNA or any of the non-structural proteins. The absence of these viral components also makes it possible to distinguish between vaccinated and infected animals using molecular diagnostic techniques, meaning that VLP vaccines would be DIVA compliant.

The production of both BTV and AHSV VLPs is based on the hypothesis that co-synthesis of proteins VP2, VP3, VP5 and VP7 will result in the spontaneous self-assembly of VLPs via various hydrophobic, electrostatic and covalent interactions [144]. The successful formation and protective efficacy of BTV VLPs produced by the recombinant baculovirus-mediated co-expression of these proteins in insect cells, has been demonstrated [145–147]. In the past, most recombinant proteins have been produced either in insect cell lines or in microbial fermentation systems, mammalian cell cultures or transgenic animals. More recently, however, there has been an increased interest in utilizing plant-based expression systems—the so-called "biopharming" or "molecular farming" approach—whereby plants are harnessed as mini-factories to produce useful pharmaceutical proteins [148–153]. The high-level plant-based expression of fully-assembled VLPs of BTV-8 has recently been reported [154,155] and the plant-produced VLPs were shown to elicit a strong antibody response in sheep, providing protective immunity against challenge with a BTV-8 field isolate.

An initial investigation regarding the production of AHSV CLPs by co-infection of insect cells with recombinant baculoviruses expressing either AHSV VP3 or VP7 was unsuccessful [156]. CLPs represent the inner and middle protein layers of the virus and lack the outer capsid protein layer composed of proteins VP5 and VP2 which are required for complete VLP formation. In a later study co-expression of AHSV viral proteins and assembly of both AHSV CLPs and VLPs was achieved by co-infection with baculovirus recombinants simultaneously expressing two AHSV capsid proteins, i.e., VP2 and VP3 or VP5 and VP7 [157]. However, the overall VLP yield was very low and precluded quantification.

Recently, the development of a plant-produced AHSV VLP vaccine was described [158,159]. Transient co-expression of the four capsid proteins of two different AHSV serotypes in the common tobacco plant, *N. benthamiana*, resulted in the efficient self-assembly of well-formed AHSV VLPs which were shown to be both safe and highly immunogenic in horses. Confidence in this novel AHSV VLP vaccine hinges largely on the fact that it is a vaccine comprised entirely of protein, free from infectious genetic material, and produced using a cost-effective plant expression system. Furthermore, the vaccine platform mimics field exposure to the naturally immunogenic AHS virion, but without endangering the immunised animal in any way and being completely free of the possibilities of reversion to virulence, or reassortment with other vaccine or wild-type viruses. Sera from horses immunized with AHSV 5 VLPs also elicited a comparative immune response towards AHSV 8, confirming reports of cross protection between these two serotypes and demonstrating the importance of pursuing further investigation into the potential and suitability of this candidate AHSV vaccine. The various vaccine strategies against African horse sickness, together with the advantages and disadvantages of each, are summarized in Figure 4.

6. Conclusions

African horse sickness is a lethal and debilitating disease of domestic equids. There is little doubt that the live attenuated vaccine that has been used in South Africa to protect horses against AHSV for the past six decades [160,161] has ensured their continued existence. However, the manufacture of this vaccine uses very old technology and production volumes that cannot meet the current demand. Furthermore, the fact that many horses still contract the disease and often die in spite of vaccination, as well the fact that this vaccine is not licensed for use outside of the African sub-continent, has led to an increasing demand for a new, safer and more cost-effective vaccine which would not only address the concerns of South African horse owners, but also meet approval in the wider international context where live vaccines for the disease would not be acceptable. Biotechnological advances over the past few decades have paved the way for new generation vaccines which lack the associated negative features of the LAV, and which could potentially serve as adequate replacement vaccines.

TYPE OF VACCINE	STRATEGY	ADVANTAGES	DISADVANTAGES
LIVE ATTENUATED VACCINES	Attenuation of live virus by serial passage in a heterologous host or cell culture. Low-level replication in the vaccinated animal permits the stimulation of virus-specific immunity.	- Mimics sub-clinical infection - Stimulates both humoral and cellular immune responses - long-lived immunity - can contribute to herd immunity - cost-effective	- reversion to virulence - gene segment re-assortment - non-DIVA compliant - variable serotype-specific immune response - requirement for high level bio-safety facilities - not licensed for use outside Africa
INACTIVATED VACCINES	Large quantities of virus are produced and inactivated by chemical or physical procedures. Infectivity is eliminated without destroying the antigenicity of the virus.	- safe - stable - no genetic re-assortment	- transient immunity - non-DIVA compliant - expensive - possible reaction to adjuvant - requirement for high level bio-safety facilities
RECOMBINANT VACCINES			
DNA Vaccines	DNA encoding viral proteins is injected as the vaccine. DNA is transcribed and translated to produce viral protein which elicits an immune response in the vaccinated host.	- safe - stable - cost-effective once gene is cloned - DIVA compliant	- low humoral response - possible incorporation into the host genome
Subunit vaccines	A specific viral protein is purified or produced via recombinant DNA technology and presented to the host's immune system	- safe - DIVA compliant	- weak immunogens - expensive - possible protein aggregation - boost inoculations required - need potent adjuvant
Poxvirus-vectored vaccines	The gene encoding an antigen target for neutralizing antibodies is cloned into a poxviral vector. Animals vaccinated with the recombinant virus mount an immune response to the antigen of interest.	- Induces both humoral and cellular immunity - more than one gene per vector possible - DIVA compliant	- costly - complicated design strategy - requirement for high level bio-safety facilities - can't be used in immune-compromised animals
Reverse genetics vaccines	Recombinant virus strains are genetically engineered to lack a functional gene, allowing either only a single replication cycle or preventing viral egress and inhibiting viraemia. Purified recombinant viruses are used as the vaccine.	- does not cause disease - low risk of genetic reassortment - DIVA compliant	- complicated design strategy - expensive
VLP vaccines	AHSV structural proteins are expressed from recombinant bacterial strains in mammalian, insect or plant expression systems. These assemble into empty viral-like particles which are administered with adjuvant as the vaccine.	- Safe - No risk of reversion to virulence - No risk of genetic re-assortment - Cost-effective in plants - Ease of scale up in plants - DIVA compliant	- Costly if produced in insect or mammalian cells - Possible reaction to adjuvant

Figure 4. A summary of the various vaccine strategies against African horse sickness.

An ideal AHSV vaccine would activate both humoral and cell-mediated immune responses and provide rapid and long-lasting protective immunity against all nine serotypes of the virus. It would block viraemia, disallow transmission by the midge vector, ensure that no risk of reversion to virulence nor re-assortment with outbreak strains was possible, and permit accurate differentiation between vaccinated and infected animals. It should be possible to safely, consistently and economically

Viruses **2019**, *11*, 844

produce sufficient doses of such a vaccine to meet the demands of both the private and rural sectors. Importantly, the vaccine should hold sufficient interest and market potential to capture the attention of the manufacturing industry.

Five types of alternative AHS vaccine platforms have been described in recent years. Two of these, the ECRA (Entry Competent Replication Abortive) [131] and DISA (Disabled Infectious Single Animal) [132] candidate vaccines, stem from research in the area of reverse genetics technology, while the third and fourth are based on modified pox viruses [117,122]. Although the results obtained in experimental trials with these vaccines look promising, issues of cost and scalability have thus far prevented any from being commercialized. The newest VLP candidates have the potential to be as efficacious as the currently used live attenuated vaccine, but without the latter's accompanying risks and shortcomings. Furthermore, they have the associated benefits of cheap and scalable production processes. However, challenge trials with live virus and further investigation into the development of VLPs of the other AHSV serotypes are required to conclusively demonstrate the protective efficacy of these plant-produced vaccines.

Patents: The authors AM and ER have filed a patent application protecting the production of chimaeric orbivirus virus-like particles in plants (PCT/IB2017/052236).

Author Contributions: S.J.D. conceived and prepared the manuscript; A.E.M., I.I.H. and E.P.R. provided consultation and read and approved the final manuscript.

Funding: This research received no additional external funding.

Acknowledgments: The author SD acknowledges the Technology and Innovation Agency (Grant number: TAHIC13-00023-CC); the Poliomyelitis Research Foundation and the Council for Scientific and Industrial Research for PhD funding.

Conflicts of Interest: The authors declare no conflict of interest.

References

1. Mellor, P.S.; Hamblin, C. African horse sickness. *Vet. Res.* **2004**, *35*, 445–466. [CrossRef] [PubMed]
2. Henning, M.W. African Horse Sickness, Perdesiekte, Pestis Equorum. In *Animal Diseases of South Africa*, 3rd ed.; Central News Agency Ltd.: Pretoria, Africa, 1956; pp. 785–808.
3. Coetzer, J.A.W.; Guthrie, A.J. African horse sickness. In *Infectious Diseases of Livestock*, 2nd ed.; Coetzer, J.A.W., Tustin, R.C., Eds.; Oxford University Press: Cape Town, Africa, 2004; pp. 1231–1246.
4. Vandenbergh, S. The story of a disease: African horsesickness and its direct influence on the necessary development of veterinary science in South Africa c. 1890s–1920s. *Historia* **2010**, *55*, 243–262.
5. Lubroth, J. African Horsesickness and the epizootic in Spain, 1987. *Equine Pract.* **1988**, *10*, 26–33.
6. Mirchamsy, H.; Hazrati, A. A review on etiology and pathogeny of African horse sickness. *Arch. Razi Inst.* **1973**, *25*, 23–46.
7. Mellor, P. African horse sickness: Transmission and epidemiology. *Vet. Res.* **1993**, *24*, 199–212. [PubMed]
8. Howell, P.G. The 1960 epizootic of African Horsesickness in the Middle East and SW Asia. *J. S. Afr. Vet. Assoc.* **1960**, *31*, 329–334.
9. De Vos, C.J.; Hoek, C.A.; Nodelijk, G. Risk of introducing African horse sickness virus into the Netherlands by international equine movements. *Prev. Vet. Med.* **2012**, *106*, 108–122. [CrossRef] [PubMed]
10. Herholz, C.; Fussel, A.E.; Timoney, P.; Schwermer, H.; Bruckner, L.; Leadon, D. Equine travellers to the Olympic Games in Hong Kong 2008: A review of worldwide challenges to equine health, with particular reference to vector-borne diseases. *Equine Vet. J.* **2008**, *40*, 87–95. [CrossRef] [PubMed]
11. Hopley, R.; Toth, B. Focus on: African horse sickness. *Vet. Rec.* **2013**, *173*, 13–14. [CrossRef]
12. M'Fadyean, J. African horse-sickness. *J. Comp. Pathol. Ther.* **1900**, *13*, 1–20. [CrossRef]
13. Howell, P.G. The isolation and identification of further antigenic types of African horsesickness virus. *Onderstepoort J. Vet. Res.* **1962**, *29*, 139–149.
14. McIntosh, B.M. Immunological types of horsesickness virus and their significance in immunization. *Onderstepoort J. Vet. Res.* **1958**, *27*, 465–536.
15. Calisher, C.H.; Mertens, P.P. Taxonomy of African horse sickness viruses. In *African Horse Sickness*; Mellor, P.S., Baylis, M., Hamblin, C., Mertens, P.P.C., Calisher, C.H., Eds.; Springer: Vienna, Austria, 1998.

16. Du Toit, R.M. The transmission of bluetongue and horse sickness by Culicoides. *Onderstepoort J. Vet. Sci. Anim. Ind.* **1944**, *19*, 7–16.

17. Mellor, P.S.; Boorman, J.; Jennings, M. The multiplication of African horse-sickness virus in two species ofCulicoides (Diptera, Ceratopogonidae). *Arch. Virol.* **1975**, *47*, 351–356. [CrossRef] [PubMed]

18. Boorman, J.; Mellor, P.S.; Penn, M.; Jennings, M. The growth of African horse-sickness virus in embryonated hen eggs and the transmission of virus byCulicoides variipennis Coquillett (Diptera, Ceratopogonidae). *Arch. Virol.* **1975**, *47*, 343–349. [CrossRef] [PubMed]

19. Carpenter, S.; Mellor, P.S.; Fall, A.G.; Garros, C.; Venter, G.J. African Horse Sickness Virus: History, Transmission, and Current Status. *Annu. Rev. Entomol.* **2017**, *62*, 343–358. [CrossRef]

20. Roy, P. Genetically engineered structure-based vaccine for bluetongue disease. *Vet. Ital.* **2004**, *40*, 594–600.

21. Barnard, B.J.H. Epidemiology of African horse sickness and the role of the zebra in South Africa. In *African Horse Sickness*; Mellor, P.S., Baylis, M., Hamblin, C., Mertens, P.P.C., Calisher, C.H., Eds.; Springer: Vienna, Austria, 1998; pp. 13–19.

22. Theiler, A. The susceptibility of the dog to african horse-sickness. *J. Comp. Pathol. Ther.* **1910**, *23*, 315–325. [CrossRef]

23. McIntosh, B.M. Horsesickness antibodies in the sera of dogs in enzootic areas. *J. S. Afr. Vet. Assoc.* **1955**, *26*, 269–272.

24. Oellermann, R.; Els, H.; Erasmus, B. Characterization of African horsesiekness virus. *Arch. Fur Die Gesamte Virusforsch.* **1970**, *29*, 163–174. [CrossRef]

25. Coetzer, J.A.W.; Erasmus, B.J. African Horse Sickness. In *Infectious Diseases of Livestock with Special Reference to Southern Africa*; Coetzer, J.A.W., Thomson, G.R., Tustin, R.C., Eds.; Oxford Uiversity Press: Oxford, UK, 1994; pp. 460–475.

26. Grubman, M.J.; Lewis, S.A. Identification and characterization of the structural and nonstructural proteins of African horsesickness virus and determination of the genome coding assignments. *Virology* **1992**, *186*, 444–451. [CrossRef]

27. Bremer, C.W. A gel electrophoretic study of the protein and nucleic acid components of African horsesickness virus. *Onderstepoort J. Vet. Res.* **1976**, *43*, 193–199. [PubMed]

28. Bremer, C.W.; Huismans, H.; Van Dijk, A.A. Characterization and cloning of the African horsesickness virus genome. *J. Gen. Virol.* **1990**, *71*, 793–799. [CrossRef]

29. Roy, P.; Mertens, P.P.; Casal, I. African horse sickness virus structure. *Comp. Immunol. Microbiol. Infect. Dis.* **1994**, *17*, 243–273. [CrossRef]

30. Zhang, X.; Patel, A.; Celma, C.C.; Yu, X.; Roy, P.; Zhou, Z.H. Atomic model of a nonenveloped virus reveals pH sensors for a coordinated process of cell entry. *Nat. Struct. Mol. Biol.* **2016**, *23*, 74–80. [CrossRef]

31. Grimes, J.M.; Burroughs, J.N.; Gouet, P.; Diprose, J.M.; Malby, R.; Zientara, S.; Mertens, P.P.C.; Stuart, D.I. The atomic structure of the bluetongue virus core. *Nature* **1998**, *395*, 470–478. [CrossRef]

32. Zhang, X.; Boyce, M.; Bhattacharya, B.; Zhang, X.; Schein, S.; Roy, P.; Zhou, Z.H. Bluetongue virus coat protein VP2 contains sialic acid-binding domains, and VP5 resembles enveloped virus fusion proteins. *Proc. Natl. Acad. Sci. USA* **2010**, *107*, 6292–6297. [CrossRef]

33. Nason, E.L.; Rothagel, R.; Mukherjee, S.K.; Kar, A.K.; Forzan, M.; Prasad, B.V.; Roy, P. Interactions between the inner and outer capsids of bluetongue virus. *J. Virol.* **2004**, *78*, 8059–8067. [CrossRef]

34. Roy, P. Bluetongue virus structure and assembly. *Curr. Opin. Virol.* **2017**, *24*, 115–123. [CrossRef]

35. Iwata, H.; Yamagaw, M.; Roy, P. Evolutionary relationships among the gnat-transmitted orbiviruses that cause African horse sickness, bluetongue, and epizootic hemorrhagic disease as evidenced by their capsid protein sequences. *Virology* **1992**, *191*, 251–261. [CrossRef]

36. Caspar, D.L.D.; Klug, A. Physical principles in the construction of regular viruses. *Cold Spring Harb. Symp. Quant. Biol.* **1962**, *27*, 1–24. [CrossRef] [PubMed]

37. Stuart, D.I.; Gouet, P.; Grimes, J.; Malby, R.; Diprose, J.; Zientara, S.; Burroughs, J.N.; Mertens, P.P. Structural studies of orbivirus particles. *Arch. Virol. Suppl.* **1998**, *14*, 235–250. [PubMed]

38. Manole, V.; Laurinmaki, P.; Van Wyngaardt, W.; Potgieter, C.A.; Wright, I.M.; Venter, G.J.; van Dijk, A.A.; Sewell, B.T.; Butcher, S.J. Structural insight into African horsesickness virus infection. *J. Virol.* **2012**, *86*, 7858–7866. [CrossRef] [PubMed]

39. Gouet, P.; Diprose, J.M.; Grimes, J.M.; Malby, R.; Burroughs, J.N.; Zientara, S.; Stuart, D.I.; Mertens, P.P. The highly ordered double-stranded RNA genome of bluetongue virus revealed by crystallography. *Cell* **1999**, *97*, 481–490. [CrossRef]

40. Stuart, D.; Grimes, J. Structural Studies on Orbivirus Proteins and Particles. In *Reoviruses: Entry, Assembly and Morphogenesis. Current Topics in Microbiology and Immunology*; Roy, P., Ed.; Springer: Berlin/Heidelberg, Germany, 2006; pp. 221–244.

41. Chuma, T.; Le Blois, H.; Sanchez-Vizcaino, J.M.; Diaz-Laviada, M.; Roy, P. Expression of the major core antigen VP7 of African horsesickness virus by a recombinant baculovirus and its use as a group-specific diagnostic reagent. *J. Gen. Virol.* **1992**, *73*, 925–931. [CrossRef] [PubMed]

42. Grimes, J.; Basak, A.K.; Roy, P.; Stuart, D. The crystal structure of bluetongue virus VP7. *Nature* **1995**, *373*, 167. [CrossRef] [PubMed]

43. Basak, A.K.; Gouet, P.; Grimes, J.; Roy, P.; Stuart, D. Crystal structure of the top domain of African horse sickness virus VP7: Comparisons with bluetongue virus VP7. *J. Virol.* **1996**, *70*, 3797–3806. [PubMed]

44. Grimes, J.M.; Jakana, J.; Ghosh, M.; Basak, A.K.; Roy, P.; Chiu, W.; Stuart, D.I.; Prasad, B.V.V. An atomic model of the outer layer of the bluetongue virus core derived from X-ray crystallography and electron cryomicroscopy. *Structure* **1997**, *5*, 885–893. [CrossRef]

45. Roy, P.; Noad, R. Bluetongue Virus Assembly and Morphogenesis. In *Reoviruses: Entry, Assembly and Morphogenesis. Current Topics in Microbiology and Immunology*; Roy, P., Ed.; Springer: Berlin/Heidelberg, Germany, 2006; pp. 87–116.

46. Limn, C.-K.; Staeuber, N.; Monastyrskaya, K.; Gouet, P.; Roy, P. Functional dissection of the major structural protein of bluetongue virus: Identification of key residues within VP7 essential for capsid assembly. *J. Virol.* **2000**, *74*, 8658–8669. [CrossRef] [PubMed]

47. Roy, P.; Hirasawa, T.; Fernandez, M.; Blinov, V.M.; Rodrique, S.-V.J.M. The complete sequence of the group-specific antigen, VP7, of African horsesickness disease virus serotype 4 reveals a close relationship to bluetongue virus. *J. Gen. Virol.* **1991**, *72*, 1237–1241. [CrossRef]

48. Burroughs, J.N.; O'Hara, R.S.; Smale, C.J.; Hamblin, C.; Walton, A.; Armstrong, R.; Mertens, P.P. Purification and properties of virus particles, infectious subviral particles, cores and VP7 crystals of African horsesickness virus serotype 9. *J. Gen. Virol.* **1994**, *75*, 1849–1857. [CrossRef] [PubMed]

49. Burrage, T.G.; Trevejo, R.; Stone-Marschat, M.; Laegreid, W.W. Neutralizing epitopes of African horsesickness virus serotype 4 are located on VP2. *Virology* **1993**, *196*, 799–803. [CrossRef] [PubMed]

50. Hewat, E.A.; Booth, T.F.; Roy, P. Structure of bluetongue virus particles by cryoelectron microscopy. *J. Struct. Biol.* **1992**, *109*, 61–69. [CrossRef]

51. Boyce, M.; Celma, C.P.; Roy, P. Bluetongue virus non-structural protein 1 is a positive regulator of viral protein synthesis. *Virol. J.* **2012**, *9*, 178. [CrossRef] [PubMed]

52. Van Staden, V.; Theron, J.; Greyling, B.; Huismans, H.; Nel, L. A comparison of the nucleotide sequences of cognate NS2 genes of three different orbiviruses. *Virology* **1991**, *185*, 500–504. [CrossRef]

53. Patel, A.; Roy, P. The molecular biology of Bluetongue virus replication. *Virus Res.* **2014**, *182*, 5–20. [CrossRef] [PubMed]

54. Quan, M.; Van Vuuren, M.; Howell, P.G.; Groenewald, D.; Guthrie, A.J. Molecular epidemiology of the African horse sickness virus S10 gene. *J. Gen. Virol.* **2008**, *89*, 1159–1168. [CrossRef] [PubMed]

55. Celma, C.C.P.; Roy, P. Interaction of calpactin light chain (S100A10/p11) and a viral NS protein is essential for intracellular trafficking of non-enveloped Bluetongue virus. *J. Virol.* **2011**, *85*, 4783–4791. [CrossRef] [PubMed]

56. Zwart, L.; Potgieter, C.A.; Clift, S.J.; van Staden, V. Characterising Non-Structural Protein NS4 of African Horse Sickness Virus. *PLoS ONE* **2015**, *10*, e0124281. [CrossRef] [PubMed]

57. Ratinier, M.; Caporale, M.; Golder, M.; Franzoni, G.; Allan, K.; Nunes, S.F.; Armezzani, A.; Bayoumy, A.; Rixon, F.; Shaw, A. Identification and characterization of a novel non-structural protein of bluetongue virus. *PLoS Pathog.* **2011**, *7*, e1002477. [CrossRef] [PubMed]

58. Hassan, S.H.; Wirblich, C.; Forzan, M.; Roy, P. Expression and functional characterization of bluetongue virus VP5 protein: Role in cellular permeabilization. *J. Virol.* **2001**, *75*, 8356–8367. [CrossRef] [PubMed]

59. Mecham, J.O.; McHolland, L.E. Measurement of bluetongue virus binding to a mammalian cell surface receptor by an in situ immune fluorescent staining technique. *J. Virol. Methods* **2010**, *165*, 112–115. [CrossRef] [PubMed]

60. Stevens, L.M.; Moffat, K.; Cooke, L.; Nomikou, K.; Mertens, P.P.C.; Jackson, T.; Darpel, K.E. A low-passage insect-cell isolate of bluetongue virus uses a macropinocytosis-like entry pathway to infect natural target cells derived from the bovine host. *J. Gen. Virol.* **2019**, *100*, 568–582. [CrossRef] [PubMed]

61. Eaton, B.T.; Crameri, G.S. The Site of Bluetongue Virus Attachment to Glycophorins from a Number of Animal Erythrocytes. *J. Gen. Virol.* **1989**, *70*, 3347–3353. [CrossRef] [PubMed]

62. Hassan, S.S.; Roy, P. Expression and Functional Characterization of Bluetongue Virus VP2 Protein: Role in Cell Entry. *J. Virol.* **1999**, *73*, 9832. [PubMed]

63. Marchi, P.R.; Rawlings, P.; Burroughs, J.N.; Wellby, M.; Mertens, P.P.C.; Mellor, P.S.; Wade-Evans, A.M. Proteolytic cleavage of VP2, an outer capsid protein of African horse sickness virus, by species-specific serum proteases enhances infectivity in Culicoides. *J. Gen. Virol.* **1995**, *76*, 2607–2611. [CrossRef] [PubMed]

64. Forzan, M.; Marsh, M.; Roy, P. Bluetongue virus entry into cells. *J. Virol.* **2007**, *81*, 4819–4827. [CrossRef]

65. Forzan, M.; Wirblich, C.; Roy, P. A capsid protein of nonenveloped Bluetongue virus exhibits membrane fusion activity. *Proc. Natl. Acad. Sci. USA* **2004**, *101*, 2100–2105. [CrossRef]

66. Boyce, M.; Wehrfritz, J.; Noad, R.; Roy, P. Purified recombinant bluetongue virus VP1 exhibits RNA replicase activity. *J. Virol.* **2004**, *78*, 3994–4002. [CrossRef]

67. Ramadevi, N.; Burroughs, N.J.; Mertens, P.P.C.; Jones, I.M.; Roy, P. Capping and methylation of mRNA by purified recombinant VP4 protein of bluetongue virus. *Proc. Natl. Acad. Sci. USA* **1998**, *95*, 13537–13542. [CrossRef]

68. Kar, A.K.; Bhattacharya, B.; Roy, P. Bluetongue virus RNA binding protein NS2 is a modulator of viral replication and assembly. *BMC Mol. Biol.* **2007**, *8*, 4. [CrossRef] [PubMed]

69. Modrof, J.; Lymperopoulos, K.; Roy, P. Phosphorylation of bluetongue virus nonstructural protein 2 is essential for formation of viral inclusion bodies. *J. Virol.* **2005**, *79*, 10023–10031. [CrossRef] [PubMed]

70. Mohl, B.-P.; Roy, P. Cellular casein kinase 2 and protein phosphatase 2A modulate replication site assembly of bluetongue virus. *J. Biol. Chem.* **2016**, *291*, 14566–14574. [CrossRef] [PubMed]

71. Beaton, A.R.; Rodriguez, J.; Reddy, Y.K.; Roy, P. The membrane trafficking protein calpactin forms a complex with bluetongue virus protein NS3 and mediates virus release. *Proc. Natl. Acad. Sci. USA* **2002**, *99*, 13154–13159. [CrossRef] [PubMed]

72. Celma, C.C.P.; Roy, P. A viral nonstructural protein regulates bluetongue virus trafficking and release. *J. Virol.* **2009**, *83*, 6806–6816. [CrossRef] [PubMed]

73. Venter, E.; Van der Merwe, C.F.; Buys, A.V.; Huismans, H.; Van Staden, V. Comparative ultrastructural characterization of African horse sickness virus-infected mammalian and insect cells reveals a novel potential virus release mechanism from insect cells. *J. Gen. Virol.* **2014**, *95*, 642–651. [CrossRef] [PubMed]

74. Mertens, P.P.C.; Burroughs, J.N.; Walton, A.; Wellby, M.P.; Fu, H.; O'Hara, R.S.; Brookes, S.M.; Mellor, P.S. Enhanced infectivity of modified bluetongue virus particles for two insect cell lines and for TwoCulicoidesVector species. *Virology* **1996**, *217*, 582–593. [CrossRef] [PubMed]

75. Ruoslahti, E. Rgd and Other Recognition Sequences for Integrins. *Annu. Rev. Cell Dev. Biol.* **1996**, *12*, 697–715. [CrossRef] [PubMed]

76. Burrage, T.G.; Laegreid, W.W. African horsesickness: Pathogenesis and immunity. *Comp. Immunol. Microbiol. Infect. Dis.* **1994**, *17*, 275–285. [CrossRef]

77. Wohlsein, P.; Pohlenz, J.F.; Davidson, F.L.; Salt, J.S.; Hamblin, C. Immunohistochemical demonstration of African horse sickness viral antigen in formalin-fixed equine tissues. *Vet. Pathol.* **1997**, *34*, 568–574. [CrossRef] [PubMed]

78. Erasmus, B. A new Approach to Polyvalent Immunization Against African Horsesickness. In Proceedings of the 4th International Conference on Equine Infectious Diseases, Lyon, France, 24–27 September 1976; Bryans, J.T., Gerber, H., Eds.; Veterinary Publications: Princeton, NJ, USA, 1978; pp. 401–403.

79. Erasmus, B.J. The Pathogenesis of African Horsesickness. In *Equine Infectious Diseases*; Karger Publishers: Basel, Switzerland, 1974; pp. 1–11.

80. Alexander, R.A. Studies on the neurotropic virus of horsesickness III: The intracerebral protection test and its application to the study of immunity. *Onderstepoort J. Vet. Sci. Anim. Ind.* **1935**, *4*, 349–377.

81. Bentley, L.; Fehrsen, J.; Jordaan, F.; Huismans, H.; Du Plessis, D.H. Identification of antigenic regions on VP2 of African horsesickness virus serotype 3 by using phage-displayed epitope libraries. *J. Gen. Virol.* **2000**, *81*, 993–1000. [CrossRef] [PubMed]

82. Mathebula, E.M.; Faber, F.E.; van Wyngaardt, W.; van Schalkwyk, A.; Pretorius, A.; Fehrsen, J. B-cell epitopes of African horse sickness virus serotype 4 recognised by immune horse sera. *Onderstepoort J. Vet. Res.* **2017**, *84*, 1–12. [CrossRef] [PubMed]

83. Martinez-Torrecuadrada, J.L.; Langeveld, J.P.; Meloen, R.H.; Casal, J.I. Definition of neutralizing sites on African horse sickness virus serotype 4 VP2 at the level of peptides. *J. Gen. Virol.* **2001**, *82*, 2415–2424. [CrossRef] [PubMed]

84. Venter, M.; Napier, G.; Huismans, H. Cloning, sequencing and expression of the gene that encodes the major neutralisation-specific antigen of African horsesickness virus serotype 9. *J. Virol. Methods* **2000**, *86*, 41–53. [CrossRef]

85. De la Poza, F.; Marin-Lopez, A.; Castillo-Olivares, J.; Calvo-Pinilla, E.; Ortego, J. Identification of CD8 T cell epitopes in VP2 and NS1 proteins of African horse sickness virus in IFNAR(-/-) mice. *Virus Res.* **2015**, *210*, 149–153. [CrossRef] [PubMed]

86. El Garch, H.; Crafford, J.E.; Amouyal, P.; Durand, P.Y.; Edlund Toulemonde, C.; Lemaitre, L.; Cozette, V.; Guthrie, A.; Minke, J.M. An African horse sickness virus serotype 4 recombinant canarypox virus vaccine elicits specific cell-mediated immune responses in horses. *Vet. Immunol. Immunopathol.* **2012**, *149*, 76–85. [CrossRef] [PubMed]

87. Pretorius, A.; Van Kleef, M.; Van Wyngaardt, W.; Heath, J. Virus-specific CD8+ T-cells detected in PBMC from horses vaccinated against African horse sickness virus. *Vet. Immunol. Immunopathol.* **2012**, *146*, 81–86. [CrossRef]

88. Weyer, C.T.; Grewar, J.D.; Burger, P.; Rossouw, E.; Lourens, C.; Joone, C.; le Grange, M.; Coetzee, P.; Venter, E.; Martin, D.P.; et al. African Horse Sickness Caused by Genome Reassortment and Reversion to Virulence of Live, Attenuated Vaccine Viruses, South Africa, 2004–2014. *Emerg. Infect. Dis.* **2016**, *22*, 2087–2096. [CrossRef]

89. Weyer, C.T. African Horse Sickness Outbreak Investigation and Disease Surveillance Using Molecular Techniques. Ph.D. Thesis, University of Pretoria, Pretoria, South Africa, 2017.

90. Sailleau, C.; Hamblin, C.; Paweska, J.; Zientara, S. Identification and differentiation of the nine African horse sickness virus serotypes by RT–PCR amplification of the serotype-specific genome segment 2. *J. Gen. Virol.* **2000**, *81*, 831–837. [CrossRef]

91. Maree, S.; Paweska, J.T. Preparation of recombinant African horse sickness virus VP7 antigen via a simple method and validation of a VP7-based indirect ELISA for the detection of group-specific IgG antibodies in horse sera. *J. Virol. Methods* **2005**, *125*, 55–65. [CrossRef] [PubMed]

92. Weyer, C.T.; Joone, C.; Lourens, C.W.; Monyai, M.S.; Koekemoer, O.; Grewar, J.D.; van Schalkwyk, A.; Majiwa, P.O.; MacLachlan, N.J.; Guthrie, A.J. Development of three triplex real-time reverse transcription PCR assays for the qualitative molecular typing of the nine serotypes of African horse sickness virus. *J. Virol. Methods* **2015**, *223*, 69–74. [CrossRef] [PubMed]

93. Agüero, M.; Gomez-Tejedor, C.; Cubillo, Á.M.; Rubio, C.; Romero, E.; Jiménez-Clavero, M.A. Real-time fluorogenic reverse transcription polymerase chain reaction assay for detection of African horse sickness virus. *J. Vet. Diagn. Investig.* **2008**, *20*, 325–328. [CrossRef] [PubMed]

94. Guthrie, A.J.; MacLachlan, N.J.; Joone, C.; Lourens, C.W.; Weyer, C.T.; Quan, M.; Monyai, M.S.; Gardner, I.A. Diagnostic accuracy of a duplex real-time reverse transcription quantitative PCR assay for detection of African horse sickness virus. *J. Virol. Methods* **2013**, *189*, 30–35. [CrossRef] [PubMed]

95. Quan, M.; Lourens, C.W.; MacLachlan, N.J.; Gardner, I.A.; Guthrie, A.J. Development and optimisation of a duplex real-time reverse transcription quantitative PCR assay targeting the VP7 and NS2 genes of African horse sickness virus. *J. Virol. Methods* **2010**, *167*, 45–52. [CrossRef] [PubMed]

96. Rodriguez-Sanchez, B.; Fernandez-Pinero, J.; Sailleau, C.; Zientara, S.; Belak, S.; Arias, M.; Sanchez-Vizcaino, J.M. Novel gel-based and real-time PCR assays for the improved detection of African horse sickness virus. *J. Virol. Methods* **2008**, *151*, 87–94. [CrossRef] [PubMed]

97. OIE; Word Organisation for Animal Health. *Manual of Diagnostic Tests and Vaccines for Terrestrial Animals*, 8th ed.; World Organisation for Animal Health: Paris, France, 2018.

98. Koekemoer, J.J.O. Serotype-specific detection of African horsesickness virus by real-time PCR and the influence of genetic variations. *J. Virol. Methods* **2008**, *154*, 104–110. [CrossRef] [PubMed]

99. Bachanek-Bankowska, K.; Maan, S.; Castillo-Olivares, J.; Manning, N.M.; Maan, N.S.; Potgieter, A.C.; Di Nardo, A.; Sutton, G.; Batten, C.; Mertens, P.P. Real time RT-PCR assays for detection and typing of African horse sickness virus. *PLoS ONE* **2014**, *9*, e93758. [CrossRef] [PubMed]

100. Meiswinkel, R.; Baylis, M.; Labuschagne, K. Stabling and the protection of horses from Culicoides bolitinos (Diptera: Ceratopogonidae), a recently identified vector of African horse sickness. *Bull. Entomol. Res.* **2000**, *90*, 509–515. [CrossRef]

101. Alexander, R.A. The immunization of horses and mules against Horse Sickness by means of the neurotropic virus of mice and guinea-pigs. *Onderstepoort J. Vet. Sci Anim Ind* **1934**, *2*, 375–391.

102. Erasmus, B.J. Cultivation of horsesickness virus in tissue culture. *Nature* **1963**, *200*, 716. [CrossRef] [PubMed]

103. Von Teichman, B.F.; Smit, T.K. Evaluation of the pathogenicity of African Horsesickness (AHS) isolates in vaccinated animals. *Vaccine* **2008**, *26*, 5014–5021. [CrossRef] [PubMed]

104. Von Teichman, B.F.; Dungu, B.; Smit, T.K. In vivo cross-protection to African horse sickness Serotypes 5 and 9 after vaccination with Serotypes 8 and 6. *Vaccine* **2010**, *28*, 6505–6517. [CrossRef] [PubMed]

105. Molini, U.; Marucchella, G.; Maseke, A.; Ronchi, G.F.; Di Ventura, M.; Salini, R.; Scacchia, M.; Pini, A. Immunization of horses with a polyvalent live-attenuated African horse sickness vaccine: Serological response and disease occurrence under field conditions. *Trials Vaccinol.* **2015**, *4*, 24–28. [CrossRef]

106. Weyer, C.T.; Grewar, J.D.; Burger, P.; Joone, C.; Lourens, C.; MacLachlan, N.J.; Guthrie, A.J. Dynamics of African horse sickness virus nucleic acid and antibody in horses following immunization with a commercial polyvalent live attenuated vaccine. *Vaccine* **2017**, *35*, 2504–2510. [CrossRef] [PubMed]

107. Mirchamsy, H.; Taslimi, H. Inactivated African horse sickness virus cell culture vaccine. *Immunology* **1968**, *14*, 81. [PubMed]

108. Weyer, C.T.; Quan, M.; Joone, C.; Lourens, C.W.; MacLachlan, N.J.; Guthrie, A.J. African horse sickness in naturally infected, immunised horses. *Equine Vet. J.* **2013**, *45*, 117–119. [CrossRef]

109. Lelli, R.; Molini, U.; Ronchi, G.F.; Rossi, E.; Franchi, P.; Ulisse, S.; Armillotta, G.; Capista, S.; Khaiseb, S.; Di Ventura, M. Inactivated and adjuvanted vaccine for the control of the African horse sickness virus serotype 9 infection: Evaluation of efficacy in horses and guinea-pig model. *Vet. Ital.* **2013**, *49*, 89–98.

110. Erasmus, B.J. Preliminary observations on the value of the guinea-pig in determining the innocuity and antigenicity of neurotropic attenuated horsesickness strains. *Onderstepoort J. Vet. Res.* **1963**, *30*, 11–22.

111. Romito, M.; Du Plessis, D.H.; Viljoen, G.J. Immune responses in a horse inoculated with the VP2 gene of African horsesickness virus. *Onderstepoort J. Vet. Res.* **1999**, *66*, 139–144.

112. Roy, P.; Bishop, D.H.; Howard, S.; Aitchison, H.; Erasmus, B. Recombinant baculovirus-synthesized African horsesickness virus (AHSV) outer-capsid protein VP2 provides protection against virulent AHSV challenge. *J. Gen. Virol.* **1996**, *77*, 2053–2057. [CrossRef] [PubMed]

113. Martinez-Torrecuadrada, J.L.; Diaz-Laviada, M.; Roy, P.; Sanchez, C.; Vela, C.; Sanchez-Vizcaino, J.M.; Casal, J.I. Full protection against African horsesickness (AHS) in horses induced by baculovirus-derived AHS virus serotype 4 VP2, VP5 and VP7. *J. Gen. Virol.* **1996**, *77*, 1211–1221. [CrossRef] [PubMed]

114. Kanai, Y.; van Rijn, P.A.; Maris-Veldhuis, M.; Kaname, Y.; Athmaram, T.N.; Roy, P. Immunogenicity of recombinant VP2 proteins of all nine serotypes of African horse sickness virus. *Vaccine* **2014**, *32*, 4932–4937. [CrossRef] [PubMed]

115. Du Plessis, M.; Cloete, M.; Aitchison, H.; Van Dijk, A.A. Protein aggregation complicates the development of baculovirus-expressed African horsesickness virus serotype 5 VP2 subunit vaccines. *Onderstepoort J. Vet. Res.* **1998**, *65*, 321–329. [PubMed]

116. Scanlen, M.; Paweska, J.T.; Verschoor, J.A.; van Dijk, A.A. The protective efficacy of a recombinant VP2-based African horsesickness subunit vaccine candidate is determined by adjuvant. *Vaccine* **2002**, *20*, 1079–1088. [CrossRef]

117. Guthrie, A.J.; Quan, M.; Lourens, C.W.; Audonnet, J.C.; Minke, J.M.; Yao, J.; He, L.; Nordgren, R.; Gardner, I.A.; Maclachlan, N.J. Protective immunization of horses with a recombinant canarypox virus vectored vaccine co-expressing genes encoding the outer capsid proteins of African horse sickness virus. *Vaccine* **2009**, *27*, 4434–4438. [CrossRef] [PubMed]

118. Calvo-Pinilla, E.; Casanova, I.; Bachanek-Bankowska, K.; Chiam, R.; Maan, S.; Nieto, J.M.; Ortego, J.; Mertens, P.P.C. A modified vaccinia Ankara virus (MVA) vaccine expressing African horse sickness virus (AHSV) VP2 protects against AHSV challenge in an IFNAR -/- mouse model. *PLoS ONE* **2011**, *6*, e16503.

119. Manning, N.M.; Bachanek-Bankowska, K.; Mertens, P.P.C.; Castillo-Olivares, J. Vaccination with recombinant Modified Vaccinia Ankara (MVA) viruses expressing single African horse sickness virus VP2 antigens induced cross-reactive virus neutralising antibodies (VNAb) in horses when administered in combination. *Vaccine* **2017**, *35*, 6024–6029. [CrossRef]

120. Alberca, B.; Bachanek-Bankowska, K.; Cabana, M.; Calvo-Pinilla, E.; Viaplana, E.; Frost, L.; Gubbins, S.; Urniza, A.; Mertens, P.; Castillo-Olivares, J. Vaccination of horses with a recombinant modified vaccinia Ankara virus (MVA) expressing African horse sickness (AHS) virus major capsid protein VP2 provides complete clinical protection against challenge. *Vaccine* **2014**, *32*, 3670–3674. [CrossRef]

121. Chiam, R.; Sharp, E.; Maan, S.; Rao, S.; Mertens, P.; Blacklaws, B.; Davis-Poynter, N.; Wood, J.; Castillo-Olivares, J. Induction of antibody responses to African horse sickness virus (AHSV) in ponies after vaccination with recombinant modified vaccinia Ankara (MVA). *PLoS ONE* **2009**, *4*, e5997. [CrossRef]

122. Calvo-Pinilla, E.; Gubbins, S.; Mertens, P.; Ortego, J.; Castillo-Olivares, J. The immunogenicity of recombinant vaccines based on modified Vaccinia Ankara (MVA) viruses expressing African horse sickness virus VP2 antigens depends on the levels of expressed VP2 protein delivered to the host. *Antivir. Res.* **2018**, *154*, 132–139. [CrossRef] [PubMed]

123. Calvo-Pinilla, E.; de la Poza, F.; Gubbins, S.; Mertens, P.P.; Ortego, J.; Castillo-Olivares, J. Vaccination of mice with a modified Vaccinia Ankara (MVA) virus expressing the African horse sickness virus (AHSV) capsid protein VP2 induces virus neutralising antibodies that confer protection against AHSV upon passive immunisation. *Virus Res.* **2014**, *180*, 23–30. [CrossRef] [PubMed]

124. Gilbert, S.C. Clinical development of Modified Vaccinia virus Ankara vaccines. *Vaccine* **2013**, *31*, 4241–4246. [CrossRef] [PubMed]

125. Cottingham, M.G.; Carroll, M.W. Recombinant MVA vaccines: Dispelling the myths. *Vaccine* **2013**, *31*, 4247–4251. [CrossRef] [PubMed]

126. Calvo-Pinilla, E.; de la Poza, F.; Gubbins, S.; Mertens, P.P.; Ortego, J.; Castillo-Olivares, J. Antiserum from mice vaccinated with modified vaccinia Ankara virus expressing African horse sickness virus (AHSV) VP2 provides protection when it is administered 48h before, or 48h after challenge. *Antivir. Res.* **2015**, *116*, 27–33. [CrossRef] [PubMed]

127. Lulla, V.; Lulla, A.; Wernike, K.; Aebischer, A.; Beer, M.; Roy, P. Assembly of Replication-Incompetent African Horse Sickness Virus Particles: Rational Design of Vaccines for All Serotypes. *J. Virol.* **2016**, *90*, 7405–7414. [CrossRef] [PubMed]

128. Vermaak, E.; Paterson, D.J.; Conradie, A.; Theron, J. Directed genetic modification of African horse sickness virus by reverse genetics. *S. Afr. J. Sci.* **2015**, *111*, 1–8. [CrossRef]

129. Van de Water, S.G.; van Gennip, R.G.; Potgieter, C.A.; Wright, I.M.; van Rijn, P.A. VP2 Exchange and NS3/NS3a Deletion in African Horse Sickness Virus (AHSV) in Development of Disabled Infectious Single Animal Vaccine Candidates for AHSV. *J. Virol.* **2015**, *89*, 8764–8772. [CrossRef]

130. Kaname, Y.; Celma, C.C.; Kanai, Y.; Roy, P. Recovery of African horse sickness virus from synthetic RNA. *J. Gen. Virol.* **2013**, *94*, 2259–2265. [CrossRef]

131. Lulla, V.; Losada, A.; Lecollinet, S.; Kerviel, A.; Lilin, T.; Sailleau, C.; Beck, C.; Zientara, S.; Roy, P. Protective efficacy of multivalent replication-abortive vaccine strains in horses against African horse sickness virus challenge. *Vaccine* **2017**, *35*, 4262–4269. [CrossRef]

132. Van Rijn, P.A.; Maris-Veldhuis, M.A.; Potgieter, C.A.; van Gennip, R.G.P. African horse sickness virus (AHSV) with a deletion of 77 amino acids in NS3/NS3a protein is not virulent and a safe promising AHS Disabled Infectious Single Animal (DISA) vaccine platform. *Vaccine* **2018**, *36*, 1925–1933. [CrossRef] [PubMed]

133. Boyce, M.; Celma, C.C.P.; Roy, P. Development of reverse genetics systems for bluetongue virus: Recovery of infectious virus from synthetic RNA transcripts. *J. Virol.* **2008**, *82*, 8339–8348. [CrossRef] [PubMed]

134. Van Rijn, P.A.; van de Water, S.G.; Feenstra, F.; van Gennip, R.G. Requirements and comparative analysis of reverse genetics for bluetongue virus (BTV) and African horse sickness virus (AHSV). *Virol. J.* **2016**, *13*, 119. [CrossRef] [PubMed]

135. Feenstra, F.; Pap, J.S.; van Rijn, P.A. Application of bluetongue Disabled Infectious Single Animal (DISA) vaccine for different serotypes by VP2 exchange or incorporation of chimeric VP2. *Vaccine* **2015**, *33*, 812–818. [CrossRef] [PubMed]

136. Conradie, A.M.; Stassen, L.; Huismans, H.; Potgieter, C.A.; Theron, J. Establishment of different plasmid only-based reverse genetics systems for the recovery of African horse sickness virus. *Virology* **2016**, *499*, 144–155. [CrossRef] [PubMed]

137. Matsuo, E.; Celma, C.C.; Roy, P. A reverse genetics system of African horse sickness virus reveals existence of primary replication. *FEBS Lett.* **2010**, *584*, 3386–3391. [CrossRef] [PubMed]

138. Celma, C.C.; Stewart, M.; Wernike, K.; Eschbaumer, M.; Gonzalez-Molleda, L.; Breard, E.; Schulz, C.; Hoffmann, B.; Haegeman, A.; De Clercq, K. Replication-deficient particles: New insights into the next generation of bluetongue virus vaccines. *J. Virol.* **2016**, *91*, e01892-16. [CrossRef]

139. Noad, R.; Roy, P. Virus-like particles as immunogens. *Trends Microbiol.* **2003**, *11*, 438–444. [CrossRef]

140. Bachmann, M.F.; Rohrer, U.H.; Kundig, T.M.; Burki, K.; Hengartner, H.; Zinkernagel, R.M. The influence of antigen organization on B cell responsiveness. *Science* **1993**, *262*, 1448. [CrossRef]

141. Lechner, F.; Jegerlehner, A.; Tissot, A.C.; Maurer, P.; Sebbel, P.; Renner, W.A.; Jennings, G.T.; Bachmann, M.F. Virus-Like Particles as a Modular System for Novel Vaccines. *Intervirology* **2002**, *45*, 212–217. [CrossRef]

142. Lua, L.H.L.; Connors, N.K.; Sainsbury, F.; Chuan, Y.P.; Wibowo, N.; Middelberg, A.P.J. Bioengineering virus-like particles as vaccines. *Biotechnol. Bioeng.* **2014**, *111*, 425–440. [CrossRef] [PubMed]

143. Grgacic, E.V.; Anderson, D.A. Virus-like particles: Passport to immune recognition. *Methods* **2006**, *40*, 60–65. [CrossRef] [PubMed]

144. Pattenden, L.K.; Middelberg, A.P.J.; Niebert, M.; Lipin, D.I. Towards the preparative and large-scale precision manufacture of virus-like particles. *Trends Biotechnol.* **2005**, *23*, 523–529. [CrossRef] [PubMed]

145. Roy, P.; French, T.; Erasmus, B. Protective efficacy of virus-like particles for bluetongue disease. *Vaccine* **1992**, *10*, 28–32. [CrossRef]

146. Roy, P.; Bishop, D.H.L.; LeBlois, H.; Erasmus, B.J. Long-lasting protection of sheep against bluetongue challenge after vaccination with virus-like particles: Evidence for homologous and partial heterologous protection. *Vaccine* **1994**, *12*, 805–811. [CrossRef]

147. Stewart, M.; Bhatia, Y.; Athmaran, T.N.; Noad, R.; Gastaldi, C.; Dubois, E.; Russo, P.; Thiéry, R.; Sailleau, C.; Bréard, E.; et al. Validation of a novel approach for the rapid production of immunogenic virus-like particles for bluetongue virus. *Vaccine* **2010**, *28*, 3047–3054. [CrossRef]

148. Lomonossoff, G.P.; D'Aoust, M.-A. Plant-produced biopharmaceuticals: A case of technical developments driving clinical deployment. *Science* **2016**, *353*, 1237–1240. [CrossRef]

149. Rybicki, E.P. Plant-made vaccines for humans and animals. *Plant. Biotechnol. J.* **2010**, *8*, 620–637. [CrossRef]

150. Fischer, R.; Schillberg, S.; Buyel, J.F.; Twyman, R.M. Commercial aspects of pharmaceutical protein production in plants. *Curr. Pharm. Des.* **2013**, *19*, 5471–5477. [CrossRef]

151. Ma, J.K.C.; Drake, P.M.W.; Christou, P. Genetic modification: The production of recombinant pharmaceutical proteins in plants. *Nat. Rev. Genet.* **2003**, *4*, 794. [CrossRef]

152. Rybicki, E. History and Promise of Plant-Made Vaccines for Animals. In *Prospects of Plant-Based Vaccines in Veterinary Medicine*; MacDonald, J., Ed.; Springer International Publishing: Cham, Germany, 2018; pp. 1–22.

153. Rybicki, E.P. Plant-based vaccines against viruses. *Virol. J.* **2014**, *11*, 205. [CrossRef] [PubMed]

154. Thuenemann, E.C.; Meyers, A.E.; Verwey, J.; Rybicki, E.P.; Lomonossoff, G.P. A method for rapid production of heteromultimeric protein complexes in plants: Assembly of protective bluetongue virus-like particles. *Plant Biotechnol. J.* **2013**, *11*, 839–846. [CrossRef] [PubMed]

155. Van Zyl, A.R.; Meyers, A.E.; Rybicki, E.P. Transient Bluetongue virus serotype 8 capsid protein expression in Nicotiana benthamiana. *Biotechnol. Rep.* **2016**, *9*, 15–24. [CrossRef] [PubMed]

156. Maree, S.; Durbach, S.; Huismans, H. Intracellular production of African horsesickness virus core-like particles by expression of the two major core proteins, VP3 and VP7, in insect cells. *J. Gen. Virol.* **1998**, *79*, 333–337. [CrossRef] [PubMed]

157. Maree, S.; Maree, F.F.; Putterill, J.F.; de Beer, T.A.; Huismans, H.; Theron, J. Synthesis of empty african horse sickness virus particles. *Virus Res.* **2016**, *213*, 184–194. [CrossRef] [PubMed]

158. Dennis, S.J.; Meyers, A.E.; Guthrie, A.J.; Hitzeroth, I.I.; Rybicki, E.P. Immunogenicity of plant-produced African horse sickness virus-like particles: Implications for a novel vaccine. *Plant Biotechnol. J.* **2018**, *16*, 442–450. [CrossRef] [PubMed]

159. Dennis, S.J.; O'Kennedy, M.M.; Rutkowska, D.; Tsekoa, T.; Lourens, C.W.; Hitzeroth, I.I.; Meyers, A.E.; Rybicki, E.P. Safety and immunogenicity of plant-produced African horse sickness virus-like particles in horses. *Vet. Res.* **2018**, *49*, 105. [CrossRef]

160. MacLachlan, N.J.; Balasuriya, U.B.; Davis, N.L.; Collier, M.; Johnston, R.E.; Ferraro, G.L.; Guthrie, A.J. Experiences with new generation vaccines against equine viral arteritis, West Nile disease and African horse sickness. *Vaccine* **2007**, *25*, 5577–5582. [CrossRef]

161. House, J.A. Recommendations for African horse sickness vaccines for use in nonendemic areas. *Rev. D'élevage Et De Médecine Vétérinaire Des. Pays Trop.* **1993**, *46*, 77–81.

Review

Viral Equine Encephalitis, a Growing Threat to the Horse Population in Europe?

Sylvie Lecollinet [1,2,*], **Stéphane Pronost** [2,3,4], **Muriel Coulpier** [1], **Cécile Beck** [1,2], **Gaelle Gonzalez** [1], **Agnès Leblond** [5] **and Pierre Tritz** [2,6,7]

[1]   UMR (Unité Mixte de Recherche) 1161 Virologie, Anses (the French Agency for Food, Environmental and Occupational Health and Safety), INRAE (French National Institute of Agricultural, Food and Environmental Research), Ecole Nationale Vétérinaire d'Alfort, Université Paris-Est, 94700 Maisons-Alfort, France; muriel.coulpier@vet-alfort.fr (M.C.); cecile.beck@anses.fr (C.B.); gaelle.gonzalez@anses.fr (G.G.)

[2]   RESPE (Réseau d'épidémio-surveillance en pathologie équine), 14280 Saint-Contest, France; Stephane.pronost@laboratoire-labeo.fr (S.P.); pitritz@wanadoo.fr (P.T.)

[3]   LABÉO, 14280 Saint-Contest, France

[4]   BIOTARGEN, UNICAEN, NORMANDIE UNIV, 14000 Caen, France

[5]   UMR EPIA (Epidémiologie des Maladies Animales et Zoonotiques), INRAE, VetAgro Sup, Université de Lyon, 69280 Marcy L'Etoile, France; agnes.leblond@vetagro-sup.fr

[6]   Clinique Vétérinaire, 19 rue de Créhange, 57380 Faulquemont, France

[7]   AVEF (Association Vétérinaire Equine Française), Committee on Infectious Diseases, 75011 Paris, France

*   Correspondence: sylvie.lecollinet@anses.fr; Tel.: +33-143967111

Received: 1 October 2019; Accepted: 17 December 2019; Published: 24 December 2019

Abstract: Neurological disorders represent an important sanitary and economic threat for the equine industry worldwide. Among nervous diseases, viral encephalitis is of growing concern, due to the emergence of arboviruses and to the high contagiosity of herpesvirus-infected horses. The nature, severity and duration of the clinical signs could be different depending on the etiological agent and its virulence. However, definite diagnosis generally requires the implementation of combinations of direct and/or indirect screening assays in specialized laboratories. The equine practitioner, involved in a mission of prevention and surveillance, plays an important role in the clinical diagnosis of viral encephalitis. The general management of the horse is essentially supportive, focused on controlling pain and inflammation within the central nervous system, preventing injuries and providing supportive care. Despite its high medical relevance and economic impact in the equine industry, vaccines are not always available and there is no specific antiviral therapy. In this review, the major virological, clinical and epidemiological features of the main neuropathogenic viruses inducing encephalitis in equids in Europe, including rabies virus (*Rhabdoviridae*), Equid herpesviruses (*Herpesviridae*), Borna disease virus (*Bornaviridae*) and West Nile virus (*Flaviviridae*), as well as exotic viruses, will be presented.

Keywords: encephalitis; arbovirus; rabies; Equid herpesviruses; Borna disease virus; West Nile virus; horses

---

1. Introduction

Neurological disorders represent an important sanitary and economic threat to the equine industry worldwide. Even mild nervous deficits can result in poor performances and long recovery of athletic horses, while severe clinical signs can induce life-threatening injuries in infected horses and may expose owners, veterinarians and care providers to significant risks [1]. Few surveys have been carried out to evaluate the burden of neurological diseases in horses and were performed almost 20 years ago. They indicated that neurological affections accounted as the fifth cause of death (8%) in adult horses, behind foaling (24%), digestive (21%), locomotor (21%), and cardiovascular (9%) causes. In two studies performed in Australia and in France, neurological diseases were first attributed to trauma (26% to 34%), congenital malformations (19% to 20%), while inflammation and infection were reported in 6% to 17% of horses with neurological conditions [2–4]. Early recognition of neurological infectious diseases may increase the chance of a positive outcome and is key to the implementation of coordinated management measures designed to prevent large-scale outbreaks when highly contagious pathogens, such as equid herpesviruses, are involved. Many neurotropic viruses affecting equines are also significant human pathogens and rapid identification of zoonotic viruses in horses is pivotal in their surveillance and in the control of corresponding human viral diseases [5].

Multiple neuropathogenic pathogens, either viruses, bacteria or protozoa, can induce an important inflammation of the central nervous system (CNS). Bacterial meningitis are common neurological infections in foals, while neuroborreliosis, listeriosis and Equine Protozoal Myeloencephalitis are rare and difficult to diagnose in equines [1]. As far as viruses are concerned, *Rabies virus*, *Equid alphaherpesvirus 1* (EHV-1), *West Nile virus* and related flaviviruses (*Japanese Encephalitis virus*, *Saint-Louis Encephalitis virus* and *Murray Valley Encephalitis virus*), *Mammalian 1 orthobornavirus* and neurotropic alphaviruses (*Eastern, Western and Venezuelan Equine Encephalitis Virus* species) are the most largely described neuropathogenic viruses (Figure 1).

Genus (Virus)	Family	Virion	Virus size	Genome organisation	Genome size
Varicellovirus (Equid alphaherpesviruses)	*Herpesviridae*	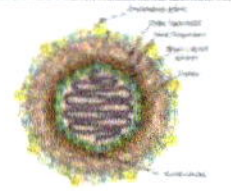	150-200 nm		DS DNA 120-240 kb 150 kb for EHV-1
Lyssavirus (rabies virus)	*Rhabdoviridae*	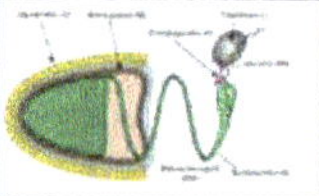	180 nm long 75 nm wide		SS RNA(-) 11 kb
Flavivirus (WNV, TBEV, LIV, JEV, Saint Louis and Murray Valley encephalitis viruses)	*Flaviviridae*	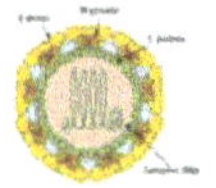	50 nm	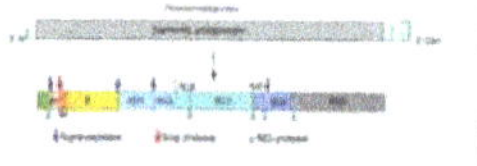	SS RNA(+) 9.7 - 12 kb
Orthobornavirus (mammalian 1 orthobornavirus)	*Bornaviridae*	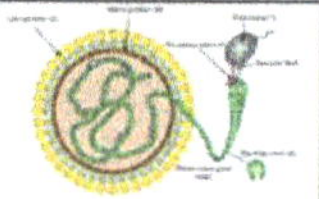	70-130 nm		SS RNA(+) 8.9 kb
Alphavirus (EEEV, VEEV, WEEV)	*Togaviridae*	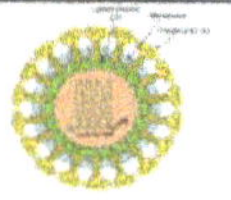	65-70 nm	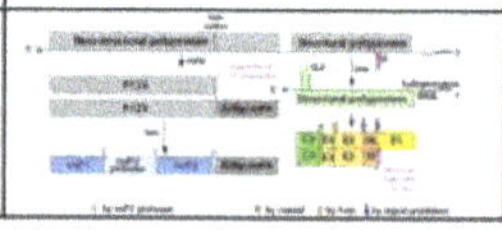	SS RNA(+) 11 - 12 kb

Figure 1. Major viruses causing encephalitis in equines. Virus classification according to ICTV 2019 nomenclature [6], structure and genome organisation are presented for viruses belonging to *Herpesviridae*, *Rhabdoviridae*, *Flaviviridae*, *Bornaviridae* and *Togaviridae* (adapted from ViralZone [7]). WNV: West Nile virus; TBEV: Tick-Borne encephalitis virus; LIV: Louping ill virus; JEV: Japanese encephalitis virus; EEEV: Eastern equine encephalitis virus; VEEV: Venezuelan equine encephalitis virus; WEEV: Western equine encephalitis virus, DS: double-stranded, SS: single-stranded.

Equine neuropathogenic viruses generally induce encephalitis or myeloencephalitis, which is an inflammation of the central nervous system (cortex, brain stem, and cerebellum) and/or of the spinal cord characterized by large or multifocal infiltrations of mononuclear cells (Figure 2d). Infected animals may experience behavioral change, as well as balance, posture and gait deficits (Figure 2a–c) [1]. Neurological examination, including testing of reflexes, evaluating postures and movements, is key in the clinical approach and allows assessing the course of disease and therefore its prognosis and response to therapeutic options. However, it is worth to note that such neurological examination and scoring is difficult to standardize—even among highly specialized practitioners [8]. In addition to posture and gait disorders, hyperthermia and sudden clinical signs peaking after 48 h of infection, guide viral encephalitis diagnosis. Usually, high fever is considered as a warming sign even if infectious diseases are not the only reason of hyperthermia and if they are not systematically detected during the horse clinical examination. Indeed, 14% to 38% of West Nile disease diagnosed in Europe and 52% of equid herpesvirus myeloencephalopathy (EHM) cases evidenced in France had hyperthermia during veterinary examination [9–11]. Cerebrospinal fluid (CSF) findings will generally be informative of a viral meningo-encephalitis, comprising an increased protein concentration, normal glucose concentration and pleomorphic leucocytosis with predominating mononuclear cells or neutrophils [12].

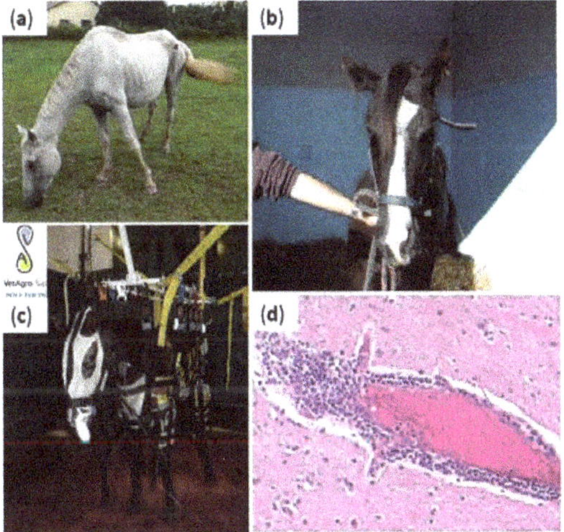

Figure 2. Clinical manifestations and lesions in viral equine encephalitis. Horses infected with equine encephalitis viruses may experience posture deficits (increasing of the lift polygon in (**a**)), cranial nerve deficits (facial paralysis in (**b**)), balance deficiencies (slings in (**c**) can be used to support paretic horses and avoid long and poor prognosis recumbency). Brain lesions are non-specific and include perivascular infiltration of inflammatory cells, observed in (**d**). Credits: Pr Agnès Leblond, VetAgroSup, and Dr Eve Laloy, French Veterinary School of Alfort.

Epidemiological parameters including knowledge on the holding conditions, horse condition and nutritional needs, prevalent pathogens in a specific region (Figure 3) will help to prioritize the hypothesis. Updated epidemiological data are strongly needed, such as the ones provided by national surveillance systems such as RESPE (Réseau d'épidémio-surveillance en pathologie équine) in France [13] or EQUINELLA in Switzerland [14]) In Europe, many viruses, including two zoonotic viruses, should be recognized promptly: rabies virus is one of the most important global zoonotic pathogen and has been eradicated from Western Europe by successful vaccination campaigns, while West Nile virus is a (re)emerging arthropod-borne virus that has recently spread in Europe to the Balkans area and northern-most countries (Germany) [15–17]. From an economic perspective, Equid herpesviruses (EHV-1 in particular) are one of the most important equine pathogens in Europe [18]. In France, EHV-1 was the principal cause of neurological infections from 2008 to 2011, with 26 cases of EHM over 214 neurological cases reported (12%) [10].

(a)

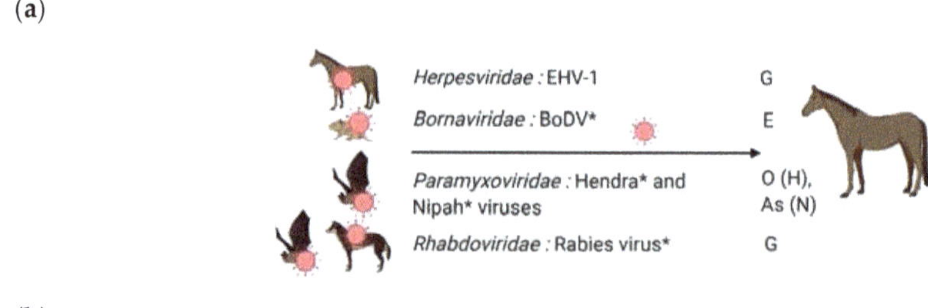

(b)

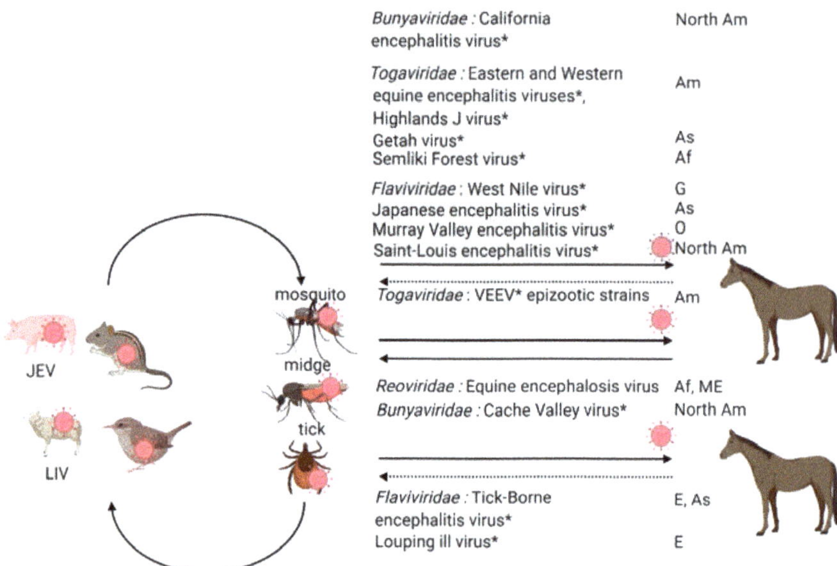

Figure 3. Encephalitis viruses in equines. Transmission mode (direct transmission in (**a**) or arthropod-borne transmission in (**b**)), zoonotic potential (zoonotic viruses are marked with an asterisk) and geographical distribution (Af for Africa, Am for America, As for Asia, E for Europe, ME for Middle East, O for Oceania and G for global) are presented. EHV-1: Equid alphaherpesvirus 1; BoDV: Borna disease virus; VEEV: Venezuelan equine encephalitis virus. Black arrows represent established virus transmission between the two partners. Doted arrows indicate limited virus transmission possibility from the infected horse to its reservoir, with the exception of midge-borne arboviruses and of the mosquito borne VEEV epizootic variants for which horses serve as amplifier hosts. The figure was prepared on BioRender [19].

Most equine neuropathogenic viruses will induce similar clinical presentations and laboratory testing, either by indirect (ELISA, seroneutralisation or other serological assays) or direct assays (PCR, virus isolation, staining of viral antigens from infected tissues) must be provided in order to confirm the etiology of the disease (Table 1). For the past few years, direct methodologies, especially for EHV-1 detection, have been promoted. However, for most arthropod-borne viruses as well as for Borna disease virus, indirect assays have sustained interest owing to low and short-lasting viremia in infected horses. Serological screenings are also widely used in epidemiological surveys for the surveillance of equine encephalitis worldwide. Since horses can be vaccinated against most equine neuropathogenic viruses (with the notable exceptions of Borna disease and some exotic equine encephalitis viruses) (Table 2), the immunization status must be known for the interpretation of the serological tests. In this review, we will present the major virological, clinical and epidemiological features of the main neuropathogenic viruses inducing encephalitis in equids in Europe, namely rabies virus (*Rhabdoviridae*), equid herpesviruses (*Herpesviridae*), Borna disease virus (*Bornaviridae*) and West Nile virus (*Flaviviridae*), as well as exotic viruses. The most relevant information on the diagnosis and prevention from equine encephalitis viruses enzootic in Europe will also be presented in this review.

Table 1. List of diagnostic assays available against equine neuropathogenic viruses enzootic in Europe. PCR = polymerase chain reaction; RT = reverse transcription; VNT = virus neutralization tests; CFT = complement fixation test; DFA = direct fluorescence assay; dRIT = direct rapid immunohistochemistry test; IHC = immunohistochemistry; IFA = indirect fluorescence assay; HIA = hemagglutination inhibition assay; MIA = multiplex immunoassay.

Virus	Diagnostic Assays	Advantages and Shortcomings
EHV-1	Direct assays: PCR, virus isolation	Direct virus detection and typing (SNP-PCR) is possible from easily accessible samples (nasal swabs and blood).
	Serology: VNT, CFT or ELISA	Due to highly prevalent and lifelong infection, diagnostic assays should be interpreted with care. Serology will be informative if serial serum samples can be obtained.
Rabies virus	Direct assays: DFA, dRIT, RT-PCR	Direct virus detection is possible only from the brain of dead animals.
BoDV	Direct assays: RT-PCR, IHC	Due to limited antibody response induced after BoDV infection, definitive diagnostic will be made only after direct virus detection from the brain of dead animals.
WNV/Flaviviruses	Indirect assays preferred: ELISA, IFA, HIA, VNT	Rapid serological screening tests (competition ELISA, IFA) are very sensitive but present a low diagnostic specificity; they should be interpreted with care and confronted with results from confirmatory serological assays (VNT, MIA).
	Direct assays: RT-PCR, virus isolation	Direct virus detection is possible from the brain of dead animals and when positive, indicates recent virus infection.

Table 2. List of vaccines licensed in Europe against equine neuropathogenic viruses. Vaccine types and recommended vaccination protocols are presented.

Virus	Vaccine Types Available in Europe	Protection Provided
EHV-1	Inactivated: BIOEQUIN® H (BIOVETA), PNEUMEQUINE® (Boehringer Ingelheim), EQUIP® EHV 1,4 (Zoetis) Live attenuated: PREVACCINOL® (MSD Animal Health), licensed in Germany	Insufficient individual protection against EHM but allows for decreased virus transmission in the vaccinated population, after 2 primes at a 1-month interval (3–4 months with the live attenuated vaccine) and boosts every 6 to 12 months.
Rabies virus	Inactivated: ENDURACELL® R MONO and VERSIGUARD® Rabies (Zoetis), NOBIVAC® Rabies (MSD), RABIGEN® mono (Virbac) and RABISIN® (Boehringer Ingelheim)	Good protection at the individual level provided after a unique prime and boosts performed every year or every 2 years.
WNV	Inactivated: EQUIP® WNV (Zoetis) Recombinant: EQUILIS® West Nile (MSD) and PROTEQ® West Nile (Boehringer Ingelheim)	Good protection at the individual level provided after 2 primes at a 1-month interval and boosts performed every year.

2. Equine Encephalitis Viruses Enzootic in Europe

Equine viruses causing encephalitis can be classified into two groups: viruses transmitted indirectly through the bites of an infected arthropod (mosquitoes, ticks, or midges) or by direct transmission (Figure 3). Direct transmission viruses are usually associated with highly contagious horse-horse contacts, secretions or excretions (with the prime example of EHV-1 infected horses) (Figure 3a) compared to arthropod-vector transmitted neuropathogenic viruses; most equines infected by neurotropic arboviruses are indeed considered as dead-end hosts, with the remarkable exception of the epizootic strains of Venezuelan equine encephalitis virus, an arbovirus only identified in America (Figure 3b) [20,21].

In this section, we will address several viral equine encephalitis inducing the most important clinical and economic consequences in Europe: rabies virus (*Rhabdoviridae*), Equid herpesviruses (*Herpesviridae*), Borna disease virus (*Bornaviridae*) and neurotropic flaviviruses transmitted by mosquitoes and ticks, e.g., West Nile, tick-borne encephalitis and Louping ill viruses (*Flaviviridae*).

2.1. Equid Herpesviruses

Highly successful pathogens of horse populations worldwide, Equid herpesviruses induce latent infections that may cause abortions, respiratory (rhinopneumonia), and neurological diseases (myeloencephalopathy). Nine herpesviruses have been described in the family Equidae, which includes horses, ponies, donkeys and zebras and two natural hosts have been identified key for EHV epidemiology: horses for equid herpesviruses 1 (EHV-1), EHV-2, EHV-3, EHV-4, and EHV-5 and donkeys for EHV-6, EHV-7, and EHV-8 [22]. Concerning EHV-9, zebras, as well as the African rhinoceros, could serve as virus reservoirs [23]. Equid alphaherpesvirus 8 (EHV-8), formerly known as asinine herpesvirus 3, was recently considered as a new threat to the horse industry; it was shown to cause abortion in horses in Ireland and was associated with one neurological case in a donkey in China [24].

Equid alphaherpesvirus 1 (EHV-1) was recognized as a neuropathogenic virus in horses in 1966 [25]. Within the last 20 years, EHM has been considered as an uncommon sequela of EHV-1 infection in horses that led the USDA to classify EHM as a re-emerging disease. If different cases of EHM caused by EHV-4 were strongly suspected, no cases have been reported in the literature up to now. Although EHV-1 and EHV-4 share a high degree of genetic and antigenic similarities, differential virus tropism (the former virus being more endotheliotropic than EHV-4) and ability to interfere with the innate immune response could explain the differences in host range and pathogenicity between EHV-1 and EHV-4, as suggested by Ma et al. [26]. Nevertheless, similar to sporadic cases of EHV-4 abortion, it cannot be ruled out that neurological cases may occur with EHV-4.

Virus: EHV-1 and 4 are members of the *Varicellovirus* genus, in the *Alphaherpesvirinae* sub-family in the family *Herpesviridae* (Figure 1). These alphaherpesviruses are characterized by lytic infection and can establish a lifelong latent infection in blood circulating and lymph node-residing lymphocytes, as well as in sensory neurons within the trigeminal ganglia, which may reactivate upon stress [27,28]. The linear double-stranded DNA genome of EHV-1 contains 80 open reading frames and is 150 kb long and consists of a unique long (UL) and unique short region (US) [29].

Transmission and epidemiology: Virus transmission occurs through direct contact between horses, fomites, infectious aerosols and/or indirectly by humans. A recent study demonstrating the survival of EHV-1 in water strongly suggests a potential new way of transmission [30]. EHV-1 causes frequent outbreaks of abortion and myeloencephalopathy worldwide, even in vaccinated horses (Figure 4). EHV-1 outbreaks have been reported for centuries and many cases are reported across Europe, in France, Great Britain and Belgium, in the United States, in New-Zealand, Australia, Chile, Argentina, Israel and United Arab Emirates [31–33]. EHM incidence has increased in most parts of the world, in Europe and North America, as well as in Oceania, Africa, and Asia [10,11,28].

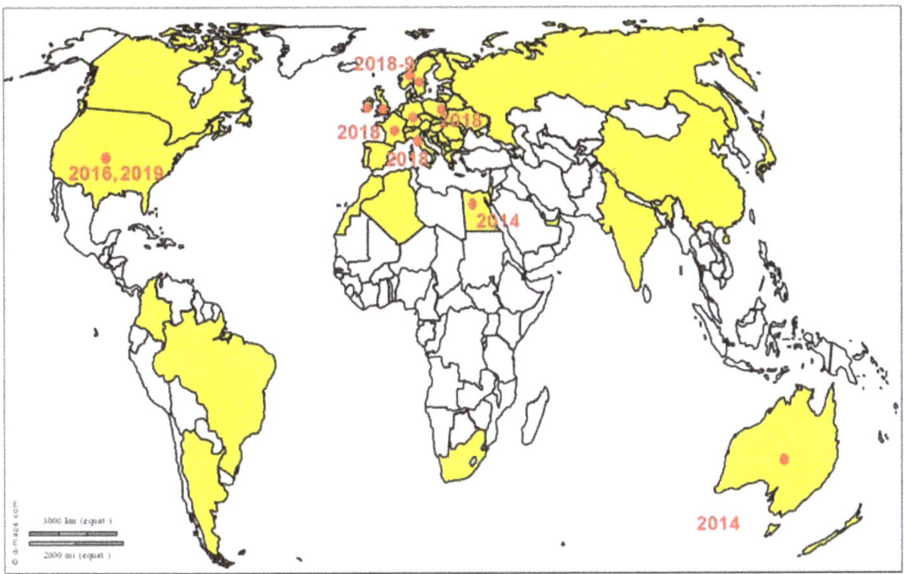

Figure 4. Distribution of EHV-1 outbreaks. Recent outbreaks in horses reported to the OIE WAHIS (World Animal Health Information System) interface [15], Promed Alerts (http://www.promedmail.org) [16] or in the scientific literature are depicted by dots (2014–2019).

Pathogenesis and clinical disease: After primary replication in the respiratory tract, EHV-1 disseminates via cell-associated viremia in peripheral blood mononuclear cells and subsequently infects the endothelial cells of the pregnant uterus or central nervous system, leading in some cases to abortion and/or neurological disorders [34]. The incubation period of the disease is 6–8 days (before neurological signs become apparent), both in experimentally and in naturally EHV-1-infected horses [35]. Histologically, the most frequently observed lesion of EHM is vasculitis in the brain and/or spinal cord, leading to brain damages by hypoxia [36,37]. Neurological signs of EHM range from temporary ataxia, paresis, loss of sensation around the tail and perineal area and urinary incontinence to complete paralysis and death [38]. Affected horses may recover completely, while recumbency often leads to a fatal outcome [39]. The increased interest of researchers in the manifestations of this disease is not only due to the lack of current scientific understanding but also to the associated economic impact [40,41]. Infections that cause severe neurological dysfunction may only involve either one or two horses [42,43] or be associated with larger outbreaks [42,44]. Neurological syndromes have also been observed in various environments open to horses: breeding farms, riding schools, racetracks and, more recently—veterinary hospitals. Furthermore, while several breeds and age groups seem to be at a lower risk of developing EHM [42,45], other factors, including EHV-1 strains, host immunity and still unknown parameters, could explain why experimental EHV-1 infections were frequently partially successful [46,47].

Interestingly, Nuggent et al. and Allen et al. showed in 2006 that a single point mutation of adenine to guanine at nucleotide position 2254 in the catalytic subunit of the gene encoding DNA polymerase (ORF30) was often associated with EHV-1 neuropathogenicity (in 83% to 86% of cases), while absent in the majority of EHV-1 abortion outbreaks [18,19]. The non synonymous A to G substitution at nucleotide position 2254 results in the replacement of asparagine (N) in position 752 (N752→D752). Experimental infections with recombinant viruses performed by Goodman et al. (2007) demonstrated that the N752 sequence variant of EHV-1 DNA Pol, when compared to the D752 variant, generated a low level of viremia in natural hosts and presented with reduced overall pathogenicity and capacity to induce neurological signs [46]. The discovery of this single polymorphism in ORF30 led to the development of a SNP-PCR (SNP—single nucleotide polymorphism) test for the detection of the two genotypes (potential neuropathogenic and non-neuropathogenic strains) [50]. Many studies performed in the field in different countries to characterize the neuropathogenic and non-neuropathogenic variants of Equid alphaherpesvirus 1, demonstrated neurological cases with A2254 variants [43,51]. This finding suggested that the current dogma that a significant percentage of EHM outbreaks are caused by a mutant strain (G2254) is too overly simplistic [52].

Diagnosis: Over the past 15 years, diagnosis tools have been improved (Table 1). PCR that allows for the direct detection of the virus has become the new standard [41]. The use of this powerful diagnostic assay has to nevertheless be considered with all the clinical information available by the practitioner, in particular the time of completion of the sampling (nasal swab and blood). Indeed, the viral load observed during EHM is generally much lower than the one observed during abortions. Given that EHV-1 latently infects leukocytes, PCR results obtained from blood samples should be interpreted with care. It is important for samples taken a few days after the observation of neurological clinical signs, not to exclude the possibility of an infection with EHV-1—even when a negative PCR result was obtained. Monitoring contact horses with PCR tests performed on nasopharyngeal swabs is then recommended. The discriminatory test between the neuropathogenic and non-neuropathogenic strains is also used in many laboratories (SNP PCR A/G2254), but regardless of the results of the SNP PCR, the practitioners will have to apply equivalent EHV-1 management measures. Virus isolation, serological testing (virus neutralization) and post-mortem examination are still informative.

Prevention and control: Practitioners need to identify quickly EHV-1 infections and to apply strict sanitary measures to stop virus spreading. Isolation and quarantine measures have to be applied according to high-risk groups. There is an urgent need to screen and separate potential virus shedders (either confirmed to be infected with EHV-1 or exposed) from non-exposed and healthy animals. Specific staff caregivers would be affected to each group and would have specific supplies (gloves, coats, calots, boots and footbath). A three weeks' quarantine starts as soon as the last reported case is declared. At the end of the infected period, an eight days' crawlspace, after carrying out cleaning and disinfection of boxes, is needed.

Several inactivated and live vaccines are available against EHV-1 and EHV-4, and both types of vaccines have been marketed in Europe (Table 2) (reviewed in [53]). Although they reduce both clinical signs of the respiratory disease and virus shedding [54], their efficacy against neurological disorders and abortion is limited [41]. In respect to these limitations, practitioners have turned toward alternative treatments by using antiviral molecules—even if no marketing authorization are available in horses. In vivo, valaciclovir, which is the prodrug of aciclovir, was tested in experimental EHV-1 infection, but showed no antiviral effects [55]. *In vitro* studies are performed around the world to identify new molecules with a strong antiviral potential [56,57].

2.2. Rabies Virus

Rabies virus is a neurotropic virus belonging to the genus *Lyssavirus*, family *Rhabdoviridae*. It is responsible for a zoonotic and inevitably fatal disease, once neurological signs have been recognized and is nevertheless considered as a neglected disease in tropical areas. Every mammal is susceptible, but some species such as dogs, jackals, coyotes, wolves, foxes, skunks, mongooses, raccoons, and bats act as reservoir hosts.

Virus: Lyssaviruses are currently classified into 17 different species: *Rabies virus* (RABV), *Lagos bat virus* (LBV), *Mokola virus* (MOKV), *Duvenhage virus* (DUVV), *European bat lyssavirus types 1* and *2* (EBLV-1 and -2), *Australian bat lyssavirus* (ABLV), *Aravan virus* (ARAV), *Khujand virus* (KHUV), *Irkut virus* (IRKV), *West Caucasian bat virus* (WCBV), *Shimoni bat virus* (SHIBV), and more recently described bat lyssaviruses (*Bokeloh bat lyssavirus* (BBLV), *Ikoma lyssavirus* (IKOV), *Gannoruwa bat lyssavirus* (GBLV) and *Lleida bat lyssavirus* (LLEBV)) [6,58–63]. Lyssaviruses are also separated into three phylogroups, based on their genetic, immunologic, and pathogenic characteristics. Phylogroup I includes RABV, DUVV, EBLV-1, EBLV-2, ABLV, ARAV, IRKV, BBLV, GBLV, and KHUV, phylogroup II includes LBV, MOKV, and SHIBV, and phylogroup III includes WCBV, IKOV, and LLEBV [59]. All lyssaviruses are capable of causing fatal acute encephalitis indistinguishable from clinical rabies in humans and other mammals. With the exception of Mokola and Ikoma lyssaviruses, every species have known bat reservoirs, leading to the speculation that lyssaviruses originated in the order Chiroptera [62]. Human clinical rabies cases have been documented for RABV, MOKV, DUVV, EBLV-1, EBLV-2, ABLV, and IRKV [63].

Lyssaviruses are enveloped, bullet-shaped viruses with a single-stranded, negative sense RNA genome of about 12 kb that encodes five viral proteins: nucleoprotein (N), phosphoprotein (P), matrix (M), glycoprotein (G), and RNA polymerase (L) (Figure 1). The RNA genome is encapsidated by the N protein, forming the ribonucleoprotein (RNP) complex, which is the functional template for transcription and replication.

Transmission and epidemiology: Rabies cases have been reported across the globe in more than 150 countries [5]. According to recent estimates by the World Health Organization, 55,000 to 60,000 human deaths due to rabies infection are expected to occur every year [64]. The majority of them occur in developing countries in Asia and Africa, with about 35,000 and 21,000 human cases, respectively, and rabies virus is usually transmitted by free roaming dogs in these areas [64,65]. Two major epidemiological cycles are reported, urban canine rabies, now largely confined to developing countries and sylvatic or wildlife rabies which predominates throughout most of Europe and North America [65]. Animal species involved in rabies virus transmission along sylvatic cycles may vary. In the United States, skunks, raccoons and bats are the wild species most often found rabid, while in Canada, these are foxes and skunks. In Europe, the red foxes and raccoon dogs serve as the main reservoir hosts and red foxes were found to account for 60% of all reported cases in central and western Europe. Bat species involved in rabies virus transmission also differ between countries: in Latin America, vampire bats are an exceptionally devastating source of infection for cattle and equids, while in North America rabid non-hematophagous bats have occasionally transmitted the disease to horses [66,67]. Horses are sensitive to both canine and bat rabies strains. Rabies is fairly rare in horses and usually less than 100 cases are reported in the United States every year: horses and mules (*Equus* spp.; 31 [6.6% of rabid animals] in 2013) [68] (Figure 5). A large number of rabies cases (172 out of 467 suspected cases) have been reported in donkeys in Sudan over a period of 10 years from 1992 to 2002 [69]. In Australia, two equine cases also arose recently in 2013, documenting the first occurrence of ABLV in animals other than bats or humans [70]. In Europe, 233 cases have been reported between 2010 and 2019, mainly in Eastern Europe (Russian Federation (49), Ukraine (41), Turkey (70), Belarus (27), Moldova (9), Romania (17), Georgia (8), Poland (2), Croatia (7), Serbia (1), and Latvia (1)), while for the same period, only one case was reported in Western Europe, in Italy (2010) [71].

Pathogenesis and clinical disease: Rabies virus is primarily transmitted to equids through the saliva of an infected animal. Contamination occurs mainly through bites or contact of a cutaneous or mucous (oral, nasal, eye mucosa) lesion with infected saliva. Rabies pathogenesis is characterized by three distinct phases. Phase 1 corresponds to the ascending or centripetal period during which the virus is transported toward the CNS. Phase 1 occurs after the bite of a rabid animal, and after a short-lasting replication in local muscle cells, the virus enters motor and sensory neurons. Paresthesia at the biting site may develop, which results in rubbing or automutilation through biting. Lyssavirus mainly shows axon-neuronal transport by binding with acetylcholine-receptors at motor end plate and multiply at the ventral horn of the spinal cord before CNS spreading [72]. The virus replicates within the CNS during phase 2, leading to clinical signs of encephalomyelitis. Phase 2 in horses is characterized by extensive virus replication in the limbic system (including hypothalamus, hippocampus, amygdala, and other nearby areas) and the spinal cord [73]. Because phase 2 is the period with the most dramatic clinical signs, most horses will be euthanized during that phase. Phase 3, also called centrifugal phase, is the period where the virus leaves the CNS and infects other organs in the body. Phase 3 is characterized by neuronal transportation of virus into highly vascularized organs, such as the salivary glands, facilitating virus transmission to new hosts and its excretion into the environment.

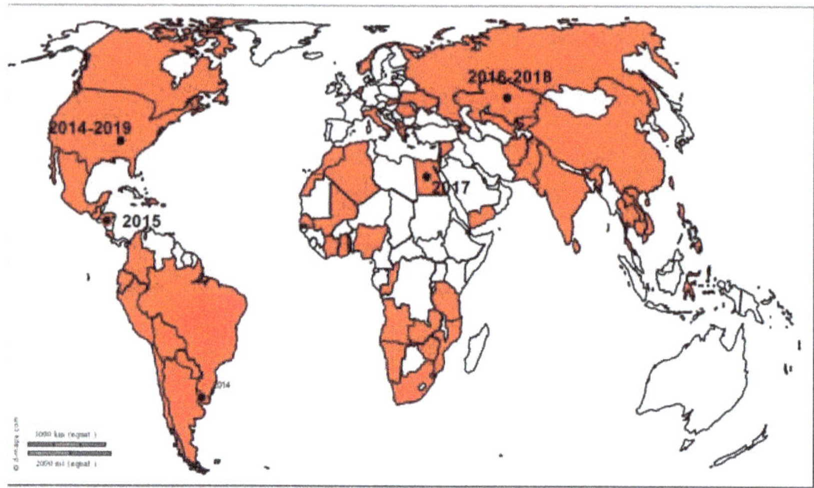

Figure 5. Distribution of rabies outbreaks. Recent outbreaks in horses reported to the OIE WAHIS interface [15], Promed Alerts [16] or in the scientific literature are depicted by dots (2014–2019).

Clinical signs of rabies are highly variable in horses and three forms are classically described according to the injured area: silent, paralytic or furious forms. The furious form is uncommon (10 to 17% of cases), while silent and paralytic forms are the most common. After a long incubation time (over 6 months), the disease progresses rapidly (within 3 to 6 days), including a sudden change in behaviour (depression to manic), itching at the biting site, loss of appetite, high fever, gait disorder, paralysis at the inoculation point, aggressivity, and hyperesthesia [74].

Diagnosis: Confirmatory diagnosis is preferably undertaken through virus identification by direct fluorescent antibody (DFA) test, direct rapid immunohistochemistry test (dRIT), or pan-lyssavirus RT-PCR assays [75]. DFA test, dRIT, and RT-PCR provide a reliable diagnosis in 98% to 100% of cases for all lyssavirus strains if an appropriate conjugate or primer/probe is used [76]. For a large number of samples, conventional and real-time PCR can provide rapid results in equipped laboratories. Histological techniques such as Seller staining (evidencing Negri bodies) are no longer recommended for diagnosis. In case of inconclusive results from primary diagnosis tests (DFA test, dRIT, or pan-Lyssavirus RT-PCR), further confirmatory tests (molecular tests, cell culture or mouse inoculation tests) on the same sample or repeated tests on additional samples are recommended. Wherever possible, virus isolation in cell culture should replace mouse inoculation tests.

Prevention and control: Rabies control has been mainly afforded through the vaccination of wild and domestic susceptible animal species [77]. Vaccination is recommended in endemic areas. Several inactivated adjuvant vaccines are commercialized in Europe and can be used in domestic mammals (see Table 2 for the list). Rabies vaccination intervals >1 year may be appropriate for previously vaccinated horses, but not in primed horses vaccinated only once [78].

2.3. Borna Disease Virus

The Borna disease virus (BDV, renamed BoDV), *Mammalian 1 orthobornavirus* according to ICTV nomenclature [6], is the prototype member of the *Bornaviridae* family, within the order *Mononegavirales* [79]. For years, it was the only member of this family, but since 2008 new bornaviruses were discovered in birds, reptiles, and mammals (reviewed in [80]). Amongst them, an avian borna virus (ABV), the Psittaciform 1 orthobornavirus, was shown to be responsible for the proventricular dilatation disease [81], and a mammal Bornavirus, the Variegated squirrel 1 bornavirus (VSBV-1), recently reappointed *Mammalian 2 orthobornavirus*, was associated with fatal encephalitis in humans [82].

This expansion of the *Bornaviridae* family and the association of the new viruses with animal and human diseases revived the interest for this family and called for new classification [80].

Virus: Bornaviruses are enveloped, non-segmented, single-stranded, negative-sense RNA viruses. Their 8.9 kb genome encodes six viral proteins, five structural (nucleoprotein N, phosphoprotein P, matrix M, surface glycoprotein G and the large structural protein L directing the replication of BoDV RNA genome) and one non-structural (X) (Figure 1). They have the particularity, amongst the Mononegavirales, to replicate within the nucleus and to be poorly released from infected cells [83].

Transmission and epidemiology: BoDV is the causative agent of the Borna disease, a rare but severe, often lethal, encephalitis, which was first described in horses during a devastating outbreak in the little city of Borna in Saxony/Germany around 1894 [84]. Later, it was reported to infect a large range of animals, including sheep, cattle, dogs, cats, shrews, ostriches, birds, macaques, and several zoo animals, suggesting that its potential host-range includes all warm-blooded animals [85–91]. BoDV infection, based on antibodies, antigen, RNA and virus detection, has been reported from horses and other animals in many countries in several continents, Europe, North America, Australia, and Asia, suggesting a worldwide distribution of the virus (reviewed in [89]). This should, however, be taken cautiously as diagnosis is not always reliable (see Diagnosis section below). Illustrating these difficulties, viral RNA detection outside central Europe, the endemic area in which the highest clinical incidence was consistently found as well as the verified classical Borna disease's cases, has been suspected to be caused by contamination [92]. Endemic area includes eastern and southern Germany, the eastern part of Switzerland and the area bordering Liechtenstein as well as the most western part of Austria (reviewed in [92,93]) and, as recently reported, upper Austria [94] (Figure 6).

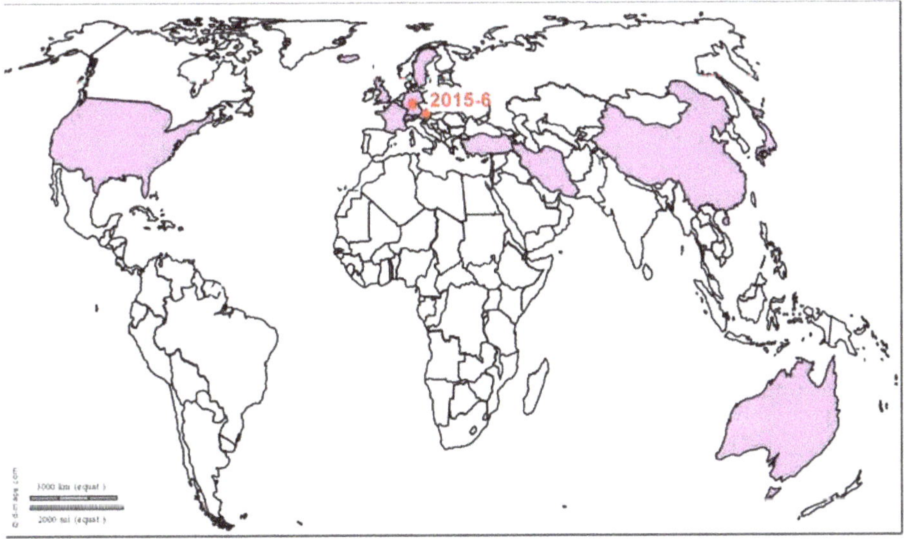

Figure 6. Distribution of Borna disease outbreaks. Recent outbreaks in horses reported to Promed Alerts [16] or in the scientific literature are depicted by dots (2014–2019).

The bicolored white-toothed shrew, *Crocidura leucodon*, is recognized as the natural reservoir host of BoDV [88,94,95]. In this host, the virus replicates in numerous tissues, neural and non-neural, without causing clinical symptoms or pathological lesions. It is secreted in saliva, urine, skin, tears, and feces [96]. Horses may be infected via the olfactory route, as the presence of BoDV antigen and RNA, as well as inflammation and edema, have been found in the olfactory bulb of naturally infected horses early in the course of the disease [97]. This is supported by successful experimental intranasal infection of rats, mice, sheep, and horses (reviewed in [93]). In horses, the virus is mostly confined into the brain and, although BoDV RNA was found in oral, nasal and conjunctival fluids of naturally infected horses, infectious virus was rarely detected [98], indicating that transmission from horse to horse in stables is unlikely. Vertical transmission may be possible as viral RNA was detected in the brain of a pregnant mare and her fetus [99] and was shown to occur experimentally in mice [100].

The question of whether BoDV is a human pathogen has been debated for years. Several studies showed it may be a cause for some psychiatric disorders [101,102], and possible responsible mechanisms have been proposed [103,104], but others have suspected that contamination occurred in initial studies [105], and have refuted any association between BoDV infection and mental illness [106]. While this remains a question, two recent studies convincingly showed that BoDV is a human pathogen, as it was associated with fatal encephalitis [107,108]. How humans have been infected remains to be elucidated but the proximity of BoDV sequences from humans, shrews and horses leads to suspect a zoonotic risk.

Pathogenesis and clinical disease: In horses, the typical course of the disease is characterized by an acute encephalitis that develops following an incubation period which lasts up to 3 months [109]. During the acute phase, neurological and neuro-behavioural signs vary but may include unusual posture (crossing legs), repetitive movement disturbance, teeth grinding, circle walking, neck stiffness, nystagmus, strabismus, myosis associated with external stimuli such as hyper excitability, aggressivity, lethargy, sleepiness and stupor. Hyperthermia, which may precede neurological signs, is not always noticed during Borna disease. During the final stage, paralysis may appear followed by seizures associated with specific movements named "push to the wall". Death occurs in 80% to 100% of cases, in 1 to 4 weeks after the onset of clinical signs. In survivors, infection is life-long, and a chronic form of the disease develops with recurrent clinical events such as depression, apathy, somnolence and scared behavior, in particular following stress [110]. Of note, some infection may be asymptomatic. Histopathological examination of infected brains revealed viral antigens mainly in neuronal nuclei and the characteristic Joest-Degen inclusion bodies, accompanied with massive infiltration of inflammatory cells [109,110].

Diagnosis: The diagnostic of BoDV infection is particularly difficult. Clinical signs are not specific and low titres of antibodies in infected horses and low viremia (the virus is confined within the brain) does not allow a reliable diagnostic from serum or cerebrospinal fluids, even when the most sensitive serological or molecular tests are used. Standardized tests, validated by inter-laboratory assays, does not exist. Although intra-vitam studies give useful indicators, only *post-mortem* analyses performed in brain, the tissue with the highest viral load, will confirm a definitive diagnostic (Table 1).

Prevention and control: A few therapeutics (amantadine sulfate) and vaccines (attenuated and inactivated candidates) have been developed against equine BoDV infection thus far, but none are available in veterinary medicine since none proved effective in controlling or preventing the disease [109].

2.4. Enzootic Flaviviruses: West Nile Virus, Tick-Borne Encephalitis Virus and Louping Ill Virus

Three neurotropic flaviviruses documented in horses suffering from meningoencephalitis are enzootic in Europe, West Nile virus (WNV), tick-borne encephalitis virus (TBEV) and Louping ill virus (LIV). Flaviviruses can be divided into three distinct groups according to their vectors: tick-borne viruses, mosquito-borne viruses and viruses with unknown vectors [111].

Viruses: Flaviviruses, belonging to the *Flaviviridae* family, are enveloped, non-segmented, single-stranded and positive-sense RNA viruses. Their genome of approximately 11 kb encodes three structural proteins (capsid C, preMembrane prM and Envelop E) and seven non-structural proteins (NS1, NS2A, NS2B, NS3, NS4A, NS4B, and NS5) involved in virus replication and counteractive of immune responses (Figure 1). Notably, flaviviruses belonging to the Japanese Encephalitis serocomplex (such as WNV, Japanese, Saint-Louis and Murray Valley encephalitis viruses) express an additional non-structural protein, NS1′, resulting from ribosomal frameshift occurring at a specific heptanucleotide motif close to the beginning of the NS2A gene [112]. Even though the precise functions of NS1′ are still largely unknown, this protein has been involved in virus neuroinvasiveness.

Transmission and Epidemiology

WNV

WNV is maintained and amplified in an enzootic cycle involving birds and mosquitoes from the *Culex* genus as vectors. WNV is transmitted to different animal species (mainly mammals but also reptiles and amphibians) through the bite of infected mosquitoes. Horses and humans are highly susceptible to WNV infection but are considered as dead-end hosts owing to limited and short viremia that does not sustain transmission to naïve mosquitoes. In both species, asymptomatic infections are the most common, but in rare cases (approximately 1 out of 140 infections in humans and up to 10% infected horses), neuroinvasive forms with meningitis, encephalitis or myelitis may occur [113,114]. WNV is the most widely distributed arbovirus that induces equine encephalitis (Figure 7). Over the last 15 years, WNV has been repeatedly reported in Europe with a high frequency in the Mediterranean region and in Eastern Europe. This virus was first described in France, Portugal and Cyprus in the 1960s [115]. After a silence of more than 30 years, WNV lineage 1 strains resurfaced in North Africa (in Morocco (1996), Algeria (1994) and Tunisia (1997)), as well as in Western and Eastern Europe (Romania (1996), Italy (1998), Russia (1999) and France (2000)) [116–118]. WNV strikingly exemplified how fast and unpredictable flaviviruses can emerge when the virus was introduced in New York City in 1999. It produced large and dramatic outbreaks in humans and horses and rapidly spread, in less than 4 years, throughout the United States of America, causing more than 30,000 cases and 1200 deaths in humans and more than 24,000 cases in the equine population for the United States only over a 10-year period [119]. Interestingly, in Europe the epidemiological scenario in 1996–2010 was quite different from the one in North America as epidemics were irregular and limited in time and space. Nevertheless, a revival of WNV activity in Europe has been associated in particular with the introduction in 2004 of a new WNV strain within lineage 2, most likely originating from Africa [120]. This WNV lineage 2 was initially identified in Hungary and then spread to the eastern part of Austria and to southern European countries including Greece in 2010 and Italy in 2011 [121]. Unprecedented WNV transmission seasons in Europe were registered in 2010, 2012, 2013, or 2015, in association with climatic and environmental conditions sustaining mosquito activity and close mosquito-bird contact rates. Nevertheless, these recent transmission seasons were in no way comparable to the exceptional transmission wave experienced in 2018. Indeed, 2018 showed a 7.2-fold increase of reported cases compared to the 2017 transmission season and a final total number of reported autochthonous infections in humans (n = 2083) higher than the cumulative number from the previous seven years (n = 1832) [17]. The highest increase compared to 2017 was observed in Bulgaria (15-fold) followed by France (13.5-fold) and Italy (10.9-fold). The number of European equine WNV outbreaks doubled in 2018 (n = 285) in comparison with earlier WNV transmission seasons (n = 97–191 in 2013–2017, with on average 145 equine cases reported annually for this period) [17]. The second remarkable pattern was the reporting in 2018 for the first time of WNV in northern Europe, with several bird species and two horses found infected by WNV lineage 2 in Germany [122]. In Australia, specific virus variants called Kunjin virus and classified into WNV lineage 1b, recently caused unprecedented epizootics of neurological disease in horses in Southeast Australia, resulting in almost 1000 cases and a 9% case fatality rate in 2011; unusual climatic conditions, as well as enhanced virus transmission by infected mosquitoes, could have contributed to the phenomenon [123].

WNV neurovirulence and neuroinvasion are typically associated with sequence variations in the flavivirus E protein [124]; strikingly, a unique mutation at position 249 in the helicase portion of NS3 (NS3249P) has been identified in virus strains that have been responsible for major WNV outbreaks during the two last decades [125] and its role in the modulation of WNV virulence and transmission is nowadays largely debated [126,127].

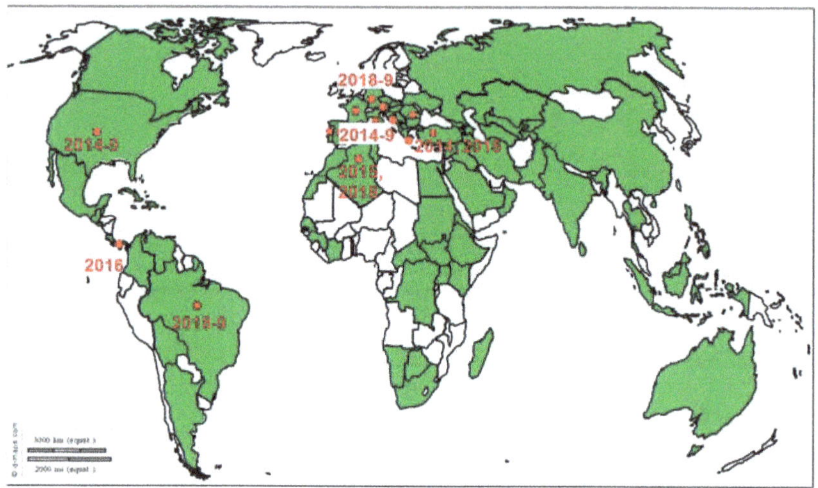

Figure 7. Distribution of WNV outbreaks. Recent outbreaks in horses reported to the OIE WAHIS interface [15], Promed Alerts [16] or in the scientific literature are depicted by dots (2014–2019).

TBEV and LIV

TBEV is the most important human tick-borne pathogen in Europe and Asia. The estimated annual incidence rate is 10,000 human infections with a case-fatality rate ranging from 1% to 20%. For TBEV, the arthropod vectors are primarily hard ticks and in Europe, the most important tick vector is *Ixodes ricinus*. Contrary to mosquitoes, which become infected only if there is sufficient viremia in the vertebrate host, ticks can become infected during a shared meal or "co-feeding", not requiring a systemic infection of the host [128]. In fact, adult and immature ticks, as well as larvae and nymphs are attached to their host for several days and can feed together on the same reservoir. Contamination between naïve (uninfected) and infected ticks can occur during this meal. Co-feeding is facilitated by the proximity of ticks and the action of saliva, which allows for the transfer of arboviruses including TBEV [129]. The main vertebrate reservoir hosts of TBEV are rodents of the genus *Myodes* and *Apodemus* although other small rodents and shrews can contribute to the natural transmission cycle. Larger animals such as sheep, goats and more rarely cattle can be additional competent hosts. Goats, sheep and cows excrete the virus in the milk. Humans, horses and game (deer, wild boar, fox) are epidemiological dead-end hosts. Human infection can occur through a bite from a TBEV- infected tick, more rarely through the ingestion of unpasteurized milk or milk products from goats and less often from infected cows or sheep [130]. Hard ticks (*Ixodes ricinus*) also transmit LIV. The vertebrate reservoir hosts are the wood mouse (*Apodemus sylvaticus*), the common shrew (*Sorex araneus*), the red grouse (*Lagopus lagopus scoticus*) and the sheep (*Ovis aries*). A co-feeding mechanism has also been reported for LIV transmission in *I. ricinus* ticks feeding on mountain hares (*Lepus timidus*) [131]. Sheep develop the disease and the virus could occasionally been detected in a range of other animal species such as goats, dogs, pigs, horses, humans, deer, llamas, alpacas, and mountain hare [132]. Finally, very uncommon cases of horse infected by TBEV or LIV have been described [133–135] and a few serological surveys in equids are available in the scientific literature. TBEV seroprevalence rates of 20% to 30% among asymptomatic horses have been reported in Austria and Germany [134,136] while lower ones have been reported in the Balkans (3% to 5% in Serbia and Slovakia) [137].

TBEV is reported in the northern hemisphere of Europe and Asia. There are three main subtypes of TBEV: European (TBEV-Eu), Siberian (TBEV-Si) and Far Eastern (TBEV-FE) circulating in Europe with TBEV-Si and TBEV-FE recently detected in the Baltic countries and in Eastern Finland [138,139]. Based on epidemiological investigations, LIV distribution area initially limited to the British Isles (particularly in Scotland, Cumbria, Wales, Devon and Ireland), but seroconversion or clinical cases in sheep due to LIV-like viruses have been reported during the last decade in Norway, Denmark, and Spain [140,141].

Pathogenesis and clinical disease: TBEV, LIV and WNV induce severe neurological syndromes, through pathogenic mechanisms that are still largely unknown. The majority of WNV and TBEV studies were done in vitro on transformed cells lines and in vivo on mouse experimental models [142]. The exact mechanism of WNV and TBEV CNS invasion is unclear, but five models have been proposed that rely on the anatomy of the blood-brain barrier (BBB): (i) entry through the BBB via infected leukocytes (a so-called Trojan horse mechanism, demonstrated for WNV) [143]; (ii) a direct passage of the hemato meningeal barrier after its integrity has been compromised by the action of cytokines (TNFα) or metalloproteinases (MMP9) inducing changes in capillary permeability (WNV) [144]; (iii) direct infection of the brain microvascular endothelial cells without effect on cells integrity (WNV and TBEV) [145]; (iiii) infection or passive transport through the epithelial cells of choroid plexuses, whose function is the production of CSF (WNV) [146]; (iiiii) infection via the neuronal pathway by the infection of olfactory neurons and/or axonal transport in the retrograde direction (mechanisms demonstrated for WNV only) [147]. This latter axonal transport promotes entry into the CNS from peripheral inoculation near nerve connections and acute flaccid paralysis of a limb. In an experimental model of infection of hamster, WNV was present 4–5 days after infection in multiple sites of the brain and spinal cord. Foci of neuronal infection were observed in the cortex, hippocampus, cerebellum, basal ganglia or the anterior horn of the spinal cord [148]. Neurons and astrocytes are infected with WNV and TBEV [142,149,150]. Following neuronal cell death, inflammatory molecules (such as IL1-β, IL6, IL8 and TNFα) have potentially toxic effects on uninfected neurons. Later, during infection, lymphoid infiltration can be observed in the infected regions with cells releasing proinflammatory cytokines destroying flavivirus-infected cells but also contributing to the pathogenesis of the virus by their cytotoxic action in the CNS [133,151].

An equine infection by WNV can be suspected when the following symptoms are noticed: hyperthermia, ataxia, hind legs paresis, muscle tremors, teeth grinding, cranial nerve deficits, dysphagia or face paralysis [9]. During 2000 and 2004 WNV epizootics in Camargue, France, ataxia was the main clinical sign observed in 64% to 72% of cases, while behavioral modifications (45% of cases) and muscle tremors (35% of cases) were less frequently reported [9]. Neurological disorders could persist within 5 to 30 days with a complete or partial remission after several months. Case fatality rates are variable, generally ranging from 20% to 57%.

Few publications describe TBEV infections in animals. Clinical signs of encephalitis induced by TBEV are rare in horses and include anxiety, decreased appetite, nervousness, and emaciation [134,135]. Generally asymptomatic in horses, the confirmation of equine TBEV infections can be challenging due to close and partially shared antigenicity with other flaviviruses such as WNV and Japanese Encephalitis Virus (JEV) [136]. Encephalitis caused by LIV in horses is very uncommon. During the outbreak reported in Ireland, horses developed neurological disorders comparable to the ones described for other arborviral infections affecting the CNS. Its principal symptoms consisted of ataxia ranging from slight incoordination to falling risk. Face and neck muscle tremors, depression, fear of light, behavioral modification such as constant chewing or mild fever could be observed [133,152].

Diagnosis: The neurological symptoms and lesions are not specific to flavivirus infection, making laboratory tests compulsory to confirm or exclude viral etiology. The diagnosis of flavivirus infection is delicate and relies on the detection of viral RNA in blood and CSF, viral isolation in cell culture and/or the detection of IgM and IgG in serum and CSF. Flavivirus infection (WNV, TBEV, LIV) diagnosis is primarily based on indirect methods. Indeed, viremia of an infected horse is quite low and fleeting and already vanished by the time horses develop neurological signs. It is therefore quite difficult to identify the pathogen by direct diagnostic assays, even with highly sensitive real-time RT-PCR assays. Serology tests such as ELISA for WNV and TBEV detection and hemagglutination inhibition assay for LIV are fast and allow for the identification of anti-flavivirus antibodies. These tests suffer from poor specificity because of antibody cross-reactions between flaviviruses [153]. Several serological tests have to be carried out in order to detect a recent infection (IgM ELISA for WNV or an antibody titer kinetics). A virus neutralisation test, with higher diagnostic specificity, practiced in a biological security 3 level laboratory allows the definitive identification of the flavivirus. WNV infection can also be diagnosed by RT-PCR on EDTA-blood samples collected during the first clinical period (marked by hyperthermia only, a few days after mosquito bites and before the neuro-invasive form of the disease) or on cerebrospinal fluid and brain on post-mortem. Samples must be shipped cold or frozen, as WNV is very sensitive to thermal and chemical inactivation.

Prevention and control: There is no specific treatment for flaviviruses and disease control primarily relies on integrated virus surveillance and on vector control, with strategies and methodologies differing between European countries [154,155]. However, for WNV, three equine vaccines have a European Marketing Authorization (Table 2). The Zoetis Equip WNV vaccine is composed of the inactivated West Nile virus, strain VM-2 New York 1999 combined with an adjuvant [156]. The other two vaccines are recombinant and adjuvanted vaccines. Proteq West Nile is composed of a canary poxvirus vector expressing the prM and E genes of WNV [157]. The canarypox vector performs an abortive replication cycle in mammalian cells where the inserted gene product (transgene) is expressed [158]. Finally, the Equilis West Nile vaccine from Intervet consists of the yellow fever virus (YFV) 17D vaccine strain where the prM and E genes of YFV have been replaced by those of WNV. This recombinant vaccine is injected under an inactivated form [159]. No vaccine against TBEV and LIV are available for horses.

3. Exotic Equine Encephalitis Viruses in Europe

Equine encephalitis viruses enzootic in Europe are not the only threat to European horses. Horses could contract the disease abroad during equine competitions or international events. Moreover, because of enhanced risks of emergence of exotic viruses and increasing animal movements owing to globalization, exotic equine encephalitis viruses could be reported in the future in new naïve territories. Expert opinion and risk assessment through modelling is strongly required to identify the viruses that are prone to emergence and the most at-risk areas in Europe and at the international level and to adapt surveillance plans [160,161].

The main exotic equine encephalitis viruses are vector-borne, either transmitted by mosquitoes or flying midges (Figure 3b). They encompass viruses belonging to three genera, flaviviruses in the *Flaviviridae* family (Japanese encephalitis virus (JEV), Saint-Louis encephalitis virus (SLEV) and Murray Valley encephalitis virus (MVEV)), alphaviruses in the *Togaviridae* family (eastern, western and venezuelan equine encephalitis viruses, EEEV, WEEV and VEEV respectively) and orbiviruses in the *Reoviridae* family (equine encephalosis virus, EEV) [21]. Most of them, with the exception of EEV, are zoonotic. During such zoonotic infections, humans and equines will generally not develop viremia elevated enough to infect naïve mosquitoes and they are considered as dead end hosts; horses can however effectively replicate epizootic strains of VEEV, I-AB and I-C variants, developing high viremia for an average of 4 days and serving as virus reservoir for mosquito transmission to animals or humans [20,21].

Pertaining to flavivirus infections, JEV, SLEV and MVEV induce similar clinical presentations to the ones reported with WNV but mainly differ in their case fatality rate and more restricted host and vector ranges, likely contributing to more limited geographical range (reviewed in [12,20,162,163]). Case fatality rates reported for JEV, SLEV and MVEV are 5% to 30%, 3% to 30%, and 15% to 20%, respectively [5,164]. JEV, SLEV and MVEV have been mainly reported from South-Eastern Asia, America (from Northern America up to Argentina) and Northern Australia/Papua New Guinea respectively (Figure 8). JEV and SLEV infections have been unfrequently reported in horses during the last decade, while MVEV equine outbreaks across south-eastern Australia have been identified in 2011 [12]. However, among these exotic flaviviruses, JEV presented the highest propensity to spread, with transmission evidenced westward to Nepal and Pakistan, as well as eastward in western Pacific regions in the 1990s (Eastern-most territories in Indonesia, Papua New Guinea and Australia in 1995 and 1998), subsequent to changes in human activities (deforestation, irrigation, expansion of pig breeding) [165]. Intriguingly, JEV genome fragments have been recently identified in birds and mosquitoes collected in Italy [166,167]. JEV actively circulates in rice paddies, rural and semi-urban areas, amplified by Ardeid birds (herons and egrets) and pigs as natural maintenance reservoirs and *Culex tritaeniorhynchus* or other mosquito species mainly from the *Culex* genus as vectors (reviewed in [20]). Finally, SLEV appears as a recently re-emerging flavivirus. Subsequent to WNV introduction in Northern America in 1999, probable virus competition for avian amplifier hosts likely contributed to initial disappearance of SLEV from the Western United States (1999–2014) [168,169]. From 2015, SLEV has been again reported in California and the re-emerging strain clusters genetically with an epidemic strain identified 10 years earlier during an unprecedented human encephalitis outbreak in Cordoba, Argentina [162,170].

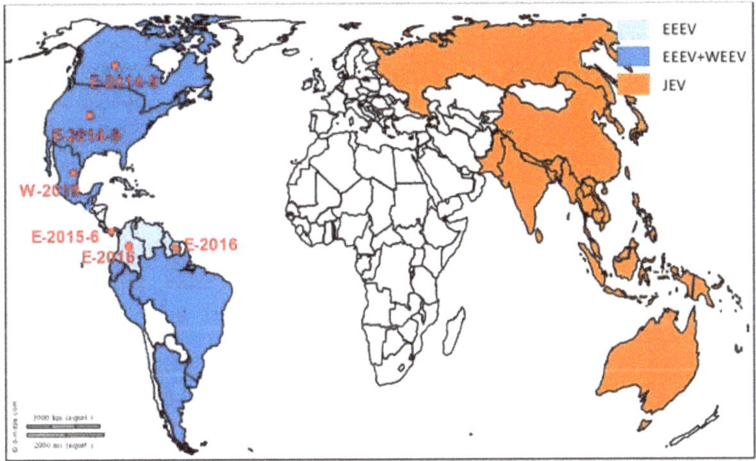

(**a**) Exotic equine encephalitis viruses—except VEEV.

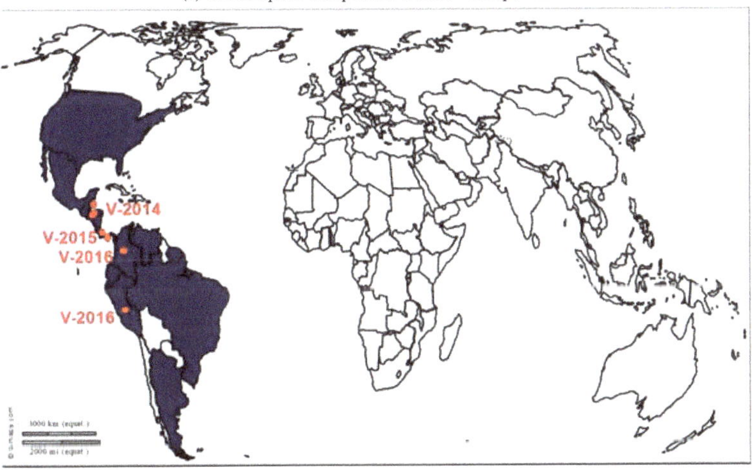

(**b**)VEEV.

Figure 8. Distribution of exotic equine encephalitis outbreaks: Japanese encephalitis, western and eastern equine encephalitis in (**a**), venezuelan equine encephalitis in (**b**). Recent outbreaks in horses reported to the OIE WAHIS interface [15], Promed Alerts [16] or in the scientific literature are depicted by dots (2014–2019).

EEEV, WEEV, and VEEV are alphaviruses regularly identified in equine encephalitis in America. These enveloped, non-segmented, single-stranded and positive-sense RNA viruses encode two polyprotein gene clusters, driving the translation of four non-structural proteins (NSP1-4) and virus structural proteins (and in particular two Envelop proteins, E1 and E2) at the 5′ and 3′ ends, respectively (Figure 1). Epidemics or epizootics of EEE, WEE, and VEE have been recognized at irregular intervals since 1831, 1847, and 1939, respectively, in different regions of America (Figure 8) [171]. The last documented human WEE case in North America occurred in 1994 and the virus has not been detected in mosquito pools since 2008, while severe epizootics involving more than 41 horses have been identified in 2019 in Mexico (Promed archive 20190406.6407111) [16,172]. The most severe forms of alphavirus equine encephalitis are associated with EEEV and VEEV epizootic variants. VEEV strains are classified into six subtypes (subtypes I-VI), with subtype I including five antigenic variants (A, B, C, D, E, and F) and epizootic strains corresponding to I-AB and C strains only. All other subtypes and other subtype I variants are endemic strains and seldom cause encephalitis in horses. Epizootic I-AB and C strains were shown to arise from E2 mutations of enzootic I-D or E variants leading to increased protein positive charges and in particular from T213R/K substitution [173] and have been described in Argentina, Peru, Columbia, Ecuador, Mexico, Trinidad, Venezuela, the United States, as well as in Panama in 2019 [16,174]. VEEV enzootic strains are mostly maintained in birds and rodents belonging to *Sigmodon*, *Zygodontomys*, *Heteromys* and *Proechimys* genus and in *Culex melanoconion* mosquitoes, while epizootic variants are transmitted by more diverse mosquito species (*Aedes*, *Psorophora* genus) [175]. Regarding EEEV, North and South American variants have been described, with North American variants being the most pathogenic for mammals [176,177]. Alternate infection of birds and *Culiseta melanura* or *morsitans* mosquitoes maintain this virus in nature, while *Aedes* bridge vectors are involved in EEEV transmission to humans and horses [20]. EEEV clinical infections in horses are highly lethal, with 70% to 90% case fatality rate reported in the literature, in comparison with 20% to 50% for WEEV and 40% to 80% for VEEV (reviewed in [20]). EEEV equine epizootics are regularly observed in the United States and in Canada, Ontario and in this respect, 2019 has sustained active transmission in North America, with more than 18 states from Eastern United States (the western limit being delineated by Minnesota and Louisiana) having reported more than 99 EEEV horse cases from early spring (end of March-April) to late autumn [16].

Equine encephalosis virus (EEV) is an orbivirus close to African horse sickness and bluetongue viruses, two arboviruses associated with unexpected emergence in Europe in 1989 and 2006 respectively [178,179]. It is a primary endotheliotropic virus and most EEV infection cases are asymptomatic or poorly symptomatic, but its clinical presentation may include ataxia, depression, hyper excitability and convulsions. Since its description in Israel in 2008–2009, it is of primary importance to improve the preparedness of European countries to EEV emergence [180].

4. Conclusions

Many neurological diseases represent an important sanitary and economic threat to the horse population worldwide. Non-vector-borne equine encephalitis viruses, such as EHV-1, rabies and BoDV, are variably reported in Europe, whereas arthropod-borne infections are usually exotic diseases. Viral encephalitis therapeutic strategies are comparable, whatever the etiologic virus. Horses with neurological conditions must be isolated in a quiet box with limited stimuli (noise, light) and an appropriate bedding providing warmth, comfort, and security. Slings can be used to support paretic horses and avoid long and poor prognosis recumbency (Figure 2). Supportive care will contribute substantially, avoiding complications and improving the prognosis. The use of DMSO (0.4–0.9 g/kg for 5–6 days) has been advocated on the basis of its free radical scavenging properties but its efficacy has not been evaluated scientifically [1]. Nonsteroidal anti-inflammatory drugs may be used to control pyrexia, inflammation and discomfort, while short-term use of glucocorticoids may be beneficial; glucocorticoids proved to be valuable in some EHM horses and it is hypothesized that the treatment reduces the supposed immune-mediated EHV-1 pathogenesis [181]. However, they were also shown to reactivate latent herpesvirus infection and to increase the level and duration of virus shedding [182]. Vaccination is controversial in the face of outbreaks and in particular inactivated vaccines take too long to generate immune responses capable to limit disease spread when outbreaks are seasonal (WNV, other vector-borne viruses in temperate areas).

Clinical signs of viral equine encephalitis are not specific and overlapping geographical areas can make virus identification very challenging. One recent and striking example of delayed identification of emerging arboviruses due to similarities in clinical presentation and cross-reactive diagnostic tools was given during WNV introduction in the United States, when WNV was initially misdiagnosed with the closely related SLEV [183]. Bearing in mind that three flaviviruses responsible for equine encephalitis are described in Europe, and that serological cross-reactivity is frequently observed in flavivirus indirect diagnosis assays, the development of multiplex approaches that allow the comparison of serological reactions against a wide range of pathogens appear to be valuable options [184,185]. Furthermore, because in about one-half of infectious equine encephalitis, no known pathogen can be evidenced [4], identification of unknown neuropathogenic viruses by classical (electron microscopy) and more recent high-throughput techniques (next generation sequencing for example) is highly desirable [186,187]. In these two recent studies, three viruses, Shuni virus, horse parvovirus-CSF and eqcopivirus, have been identified as potential causes of neurologic disease in horses through unbiased detection from different tissues or body fluids; the demonstration of infectious virus from the brain of sick horses establish Shuni virus as a novel equine neuropathogenic virus [186], while for the other two viruses for which genomic DNA was detected in CSF and/or plasma [187], comparison of virus prevalence in the CSF of healthy horses (case-control study) would be required before a conclusion on the aetiology of equine encephalitis can be reached.

Arboviruses are the most important cause of encephalitis in horses and many of these viruses are also significant human pathogens. Some of these arboviruses have recently emerged or resurged, such as WNV, JEV, SLEV, EEEV or EEV and an increased rate of emergence of vector-borne diseases can be inferred from recent studies [188]. A high diversity of mosquito species have been reported in Europe (mainly from *Aedes*, *Culex* and *Culiseta* genera), and highly invasive *Aedes albopictus* and *Ae. japonicus* have rapidly established in several European countries over the last decade [189,190]. Vector competence of native and invasive European mosquito species for equine encephalitis viruses, other than WNV and JEV, has been unfrequently evaluated [191–193]. Consequently, identification of European regions at risk for the spread of exotic equine encephalitis viruses is difficult and mainly relies on information on mosquito and animal hosts density and on records of opportunistic mosquito species [160]. On-time control of vector-borne infections relies on the use of sentinel systems, including horses or sentinel chicken flocks for example, to provide warning of virus activity and initiate mosquito control measures [155].

Author Contributions: Writing—original draft preparation, S.L., S.P., M.C., C.B., G.G., A.L., P.T.; writing—review, S.L., S.P., M.C., C.B., G.G., A.L., P.T.; writing—editing, S.L. All authors have read and agreed to the published version of the manuscript.

Funding: This research on equine encephalitis viruses received funding from the EU Commission, Directorate-General for Health and Food Safety (DG SANTE). DG SANTE appointed Anses (affiliation 1) as the EU-RL on equine diseases.

Acknowledgments: We are grateful to E. Laloy, French Veterinary School of Alfort for providing histology cliché.

Conflicts of Interest: The authors declare no conflict of interest.

References

1. Pellegrini-Masini, A.; Livesey, L.C. Meningitis and encephalomyelitis in horses. *Vet. Clin. N. Am. Equine Pract.* **2006**, *22*, 553–589. [CrossRef] [PubMed]
2. Leblond, A.; Villard, I.; Leblond, L.; Sabatier, P.; Sasco, A.J. A retrospective evaluation of the causes of death of 448 insured French horses in 1995. *Vet. Res. Commun.* **2000**, *24*, 85–102. [CrossRef] [PubMed]

3. Tyler, C.M.; Davis, R.E.; Begg, A.P.; Hutchins, D.R.; Hodgson, D.R. A survey of neurological diseases in horses. *Aust. Vet. J.* **1993**, *70*, 445–449. [CrossRef] [PubMed]

4. Laugier, C.T.; Tapprest, J. Fréquence de la pathologie nerveuse et de ses différentes causes dans un effectif de 4319 chevaux autopsiés. *Bull. Epidémiologique St. Anim. Et Aliment. Spécial Équidé* **2012**, *19*, 9.

5. Kumar, B.; Manuja, A.; Gulati, B.R.; Virmani, N.; Tripathi, B.N. Zoonotic Viral Diseases of Equines and Their Impact on Human and Animal Health. *Open Virol. J.* **2018**, *12*, 80–98. [CrossRef]

6. Walker, P.J.; Siddell, S.G.; Lefkowitz, E.J.; Mushegian, A.R.; Dempsey, D.M.; Dutilh, B.E.; Harrach, B.; Harrison, R.L.; Hendrickson, R.C.; Junglen, S.; et al. Changes to virus taxonomy and the International Code of Virus Classification and Nomenclature ratified by the International Committee on Taxonomy of Viruses (2019). *Arch. Virol.* **2019**, *164*, 2417–2429. [CrossRef]

7. Viralzone, SIB Swiss Institute of Bioinformatics. Available online: www.expasy.org/viralzone (accessed on 15 December 2019).

8. Mayhew, I.G.; de Lahunta, A.; Whitlock, R.H.; Krook, L.; Tasker, J.B. Spinal cord disease in the horse. *Cornell. Vet.* **1978**, *68*, 1–207.

9. Porter, R.S.; Leblond, A.; Lecollinet, S.; Tritz, P.; Cantile, C.; Kutasi, O.; Zientara, S.; Pradier, S.; van Galen, G.; Speybroek, N.; et al. Clinical diagnosis of West Nile Fever in Equids by classification and regression tree (CART) analysis and comparative study of clinical appearance in three European countries. *Transbound. Emerg. Dis.* **2011**, *58*, 197–205. [CrossRef]

10. Van Galen, G.; Leblond, A.; Tritz, P.; Martinelle, L.; Pronost, S.; Saegerman, C. A retrospective study on equine herpesvirus type-1 associated myeloencephalopathy in France (2008–2011). *Vet. Microbiol.* **2015**, *179*, 304–309. [CrossRef]

11. Pronost, S.; Legrand, L.; Pitel, P.H.; Wegge, B.; Lissens, J.; Freymuth, F.; Richard, E.; Fortier, G. Outbreak of equine herpesvirus myeloencephalopathy in France: A clinical and molecular investigation. *Transbound. Emerg. Dis.* **2012**, *59*, 256–263. [CrossRef]

12. Knox, J.; Cowan, R.U.; Doyle, J.S.; Ligtermoet, M.K.; Archer, J.S.; Burrow, J.N.; Tong, S.Y.; Currie, B.J.; Mackenzie, J.S.; Smith, D.W.; et al. Murray Valley encephalitis: A review of clinical features, diagnosis and treatment. *Med. J. Aust.* **2012**, *196*, 322–326. [CrossRef]

13. RESPE. Available online: http://www.respe.net (accessed on 15 December 2019).

14. EQUINELLA. Available online: http://www.equinella.ch (accessed on 15 December 2019).

15. World Organisation for Animal Health (OIE), Animal Health Information. Available online: https://www.oie.int/wahis_2/public/wahid.php/Diseaseinformation/Immsummary (accessed on 15 December 2019).

16. Promed, International Society for Infectious Diseases. Available online: https://promedmail.org/ (accessed on 15 December 2019).

17. ECDC. Epidemiological Update: West Nile Virus Transmission Season in Europe. Available online: https://www.ecdc.europa.eu/en/news-events/epidemiological-update-west-nile-virus-transmission-season-europe-2018 (accessed on 28 November 2019).

18. Dunowska, M. A review of equid herpesvirus 1 for the veterinary practitioner. Part B: Pathogenesis and epidemiology. *N. Z. Vet. J.* **2014**, *62*, 179–188. [CrossRef] [PubMed]

19. BioRender. Available online: https//app.biorender.com/ (accessed on 15 December 2019).

20. Weaver, S.C.; Reisen, W.K. Present and future arboviral threats. *Antiviral Res.* **2010**, *85*, 328–345. [CrossRef] [PubMed]

21. Chapman, G.E.; Baylis, M.; Archer, D.; Daly, J.M. The challenges posed by equine arboviruses. *Equine Vet. J.* **2018**, *50*, 436–445. [CrossRef] [PubMed]

22. Davison, A.J.; Eberle, R.; Ehlers, B.; Hayward, G.S.; McGeoch, D.J.; Minson, A.C.; Pellett, P.E.; Roizman, B.; Studdert, M.J.; Thiry, E. The order Herpesvirales. *Arch. Virol.* **2009**, *154*, 171–177. [CrossRef]

23. Abdelgawad, A.; Damiani, A.; Ho, S.Y.; Strauss, G.; Szentiks, C.A.; East, M.L.; Osterrieder, N.; Greenwood, A.D. Zebra Alphaherpesviruses (EHV-1 and EHV-9): Genetic Diversity, Latency and Co-Infections. *Viruses* **2016**, *8*, 262. [CrossRef]

24. Garvey, M.; Suarez, N.M.; Kerr, K.; Hector, R.; Moloney-Quinn, L.; Arkins, S.; Davison, A.J.; Cullinane, A. Equid herpesvirus 8: Complete genome sequence and association with abortion in mares. *PLoS ONE* **2018**, *13*, e0192301. [CrossRef]

25. Saxegaard, F. Isolation and identification of equine rhinopneumonitis virus (equine abortion virus) from cases of abortion and paralysis. *Nord. Vet. Med.* **1966**, *18*, 504–516.

26. Ma, G.; Azab, W.; Osterrieder, N. Equine herpesviruses type 1 (EHV-1) and 4 (EHV-4)—Masters of co-evolution and a constant threat to equids and beyond. *Vet. Microbiol.* **2013**, *167*, 123–134. [CrossRef]

27. Chesters, P.M.; Allsop, R.; Purewal, A.; Edington, N. Detection of latency-associated transcripts of equid herpesvirus 1 in equine leukocytes but not in trigeminal ganglia. *J. Virol.* **1997**, *71*, 3437–3443.

28. Oladunni, F.S.; Horohov, D.W.; Chambers, T.M. EHV-1: A constant threat to the horse industry. *Front. Microbiol.* **2019**, *10*, 2668. [CrossRef]

29. Telford, E.A.; Watson, M.S.; McBride, K.; Davison, A.J. The DNA sequence of equine herpesvirus-1. *Virology* **1992**, *189*, 304–316. [CrossRef]

30. Dayaram, A.; Franz, M.; Schattschneider, A.; Damiani, A.M.; Bischofberger, S.; Osterrieder, N.; Greenwood, A.D. Long term stability and infectivity of herpesviruses in water. *Sci. Rep.* **2017**, *7*, 46559. [CrossRef] [PubMed]

31. Foote, C.E.; Love, D.N.; Gilkerson, J.R.; Whalley, J.M. Detection of EHV-1 and EHV-4 DNA in unweaned Thoroughbred foals from vaccinated mares on a large stud farm. *Equine Vet. J.* **2004**, *36*, 341–345. [CrossRef] [PubMed]

32. Kydd, J.H.; Lunn, D.P.; Osterrieder, K. Report of the Fourth International Havemeyer Workshop on Equid Herpesviruses (EHV) EHV-1, EHV-2 and EHV-5. *Equine Vet. J.* **2019**, *51*, 565–568. [CrossRef] [PubMed]

33. Vissani, M.A.; Becerra, M.L.; Olguin Perglione, C.; Tordoya, M.S.; Mino, S.; Barrandeguy, M. Neuropathogenic and non-neuropathogenic genotypes of Equid Herpesvirus type 1 in Argentina. *Vet. Microbiol.* **2009**, *139*, 361–364. [CrossRef]

34. Paillot, R.C.; Case, R.; Ross, J.; Newton, R.; Nugent, J. Equine Herpes Virus-1: Virus, Immunity and Vaccines. *TOVSJ* **2008**, *2*, 68–91. [CrossRef]

35. Crabb, B.S.; Studdert, M.J. Equine herpesviruses 4 (equine rhinopneumonitis virus) and 1 (equine abortion virus). *Adv. Virus Res.* **1995**, *45*, 153–190.

36. Edington, N.; Smyth, B.; Griffiths, L. The role of endothelial cell infection in the endometrium, placenta and foetus of equid herpesvirus 1 (EHV-1) abortions. *J. Comp. Pathol.* **1991**, *104*, 379–387. [CrossRef]

37. Whitwell, K.E.; Blunden, A.S. Pathological findings in horses dying during an outbreak of the paralytic form of Equid herpesvirus type 1 (EHV-1) infection. *Equine Vet. J.* **1992**, *24*, 13–19. [CrossRef]

38. Slater, J.D.; Lunn, D.P.; Horohov, D.W.; Antczak, D.F.; Babiuk, L.; Breathnach, C.; Chang, Y.W.; Davis-Poynter, N.; Edington, N.; Ellis, S.; et al. Report of the equine herpesvirus-1 Havermeyer Workshop, San Gimignano, Tuscany, June 2004. *Vet. Immunol. Immunopathol.* **2006**, *111*, 3–13. [CrossRef]

39. Van Maanen, C. Equine herpesvirus 1 and 4 infections: An update. *Vet. Q.* **2002**, *24*, 58–78. [CrossRef] [PubMed]

40. Kydd, J.H.; Slater, J.; Osterrieder, N.; Antczak, D.F.; Lunn, D.P. Report of the Second Havemeyer EHV-1 Workshop, Steamboat Springs, Colorado, USA, September 2008. *Equine Vet. J.* **2010**, *42*, 572–575. [CrossRef] [PubMed]

41. Lunn, D.P.; Davis-Poynter, N.; Flaminio, M.J.; Horohov, D.W.; Osterrieder, K.; Pusterla, N.; Townsend, H.G. Equine herpesvirus-1 consensus statement. *J. Vet. Intern. Med.* **2009**, *23*, 450–461. [CrossRef] [PubMed]

42. Goehring, L.S.; van Winden, S.C.; van Maanen, C.; Sloet van Oldruitenborgh-Oosterbaan, M.M. Equine herpesvirus type 1-associated myeloencephalopathy in The Netherlands: A four-year retrospective study (1999–2003). *J. Vet. Intern. Med.* **2006**, *20*, 601–607. [CrossRef] [PubMed]

43. Pronost, S.; Leon, A.; Legrand, L.; Fortier, C.; Miszczak, F.; Freymuth, F.; Fortier, G. Neuropathogenic and non-neuropathogenic variants of equine herpesvirus 1 in France. *Vet. Microbiol.* **2010**, *145*, 329–333. [CrossRef] [PubMed]

44. Sutton, G.; Garvey, M.; Cullinane, A.; Jourdan, M.; Fortier, C.; Moreau, P.; Foursin, M.; Gryspeerdt, A.; Maisonnier, V.; Marcillaud-Pitel, C.; et al. Molecular Surveillance of EHV-1 Strains Circulating in France during and after the Major 2009 Outbreak in Normandy Involving Respiratory Infection, Neurological Disorder, and Abortion. *Viruses* **2019**, *11*, 916. [CrossRef]

45. Allen, G.P. Risk factors for development of neurologic disease after experimental exposure to equine herpesvirus-1 in horses. *Am. J. Vet. Res.* **2008**, *69*, 1595–1600. [CrossRef]

46. Goodman, L.B.; Loregian, A.; Perkins, G.A.; Nugent, J.; Buckles, E.L.; Mercorelli, B.; Kydd, J.H.; Palu, G.; Smith, K.C.; Osterrieder, N.; et al. A point mutation in a herpesvirus polymerase determines neuropathogenicity. *PLoS Pathog.* **2007**, *3*, e160. [CrossRef]

47. Goehring, L.S.; van Maanen, C.; Berendsen, M.; Cullinane, A.; de Groot, R.J.; Rottier, P.J.; Wesselingh, J.J.; Sloet van Oldruitenborgh-Oosterbaan, M.M. Experimental infection with neuropathogenic equid herpesvirus type 1 (EHV-1) in adult horses. *Vet. J.* **2010**, *186*, 180–187. [CrossRef]

48. Nugent, J.; Birch-Machin, I.; Smith, K.C.; Mumford, J.A.; Swann, Z.; Newton, J.R.; Bowden, R.J.; Allen, G.P.; Davis-Poynter, N. Analysis of equid herpesvirus 1 strain variation reveals a point mutation of the DNA polymerase strongly associated with neuropathogenic versus nonneuropathogenic disease outbreaks. *J. Virol.* **2006**, *80*, 4047–4060. [CrossRef]

49. Allen, G.P.; Breathnach, C.C. Quantification by real-time PCR of the magnitude and duration of leucocyte-associated viraemia in horses infected with neuropathogenic vs. non-neuropathogenic strains of EHV-1. *Equine Vet. J.* **2006**, *38*, 252–257. [CrossRef] [PubMed]

50. Allen, G.P. Development of a real-time polymerase chain reaction assay for rapid diagnosis of neuropathogenic strains of equine herpesvirus-1. *J. Vet. Diagn. Investig.* **2007**, *19*, 69–72. [CrossRef] [PubMed]

51. Perkins, G.A.; Goodman, L.B.; Tsujimura, K.; Van de Walle, G.R.; Kim, S.G.; Dubovi, E.J.; Osterrieder, N. Investigation of the prevalence of neurologic equine herpes virus type 1 (EHV-1) in a 23-year retrospective analysis (1984–2007). *Vet. Microbiol.* **2009**, *139*, 375–378. [CrossRef] [PubMed]

52. Pronost, S.; Cook, R.F.; Fortier, G.; Timoney, P.J.; Balasuriya, U.B. Relationship between equine herpesvirus-1 myeloencephalopathy and viral genotype. *Equine Vet. J.* **2010**, *42*, 672–674. [CrossRef] [PubMed]

53. Patel, J.R.; Heldens, J. Equine herpesviruses 1 (EHV-1) and 4 (EHV-4)—Epidemiology, disease and immunoprophylaxis: A brief review. *Vet. J.* **2005**, *170*, 14–23. [CrossRef] [PubMed]

54. Goodman, L.B.; Wagner, B.; Flaminio, M.J.; Sussman, K.H.; Metzger, S.M.; Holland, R.; Osterrieder, N. Comparison of the efficacy of inactivated combination and modified-live virus vaccines against challenge infection with neuropathogenic equine herpesvirus type 1 (EHV-1). *Vaccine* **2006**, *24*, 3636–3645. [CrossRef] [PubMed]

55. Garre, B.; Gryspeerdt, A.; Croubels, S.; De Backer, P.; Nauwynck, H. Evaluation of orally administered valacyclovir in experimentally EHV1-infected ponies. *Vet. Microbiol.* **2009**, *135*, 214–221. [CrossRef]

56. Thieulent, C.J.; Hue, E.S.; Fortier, C.I.; Dallemagne, P.; Zientara, S.; Munier-Lehmann, H.; Hans, A.; Fortier, G.D.; Pitel, P.H.; Vidalain, P.O.; et al. Screening and evaluation of antiviral compounds against Equid alpha-herpesviruses using an impedance-based cellular assay. *Virology* **2019**, *526*, 105–116. [CrossRef]

57. Vissani, M.A.; Thiry, E.; Dal Pozzo, F.; Barrandeguy, M. Antiviral agents against equid alphaherpesviruses: Current status and perspectives. *Vet. J.* **2016**, *207*, 38–44. [CrossRef]

58. Nolden, T.; Banyard, A.C.; Finke, S.; Fooks, A.R.; Hanke, D.; Hoper, D.; Horton, D.L.; Mettenleiter, T.C.; Muller, T.; Teifke, J.P.; et al. Comparative studies on the genetic, antigenic and pathogenic characteristics of Bokeloh bat lyssavirus. *J. Gen. Virol.* **2014**, *95*, 1647–1653. [CrossRef]

59. Malerczyk, C.; Freuling, C.; Gniel, D.; Giesen, A.; Selhorst, T.; Muller, T. Cross-neutralization of antibodies induced by vaccination with Purified Chick Embryo Cell Vaccine (PCECV) against different Lyssavirus species. *Hum. Vaccin. Immunother.* **2014**, *10*, 2799–2804. [CrossRef] [PubMed]

60. Gunawardena, P.S.; Marston, D.A.; Ellis, R.J.; Wise, E.L.; Karawita, A.C.; Breed, A.C.; McElhinney, L.M.; Johnson, N.; Banyard, A.C.; Fooks, A.R. Lyssavirus in Indian Flying Foxes, Sri Lanka. *Emerg. Infect. Dis.* **2016**, *22*, 1456–1459. [CrossRef] [PubMed]

61. Hayman, D.T.; Fooks, A.R.; Marston, D.A.; Garcia, R.J. The Global Phylogeography of Lyssaviruses-Challenging the 'Out of Africa' Hypothesis. *PLoS Negl. Trop. Dis.* **2016**, *10*, e0005266. [CrossRef] [PubMed]

62. Rupprecht, C.E.; Turmelle, A.; Kuzmin, I.V. A perspective on lyssavirus emergence and perpetuation. *Curr. Opin. Virol.* **2011**, *1*, 662–670. [CrossRef] [PubMed]

63. Johnson, N.; Vos, A.; Freuling, C.; Tordo, N.; Fooks, A.R.; Muller, T. Human rabies due to lyssavirus infection of bat origin. *Vet. Microbiol.* **2010**, *142*, 151–159. [CrossRef] [PubMed]

64. WHO. *WHO Expert Consultation on Rabies: Third Report*; WHO Technical Report Series; World Health Organization: Geneva, Switzerland, 2018; pp. 1–183.

65. Barecha, C.B.G.; Girzaw, F.; Kandi, V.; Pal, M. Epidemiology and Public Health Significance of Rabies. *Perspect. Med. Res.* **2017**, *5*, 55–67.

66. Green, S.L. Equine rabies. *Vet. Clin. N. Am. Equine Pract.* **1993**, *9*, 337–347. [CrossRef]

67. Sato, G.; Itou, T.; Shoji, Y.; Miura, Y.; Mikami, T.; Ito, M.; Kurane, I.; Samara, S.I.; Carvalho, A.A.; Nociti, D.P.; et al. Genetic and phylogenetic analysis of glycoprotein of rabies virus isolated from several species in Brazil. *J. Vet. Med. Sci.* **2004**, *66*, 747–753. [CrossRef]

68. Dyer, J.L.; Yager, P.; Orciari, L.; Greenberg, L.; Wallace, R.; Hanlon, C.A.; Blanton, J.D. Rabies surveillance in the United States during 2013. *J. Am. Vet. Med. Assoc.* **2014**, *245*, 1111–1123. [CrossRef]

69. Ali, Y.; Intisar, S.; Wegdan, H.; Ali, E. Epidemiology of Rabies in Sudan. *J. Anim. Vet. Adv.* **2006**, *5*, 266–270.

70. Weir, D.L.; Annand, E.J.; Reid, P.A.; Broder, C.C. Recent observations on Australian bat lyssavirus tropism and viral entry. *Viruses* **2014**, *6*, 909–926. [CrossRef] [PubMed]

71. WHO. Rabies Bulletin Europe. 2010–2019. Available online: https://www.who-rabies-bulletin.org/ (accessed on 15 December 2019).

72. Rech, R.; Barros, C. Neurologic Diseases in Horses. *Vet. Clin. N. Am. Equine Pract.* **2015**, *31*, 281–306. [CrossRef] [PubMed]

73. Kumar, R.; Patil, R.D. Cryptic etiopathological conditions of equine nervous system with special emphasis on viral diseases. *Vet. World* **2017**, *10*, 1427–1438. [CrossRef] [PubMed]

74. Meyer, E.E.; Morris, P.G.; Elcock, L.H.; Weil, J. Hindlimb hyperesthesia associated with rabies in two horses. *J. Am. Vet. Med. Assoc.* **1986**, *188*, 629–632.

75. OIE. *Manual of Diagnostic Tests and Vaccines for Terrestrial Animals 2019, Chapter 3.1.17*; OIE: Paris, France, 2019; pp. 1–35.

76. Appolinario, C.; Allendorf, S.D.; Vicente, A.F.; Ribeiro, B.D.; Fonseca, C.R.; Antunes, J.M.; Peres, M.G.; Kotait, I.; Carrieri, M.L.; Megid, J. Fluorescent antibody test, quantitative polymerase chain reaction pattern and clinical aspects of rabies virus strains isolated from main reservoirs in Brazil. *Braz. J. Infect. Dis.* **2015**, *19*, 479–485. [CrossRef]

77. Muller, F.T.; Freuling, C.M. Rabies control in Europe: An overview of past, current and future strategies. *Rev. Sci. Tech.* **2018**, *37*, 409–419. [CrossRef]

78. Harvey, A.M.; Watson, J.L.; Brault, S.A.; Edman, J.M.; Moore, S.M.; Kass, P.H.; Wilson, W.D. Duration of serum antibody response to rabies vaccination in horses. *J. Am. Vet. Med. Assoc.* **2016**, *249*, 411–418. [CrossRef]

79. De la Torre, J.C. Molecular biology of borna disease virus: Prototype of a new group of animal viruses. *J. Virol.* **1994**, *68*, 7669–7675.

80. Kuhn, J.H.; Durrwald, R.; Bao, Y.; Briese, T.; Carbone, K.; Clawson, A.N.; deRisi, J.L.; Garten, W.; Jahrling, P.B.; Kolodziejek, J.; et al. Taxonomic reorganization of the family Bornaviridae. *Arch. Virol.* **2015**, *160*, 621–632. [CrossRef]

81. Honkavuori, K.S.; Shivaprasad, H.L.; Williams, B.L.; Quan, P.L.; Hornig, M.; Street, C.; Palacios, G.; Hutchison, S.K.; Franca, M.; Egholm, M.; et al. Novel borna virus in psittacine birds with proventricular dilatation disease. *Emerg. Infect. Dis.* **2008**, *14*, 1883–1886. [CrossRef]

82. Hoffmann, B.; Tappe, D.; Hoper, D.; Herden, C.; Boldt, A.; Mawrin, C.; Niederstrasser, O.; Muller, T.; Jenckel, M.; van der Grinten, E.; et al. A Variegated Squirrel Bornavirus Associated with Fatal Human Encephalitis. *N. Engl. J. Med.* **2015**, *373*, 154–162. [CrossRef] [PubMed]

83. Gonzalez-Dunia, D.; Cubitt, B.; de la Torre, J.C. Mechanism of Borna disease virus entry into cells. *J. Virol.* **1998**, *72*, 783–788. [PubMed]

84. Rott, R.; Becht, H. Natural and experimental Borna disease in animals. *Curr. Top. Microbiol. Immunol.* **1995**, *190*, 17–30. [CrossRef] [PubMed]

85. Berg, M.; Johansson, M.; Montell, H.; Berg, A.L. Wild birds as a possible natural reservoir of Borna disease virus. *Epidemiol. Infect.* **2001**, *127*, 173–178. [CrossRef] [PubMed]

86. Bode, L.; Durrwald, R.; Ludwig, H. Borna virus infections in cattle associated with fatal neurological disease. *Vet. Rec.* **1994**, *135*, 283–284. [CrossRef] [PubMed]

87. Hagiwara, K.; Tsuge, Y.; Asakawa, M.; Kabaya, H.; Okamoto, M.; Miyasho, T.; Taniyama, H.; Ishihara, C.; de la Torre, J.C.; Ikuta, K. Borna disease virus RNA detected in Japanese macaques (Macaca fuscata). *Primates* **2008**, *49*, 57–64. [CrossRef]

88. Hilbe, M.; Herrsche, R.; Kolodziejek, J.; Nowotny, N.; Zlinszky, K.; Ehrensperger, F. Shrews as reservoir hosts of borna disease virus. *Emerg. Infect. Dis.* **2006**, *12*, 675–677. [CrossRef] [PubMed]

89. Kinnunen, P.M.; Billich, C.; Ek-Kommonen, C.; Henttonen, H.; Kallio, R.K.; Niemimaa, J.; Palva, A.; Staeheli, P.; Vaheri, A.; Vapalahti, O. Serological evidence for Borna disease virus infection in humans, wild rodents and other vertebrates in Finland. *J. Clin. Virol.* **2007**, *38*, 64–69. [CrossRef]

90. Lundgren, A.L.; Zimmermann, W.; Bode, L.; Czech, G.; Gosztonyi, G.; Lindberg, R.; Ludwig, H. Staggering disease in cats: Isolation and characterization of the feline Borna disease virus. *J. Gen. Virol.* **1995**, *76*, 2215–2222. [CrossRef]

91. Malkinson, M.; Weisman, Y.; Ashash, E.; Bode, L.; Ludwig, H. Borna disease in ostriches. *Vet. Rec.* **1993**, *133*, 304. [CrossRef]

92. Durrwald, R.; Kolodziejek, J.; Muluneh, A.; Herzog, S.; Nowotny, N. Epidemiological pattern of classical Borna disease and regional genetic clustering of Borna disease viruses point towards the existence of to-date unknown endemic reservoir host populations. *Microbes Infect.* **2006**, *8*, 917–929. [CrossRef] [PubMed]

93. Staeheli, P.; Sauder, C.; Hausmann, J.; Ehrensperger, F.; Schwemmle, M. Epidemiology of Borna disease virus. *J. Gen. Virol.* **2000**, *81*, 2123–2135. [CrossRef] [PubMed]

94. Weissenbock, H.; Bago, Z.; Kolodziejek, J.; Hager, B.; Palmetzhofer, G.; Durrwald, R.; Nowotny, N. Infections of horses and shrews with Bornaviruses in Upper Austria: A novel endemic area of Borna disease. *Emerg. Microbes Infect.* **2017**, *6*, e52. [CrossRef] [PubMed]

95. Bourg, M.; Herzog, S.; Encarnacao, J.A.; Nobach, D.; Lange-Herbst, H.; Eickmann, M.; Herden, C. Bicolored white-toothed shrews as reservoir for borna disease virus, Bavaria, Germany. *Emerg. Infect. Dis.* **2013**, *19*, 2064–2066. [CrossRef] [PubMed]

96. Nobach, D.; Bourg, M.; Herzog, S.; Lange-Herbst, H.; Encarnacao, J.A.; Eickmann, M.; Herden, C. Shedding of Infectious Borna Disease Virus-1 in Living Bicolored White-Toothed Shrews. *PLoS ONE* **2015**, *10*, e0137018. [CrossRef]

97. Ludwig, H.; Bode, L.; Gosztonyi, G. Borna disease: A persistent virus infection of the central nervous system. *Prog. Med. Virol.* **1988**, *35*, 107–151.

98. Herzog, S.; Pfeuffer, I.; Haberzettl, K.; Feldmann, H.; Frese, K.; Bechter, K.; Richt, J.A. Molecular characterization of Borna disease virus from naturally infected animals and possible links to human disorders. *Arch. Virol. Suppl.* **1997**, *13*, 183–190.

99. Hagiwara, K.; Kamitani, W.; Takamura, S.; Taniyama, H.; Nakaya, T.; Tanaka, H.; Kirisawa, R.; Iwai, H.; Ikuta, K. Detection of Borna disease virus in a pregnant mare and her fetus. *Vet. Microbiol.* **2000**, *72*, 207–216. [CrossRef]

100. Okamoto, M.; Hagiwara, K.; Kamitani, W.; Sako, T.; Hirayama, K.; Kirisawa, R.; Tsuji, M.; Ishihara, C.; Iwai, H.; Kobayashi, T.; et al. Experimental vertical transmission of Borna disease virus in the mouse. *Arch. Virol.* **2003**, *148*, 1557–1568. [CrossRef]

101. Chalmers, R.M.; Thomas, D.R.; Salmon, R.L. Borna disease virus and the evidence for human pathogenicity: A systematic review. *QJM* **2005**, *98*, 255–274. [CrossRef]

102. Nakamura, Y.; Takahashi, H.; Shoya, Y.; Nakaya, T.; Watanabe, M.; Tomonaga, K.; Iwahashi, K.; Ameno, K.; Momiyama, N.; Taniyama, H.; et al. Isolation of Borna disease virus from human brain tissue. *J. Virol.* **2000**, *74*, 4601–4611. [CrossRef] [PubMed]

103. Brnic, D.; Stevanovic, V.; Cochet, M.; Agier, C.; Richardson, J.; Montero-Menei, C.N.; Milhavet, O.; Eloit, M.; Coulpier, M. Borna disease virus infects human neural progenitor cells and impairs neurogenesis. *J. Virol.* **2012**, *86*, 2512–2522. [CrossRef] [PubMed]

104. Scordel, C.; Huttin, A.; Cochet-Bernoin, M.; Szelechowski, M.; Poulet, A.; Richardson, J.; Benchoua, A.; Gonzalez-Dunia, D.; Eloit, M.; Coulpier, M. Borna disease virus phosphoprotein impairs the developmental program controlling neurogenesis and reduces human GABAergic neurogenesis. *PLoS Pathog.* **2015**, *11*, e1004859. [CrossRef] [PubMed]

105. Schwemmle, M.; Jehle, C.; Formella, S.; Staeheli, P. Sequence similarities between human bornavirus isolates and laboratory strains question human origin. *Lancet* **1999**, *354*, 1973–1974. [CrossRef]

106. Hornig, M.; Briese, T.; Licinio, J.; Khabbaz, R.F.; Altshuler, L.L.; Potkin, S.G.; Schwemmle, M.; Siemetzki, U.; Mintz, J.; Honkavuori, K.; et al. Absence of evidence for bornavirus infection in schizophrenia, bipolar disorder and major depressive disorder. *Mol. Psychiatry* **2012**, *17*, 486–493. [CrossRef] [PubMed]

107. Korn, K.; Coras, R.; Bobinger, T.; Herzog, S.M.; Lucking, H.; Stohr, R.; Huttner, H.B.; Hartmann, A.; Ensser, A. Fatal Encephalitis Associated with Borna Disease Virus 1. *N. Engl. J. Med.* **2018**, *379*, 1375–1377. [CrossRef]

108. Schlottau, K.; Forth, L.; Angstwurm, K.; Hoper, D.; Zecher, D.; Liesche, F.; Hoffmann, B.; Kegel, V.; Seehofer, D.; Platen, S.; et al. Fatal Encephalitic Borna Disease Virus 1 in Solid-Organ Transplant Recipients. *N. Engl. J. Med.* **2018**, *379*, 1377–1379. [CrossRef]

109. Richt, J.A.; Grabner, A.; Herzog, S. Borna disease in horses. *Vet. Clin. N. Am. Equine Pract.* **2000**, *16*, 579–595. [CrossRef]

110. Ludwig, H.; Bode, L. Borna disease virus: New aspects on infection, disease, diagnosis and epidemiology. *Rev. Sci. Tech.* **2000**, *19*, 259–288. [CrossRef]

111. Beck, C.; Jimenez-Clavero, M.A.; Leblond, A.; Durand, B.; Nowotny, N.; Leparc-Goffart, I.; Zientara, S.; Jourdain, E.; Lecollinet, S. Flaviviruses in Europe: Complex circulation patterns and their consequences for the diagnosis and control of West Nile disease. *Int. J. Environ. Res. Public Health* **2013**, *10*, 6049–6083. [CrossRef]
112. Melian, E.B.; Hinzman, E.; Nagasaki, T.; Firth, A.E.; Wills, N.M.; Nouwens, A.S.; Blitvich, B.J.; Leung, J.; Funk, A.; Atkins, J.F.; et al. NS1' of flaviviruses in the Japanese encephalitis virus serogroup is a product of ribosomal frameshifting and plays a role in viral neuroinvasiveness. *J. Virol.* **2010**, *84*, 1641–1647. [CrossRef] [PubMed]
113. Pradier, S.; Lecollinet, S.; Leblond, A. West Nile virus epidemiology and factors triggering change in its distribution in Europe. *Rev. Sci. Tech.* **2012**, *31*, 829–844. [CrossRef] [PubMed]
114. Gardner, I.A.; Wong, S.J.; Ferraro, G.L.; Balasuriya, U.B.; Hullinger, P.J.; Wilson, W.D.; Shi, P.Y.; MacLachlan, N.J. Incidence and effects of West Nile virus infection in vaccinated and unvaccinated horses in California. *Vet. Res.* **2007**, *38*, 109–116. [CrossRef] [PubMed]
115. Zeller, H.G.; Schuffenecker, I. West Nile virus: An overview of its spread in Europe and the Mediterranean basin in contrast to its spread in the Americas. *Eur. J. Clin. Microbiol. Infect. Dis.* **2004**, *23*, 147–156. [CrossRef] [PubMed]
116. Autorino, G.L.; Battisti, A.; Deubel, V.; Ferrari, G.; Forletta, R.; Giovannini, A.; Lelli, R.; Murri, S.; Scicluna, M.T. West Nile virus epidemic in horses, Tuscany region, Italy. *Emerg. Infect. Dis.* **2002**, *8*, 1372–1378. [CrossRef]
117. Ceianu, C.S.; Ungureanu, A.; Nicolescu, G.; Cernescu, C.; Nitescu, L.; Tardei, G.; Petrescu, A.; Pitigoi, D.; Martin, D.; Ciulacu-Purcarea, V.; et al. West nile virus surveillance in Romania: 1997–2000. *Viral Immunol.* **2001**, *14*, 251–262. [CrossRef]
118. Murgue, B.; Murri, S.; Triki, H.; Deubel, V.; Zeller, H.G. West Nile in the Mediterranean basin: 1950–2000. *Ann. N. Y. Acad. Sci.* **2001**, *951*, 117–126. [CrossRef]
119. Chancey, C.; Grinev, A.; Volkova, E.; Rios, M. The global ecology and epidemiology of West Nile virus. *BioMed Res. Int.* **2015**, *2015*, 376230. [CrossRef]
120. Bakonyi, T.; Hubalek, Z.; Rudolf, I.; Nowotny, N. Novel flavivirus or new lineage of West Nile virus, central Europe. *Emerg. Infect. Dis.* **2005**, *11*, 225–231. [CrossRef]
121. Hernandez-Triana, L.M.; Jeffries, C.L.; Mansfield, K.L.; Carnell, G.; Fooks, A.R.; Johnson, N. Emergence of west nile virus lineage 2 in europe: A review on the introduction and spread of a mosquito-borne disease. *Front. Public Health* **2014**, *2*, 271. [CrossRef]
122. Michel, F.; Sieg, M.; Fischer, D.; Keller, M.; Eiden, M.; Reuschel, M.; Schmidt, V.; Schwehn, R.; Rinder, M.; Urbaniak, S.; et al. Evidence for West Nile Virus and Usutu Virus Infections in Wild and Resident Birds in Germany, 2017 and 2018. *Viruses* **2019**, *11*, 674. [CrossRef] [PubMed]
123. Van den Hurk, A.F.; Hall-Mendelin, S.; Webb, C.E.; Tan, C.S.; Frentiu, F.D.; Prow, N.A.; Hall, R.A. Role of enhanced vector transmission of a new West Nile virus strain in an outbreak of equine disease in Australia in 2011. *Parasites Vectors* **2014**, *7*, 586. [CrossRef] [PubMed]
124. Kaiser, J.A.; Wang, T.; Barrett, A.D. Virulence determinants of West Nile virus: How can these be used for vaccine design? *Future Virol.* **2017**, *12*, 283–295. [CrossRef] [PubMed]
125. Langevin, S.A.; Bowen, R.A.; Reisen, W.K.; Andrade, C.C.; Ramey, W.N.; Maharaj, P.D.; Anishchenko, M.; Kenney, J.L.; Duggal, N.K.; Romo, H.; et al. Host competence and helicase activity differences exhibited by West Nile viral variants expressing NS3-249 amino acid polymorphisms. *PLoS ONE* **2014**, *9*, e100802. [CrossRef]
126. Brault, A.C.; Huang, C.Y.; Langevin, S.A.; Kinney, R.M.; Bowen, R.A.; Ramey, W.N.; Panella, N.A.; Holmes, E.C.; Powers, A.M.; Miller, B.R. A single positively selected West Nile viral mutation confers increased virogenesis in American crows. *Nat. Genet.* **2007**, *39*, 1162–1166. [CrossRef]
127. Dridi, M.; Van Den Berg, T.; Lecollinet, S.; Lambrecht, B. Evaluation of the pathogenicity of West Nile virus (WNV) lineage 2 strains in a SPF chicken model of infection: NS3-249Pro mutation is neither sufficient nor necessary for conferring virulence. *Vet. Res.* **2015**, *46*, 130. [CrossRef]
128. Randolph, S.E.; Gern, L.; Nuttall, P.A. Co-feeding ticks: Epidemiological significance for tick-borne pathogen transmission. *Parasitol. Today* **1996**, *12*, 472–479. [CrossRef]
129. Labuda, M.; Jones, L.D.; Williams, T.; Nuttall, P.A. Enhancement of tick-borne encephalitis virus transmission by tick salivary gland extracts. *Med. Vet. Entomol.* **1993**, *7*, 193–196. [CrossRef]

130. Holzmann, H.; Aberle, S.W.; Stiasny, K.; Werner, P.; Mischak, A.; Zainer, B.; Netzer, M.; Koppi, S.; Bechter, E.; Heinz, F.X. Tick-borne encephalitis from eating goat cheese in a mountain region of Austria. *Emerg. Infect. Dis.* **2009**, *15*, 1671–1673. [CrossRef]

131. Jones, L.D.; Gaunt, M.; Hails, R.S.; Laurenson, K.; Hudson, P.J.; Reid, H.; Henbest, P.; Gould, E.A. Transmission of louping ill virus between infected and uninfected ticks co-feeding on mountain hares. *Med. Vet. Entomol.* **1997**, *11*, 172–176. [CrossRef]

132. Jeffries, C.L.; Mansfield, K.L.; Phipps, L.P.; Wakeley, P.R.; Mearns, R.; Schock, A.; Bell, S.; Breed, A.C.; Fooks, A.R.; Johnson, N. Louping ill virus: An endemic tick-borne disease of Great Britain. *J. Gen. Virol.* **2014**, *95*, 1005–1014. [CrossRef] [PubMed]

133. Timoney, P.J.; Donnelly, W.J.; Clements, L.O.; Fenlon, M. Encephalitis caused by louping ill virus in a group of horses in Ireland. *Equine Vet. J.* **1976**, *8*, 113–117. [CrossRef] [PubMed]

134. Klaus, C.; Horugel, U.; Hoffmann, B.; Beer, M. Tick-borne encephalitis virus (TBEV) infection in horses: Clinical and laboratory findings and epidemiological investigations. *Vet. Microbiol.* **2013**, *163*, 368–372. [CrossRef] [PubMed]

135. Waldvogel, A.; Matile, H.; Wegmann, C.; Wyler, R.; Kunz, C. Tick-borne encephalitis in the horse. *Schweiz. Arch. Tierheilkd.* **1981**, *123*, 227–233. [PubMed]

136. Rushton, J.O.; Lecollinet, S.; Hubalek, Z.; Svobodova, P.; Lussy, H.; Nowotny, N. Tick-borne encephalitis virus in horses, Austria, 2011. *Emerg. Infect. Dis.* **2013**, *19*, 635–637. [CrossRef] [PubMed]

137. Csank, T.; Drzewniokova, P.; Korytar, L.; Major, P.; Gyuranecz, M.; Pistl, J.; Bakonyi, T. A Serosurvey of Flavivirus Infection in Horses and Birds in Slovakia. *Vector Borne Zoonotic Dis.* **2018**, *18*, 206–213. [CrossRef] [PubMed]

138. Suss, J. Tick-borne encephalitis 2010: Epidemiology, risk areas, and virus strains in Europe and Asia-An overview. *Ticks Tick-Borne Dis.* **2011**, *2*, 2–15. [CrossRef]

139. Jaaskelainen, A.E.; Sironen, T.; Murueva, G.B.; Subbotina, N.; Alekseev, A.N.; Castren, J.; Alitalo, I.; Vaheri, A.; Vapalahti, O. Tick-borne encephalitis virus in ticks in Finland, Russian Karelia and Buryatia. *J. Gen. Virol.* **2010**, *91*, 2706–2712. [CrossRef]

140. Balseiro, A.; Royo, L.J.; Martinez, C.P.; Fernandez de Mera, I.G.; Hofle, U.; Polledo, L.; Marreros, N.; Casais, R.; Marin, J.F. Louping ill in goats, Spain, 2011. *Emerg. Infect. Dis.* **2012**, *18*, 976–978. [CrossRef]

141. Ytrehus, B.; Vainio, K.; Dudman, S.G.; Gilray, J.; Willoughby, K. Tick-borne encephalitis virus and louping-ill virus may co-circulate in Southern Norway. *Vector Borne Zoonotic Dis. (Larchmt. N.Y.)* **2013**, *13*, 762–768. [CrossRef]

142. Mustafa, Y.M.; Meuren, L.M.; Coelho, S.V.A.; de Arruda, L.B. Pathways Exploited by Flaviviruses to Counteract the Blood-Brain Barrier and Invade the Central Nervous System. *Front. Microbiol.* **2019**, *10*, 525. [CrossRef] [PubMed]

143. Verma, S.; Lo, Y.; Chapagain, M.; Lum, S.; Kumar, M.; Gurjav, U.; Luo, H.; Nakatsuka, A.; Nerurkar, V.R. West Nile virus infection modulates human brain microvascular endothelial cells tight junction proteins and cell adhesion molecules: Transmigration across the in vitro blood-brain barrier. *Virology* **2009**, *385*, 425–433. [CrossRef] [PubMed]

144. Wang, P.; Dai, J.; Bai, F.; Kong, K.F.; Wong, S.J.; Montgomery, R.R.; Madri, J.A.; Fikrig, E. Matrix metalloproteinase 9 facilitates West Nile virus entry into the brain. *J. Virol.* **2008**, *82*, 8978–8985. [CrossRef] [PubMed]

145. Palus, M.; Vancova, M.; Sirmarova, J.; Elsterova, J.; Perner, J.; Ruzek, D. Tick-borne encephalitis virus infects human brain microvascular endothelial cells without compromising blood-brain barrier integrity. *Virology* **2017**, *507*, 110–122. [CrossRef]

146. Cho, H.; Diamond, M.S. Immune responses to West Nile virus infection in the central nervous system. *Viruses* **2012**, *4*, 3812–3830. [CrossRef]

147. Samuel, M.A.; Wang, H.; Siddharthan, V.; Morrey, J.D.; Diamond, M.S. Axonal transport mediates West Nile virus entry into the central nervous system and induces acute flaccid paralysis. *Proc. Natl. Acad. Sci. USA* **2007**, *104*, 17140–17145. [CrossRef]

148. Xiao, S.Y.; Guzman, H.; Zhang, H.; Travassos da Rosa, A.P.; Tesh, R.B. West Nile virus infection in the golden hamster (Mesocricetus auratus): A model for West Nile encephalitis. *Emerg. Infect. Dis.* **2001**, *7*, 714–721. [CrossRef]

149. Potokar, M.; Jorgacevski, J.; Zorec, R. Astrocytes in Flavivirus Infections. *Int. J. Mol. Sci.* **2019**, *20*, 691. [CrossRef]
150. Palus, M.; Bily, T.; Elsterova, J.; Langhansova, H.; Salat, J.; Vancova, M.; Ruzek, D. Infection and injury of human astrocytes by tick-borne encephalitis virus. *J. Gen. Virol.* **2014**, *95*, 2411–2426. [CrossRef]
151. Donadieu, E.; Lowenski, S.; Servely, J.L.; Laloy, E.; Lilin, T.; Nowotny, N.; Richardson, J.; Zientara, S.; Lecollinet, S.; Coulpier, M. Comparison of the neuropathology induced by two West Nile virus strains. *PLoS ONE* **2013**, *8*, e84473. [CrossRef]
152. Hyde, J.; Nettleton, P.; Marriott, L.; Willoughby, K. Louping ill in horses. *Vet. Rec.* **2007**, *160*, 532. [CrossRef] [PubMed]
153. Beck, C.; Lowenski, S.; Durand, B.; Bahuon, C.; Zientara, S.; Lecollinet, S. Improved reliability of serological tools for the diagnosis of West Nile fever in horses within Europe. *PLoS Negl. Trop. Dis.* **2017**, *11*, e0005936. [CrossRef] [PubMed]
154. Chaskopoulou, A.; L'Ambert, G.; Petric, D.; Bellini, R.; Zgomba, M.; Groen, T.A.; Marrama, L.; Bicout, D.J. Ecology of West Nile virus across four European countries: Review of weather profiles, vector population dynamics and vector control response. *Parasites Vectors* **2016**, *9*, 482. [CrossRef]
155. Gossner, C.M.; Marrama, L.; Carson, M.; Allerberger, F.; Calistri, P.; Dilaveris, D.; Lecollinet, S.; Morgan, D.; Nowotny, N.; Paty, M.C.; et al. West Nile virus surveillance in Europe: Moving towards an integrated animal-human-vector approach. *Euro Surveill.* **2017**, *22*. [CrossRef] [PubMed]
156. European Medicines Agency. *EPAR Product Information: Equip WNV 2015*. Available online: http://www.ema.europa.eu/docs/en_GB/document_library/EPAR_-_Product_Information/veterinary/000137/WC500063683.pdf (accessed on 12 September 2019).
157. Minke, J.M.; Siger, L.; Karaca, K.; Austgen, L.; Gordy, P.; Bowen, R.; Renshaw, R.W.; Loosmore, S.; Audonnet, J.C.; Nordgren, B. Recombinant canarypoxvirus vaccine carrying the prM/E genes of West Nile virus protects horses against a West Nile virus-mosquito challenge. *Arch. Virol. Suppl.* **2004**, *18*, 221–230.
158. El Garch, H.; Minke, J.M.; Rehder, J.; Richard, S.; Edlund Toulemonde, C.; Dinic, S.; Andreoni, C.; Audonnet, J.C.; Nordgren, R.; Juillard, V. A West Nile virus (WNV) recombinant canarypox virus vaccine elicits WNV-specific neutralizing antibodies and cell-mediated immune responses in the horse. *Vet. Immunol. Immunopathol.* **2008**, *123*, 230–239. [CrossRef]
159. Guy, B.; Guirakhoo, F.; Barban, V.; Higgs, S.; Monath, T.P.; Lang, J. Preclinical and clinical development of YFV 17D-based chimeric vaccines against dengue, West Nile and Japanese encephalitis viruses. *Vaccine* **2010**, *28*, 632–649. [CrossRef]
160. Durand, B.; Lecollinet, S.; Beck, C.; Martinez-Lopez, B.; Balenghien, T.; Chevalier, V. Identification of hotspots in the European union for the introduction of four zoonotic arboviroses by live animal trade. *PLoS ONE* **2013**, *8*, e70000. [CrossRef]
161. Pfeffer, M.; Dobler, G. Emergence of zoonotic arboviruses by animal trade and migration. *Parasites Vectors* **2010**, *3*, 35. [CrossRef]
162. Diaz, A.; Coffey, L.L.; Burkett-Cadena, N.; Day, J.F. Reemergence of St. Louis Encephalitis Virus in the Americas. *Emerg. Infect. Dis.* **2018**, *24*. [CrossRef]
163. Mansfield, K.L.; Hernandez-Triana, L.M.; Banyard, A.C.; Fooks, A.R.; Johnson, N. Japanese encephalitis virus infection, diagnosis and control in domestic animals. *Vet. Microbiol.* **2017**, *201*, 85–92. [CrossRef] [PubMed]
164. Prow, N.A.; Tan, C.S.; Wang, W.; Hobson-Peters, J.; Kidd, L.; Barton, A.; Wright, J.; Hall, R.A.; Bielefeldt-Ohmann, H. Natural exposure of horses to mosquito-borne flaviviruses in south-east Queensland, Australia. *Int. J. Environ. Res. Public Health* **2013**, *10*, 4432–4443. [CrossRef] [PubMed]
165. Morita, K.; Nabeshima, T.; Buerano, C.C. Japanese encephalitis. *Rev. Sci. Tech.* **2015**, *34*, 441–452. [CrossRef] [PubMed]
166. Ravanini, P.; Huhtamo, E.; Ilaria, V.; Crobu, M.G.; Nicosia, A.M.; Servino, L.; Rivasi, F.; Allegrini, S.; Miglio, U.; Magri, A.; et al. Japanese encephalitis virus RNA detected in Culex pipiens mosquitoes in Italy. *Euro Surveill.* **2012**, *17*. [CrossRef] [PubMed]
167. Preziuso, S.; Mari, S.; Mariotti, F.; Rossi, G. Detection of Japanese Encephalitis Virus in bone marrow of healthy young wild birds collected in 1997-2000 in Central Italy. *Zoonoses Public Health* **2018**, *65*, 798–804. [CrossRef]

168. Reisen, W.K.; Lothrop, H.D.; Wheeler, S.S.; Kennsington, M.; Gutierrez, A.; Fang, Y.; Garcia, S.; Lothrop, B. Persistent West Nile virus transmission and the apparent displacement St. Louis encephalitis virus in southeastern California, 2003–2006. *J. Med. Entomol.* **2008**, *45*, 494–508. [CrossRef]

169. Fang, Y.; Reisen, W.K. Previous infection with West Nile or St. Louis encephalitis viruses provides cross protection during reinfection in house finches. *Am. J. Trop. Med. Hyg.* **2006**, *75*, 480–485. [CrossRef]

170. Spinsanti, L.I.; Diaz, L.A.; Glatstein, N.; Arselan, S.; Morales, M.A.; Farias, A.A.; Fabbri, C.; Aguilar, J.J.; Re, V.; Frias, M.; et al. Human outbreak of St. Louis encephalitis detected in Argentina, 2005. *J. Clin. Virol.* **2008**, *42*, 27–33. [CrossRef]

171. Rico-Hesse, R. Venezuelan equine encephalomyelitis. *Vet. Clin. N. Am. Equine Pract.* **2000**, *16*, 553–563. [CrossRef]

172. Bergren, N.A.; Auguste, A.J.; Forrester, N.L.; Negi, S.S.; Braun, W.A.; Weaver, S.C. Western equine encephalitis virus: Evolutionary analysis of a declining alphavirus based on complete genome sequences. *J. Virol.* **2014**, *88*, 9260–9267. [CrossRef]

173. Brault, A.C.; Powers, A.M.; Holmes, E.C.; Woelk, C.H.; Weaver, S.C. Positively charged amino acid substitutions in the e2 envelope glycoprotein are associated with the emergence of venezuelan equine encephalitis virus. *J. Virol.* **2002**, *76*, 1718–1730. [CrossRef] [PubMed]

174. Arechiga-Ceballos, N.; Aguilar-Setien, A. Alphaviral equine encephalomyelitis (Eastern, Western and Venezuelan). *Rev. Sci. Tech.* **2015**, *34*, 491–501. [CrossRef] [PubMed]

175. Deardorff, E.R.; Forrester, N.L.; Travassos-da-Rosa, A.P.; Estrada-Franco, J.G.; Navarro-Lopez, R.; Tesh, R.B.; Weaver, S.C. Experimental infection of potential reservoir hosts with Venezuelan equine encephalitis virus, Mexico. *Emerg. Infect. Dis.* **2009**, *15*, 519–525. [CrossRef] [PubMed]

176. Aguilar, P.V.; Paessler, S.; Carrara, A.S.; Baron, S.; Poast, J.; Wang, E.; Moncayo, A.C.; Anishchenko, M.; Watts, D.; Tesh, R.B.; et al. Variation in interferon sensitivity and induction among strains of eastern equine encephalitis virus. *J. Virol.* **2005**, *79*, 11300–11310. [CrossRef] [PubMed]

177. Weaver, S.C.; Powers, A.M.; Brault, A.C.; Barrett, A.D. Molecular epidemiological studies of veterinary arboviral encephalitides. *Vet. J.* **1999**, *157*, 123–138. [CrossRef] [PubMed]

178. Zientara, S.; Weyer, C.T.; Lecollinet, S. African horse sickness. *Rev. Sci. Tech.* **2015**, *34*, 315–327. [CrossRef] [PubMed]

179. Wilson, A.; Mellor, P. Bluetongue in Europe: Vectors, epidemiology and climate change. *Parasitol. Res.* **2008**, *103*, S69–S77. [CrossRef]

180. Aharonson-Raz, K.; Steinman, A.; Bumbarov, V.; Maan, S.; Maan, N.S.; Nomikou, K.; Batten, C.; Potgieter, C.; Gottlieb, Y.; Mertens, P.; et al. Isolation and phylogenetic grouping of equine encephalosis virus in Israel. *Emerg. Infect. Dis.* **2011**, *17*, 1883–1886. [CrossRef]

181. Reed, S.M.; Toribio, R.E. Equine herpesvirus 1 and 4. *Vet. Clin. N. Am. Equine Pract.* **2004**, *20*, 631–642. [CrossRef]

182. Slater, J.D.; Borchers, K.; Thackray, A.M.; Field, H.J. The trigeminal ganglion is a location for equine herpesvirus 1 latency and reactivation in the horse. *J. Gen. Virol.* **1994**, *75*, 2007–2016. [CrossRef]

183. Roehrig, J.T. West nile virus in the United States—A historical perspective. *Viruses* **2013**, *5*, 3088–3108. [CrossRef] [PubMed]

184. Beck, C.; Despres, P.; Paulous, S.; Vanhomwegen, J.; Lowenski, S.; Nowotny, N.; Durand, B.; Garnier, A.; Blaise-Boisseau, S.; Guitton, E.; et al. A High-Performance Multiplex Immunoassay for Serodiagnosis of Flavivirus-Associated Neurological Diseases in Horses. *BioMed Res. Int.* **2015**, *2015*, 678084. [CrossRef] [PubMed]

185. Cleton, N.B.; van Maanen, K.; Bergervoet, S.A.; Bon, N.; Beck, C.; Godeke, G.J.; Lecollinet, S.; Bowen, R.; Lelli, D.; Nowotny, N.; et al. A Serological Protein Microarray for Detection of Multiple Cross-Reactive Flavivirus Infections in Horses for Veterinary and Public Health Surveillance. *Transbound. Emerg. Dis.* **2017**, *64*, 1801–1812. [CrossRef] [PubMed]

186. Van Eeden, C.; Williams, J.H.; Gerdes, T.G.; van Wilpe, E.; Viljoen, A.; Swanepoel, R.; Venter, M. Shuni virus as cause of neurologic disease in horses. *Emerg. Infect. Dis.* **2012**, *18*, 318–321. [CrossRef]

187. Altan, E.; Li, Y.; Sabino-Santos, G., Jr.; Sawaswong, V.; Barnum, S.; Pusterla, N.; Deng, X.; Delwart, E. Viruses in Horses with Neurologic and Respiratory Diseases. *Viruses* **2019**, *11*, 942. [CrossRef]

188. Cohen, M.L. Changing patterns of infectious disease. *Nature* **2000**, *406*, 762–767. [CrossRef]

189. Martinet, J.P.; Ferte, H.; Failloux, A.B.; Schaffner, F.; Depaquit, J. Mosquitoes of North-Western Europe as Potential Vectors of Arboviruses: A Review. *Viruses* **2019**, *11*, 1059. [CrossRef]

190. Cunze, S.; Koch, L.K.; Kochmann, J.; Klimpel, S. Aedes albopictus and Aedes japonicus—Two invasive mosquito species with different temperature niches in Europe. *Parasites Vectors* **2016**, *9*, 573. [CrossRef]

191. Vogels, C.B.; Goertz, G.P.; Pijlman, G.P.; Koenraadt, C.J. Vector competence of European mosquitoes for West Nile virus. *Emerg. Microbes Infect.* **2017**, *6*, e96. [CrossRef]

192. De Wispelaere, M.; Despres, P.; Choumet, V. European Aedes albopictus and Culex pipiens Are Competent Vectors for Japanese Encephalitis Virus. *PLoS Negl. Trop. Dis.* **2017**, *11*, e0005294. [CrossRef]

193. Huber, K.; Jansen, S.; Leggewie, M.; Badusche, M.; Schmidt-Chanasit, J.; Becker, N.; Tannich, E.; Becker, S.C. Aedes japonicus japonicus (Diptera: Culicidae) from Germany have vector competence for Japan encephalitis virus but are refractory to infection with West Nile virus. *Parasitol. Res.* **2014**, *113*, 3195–3199. [CrossRef] [PubMed]

Communication

Equid alphaherpesvirus 1 from Italian Horses: Evaluation of the Variability of the *ORF30*, *ORF33*, *ORF34* and *ORF68* Genes

Silvia Preziuso [1,*], Micaela Sgorbini [2], Paola Marmorini [3] and Vincenzo Cuteri [1]

1   School of Biosciences and Veterinary Medicine, University of Camerino, Via Circonvallazione 93/95, 62024 Matelica (MC), Italy
2   Department of Veterinary Sciences, University of Pisa, San Piero a Grado, 56122 Pisa, Italy
3   Private Practitioner, 52100 Arezzo, Italy
*   Correspondence: silvia.preziuso@unicam.it

Received: 28 July 2019; Accepted: 11 September 2019; Published: 13 September 2019

Abstract: *Equid alphaherpesvirus 1* (EHV-1) is an important pathogen of horses. It is spread worldwide and causes significant economic losses. The *ORF33* gene has a conserved region that is often used as target in diagnostic PCR protocols. Single nucleotide point (SNP) mutations in *ORF30* are usually used to distinguish between neuropathogenic and non-neuropathogenic genotypes. An *ORF68* SNP-based scheme has been used for grouping different isolates. Recently, the highest number of variable sites in EHV-1 from the UK has been found in ORF34. In this study, EHV-1 positive samples from Italian horses with a history of abortion were investigated by amplifying and sequencing the *ORF30*, *ORF33*, *ORF34* and *ORF68* genes. Most animals were infected by the neuropathogenic type A2254G. A 118 bp deletion was found at nucleotide positions 701–818 of the *ORF68* gene, making impossible to assign the samples to a known group. Sequencing of the *ORF34* gene with a newly designed nested PCR showed new SNPs. Analysis of these sequences and of those obtained from genetic databases allowed the identification of at least 12 groups. These data add depth to the knowledge of EHV-1 genotypes circulating in Italy.

Keywords: *Equid alphaherpesvirus 1*; horse; PCR; sequencing; *ORF30*; *ORF33*; *ORF34*; *ORF68*

---

1. Introduction

Equid alphaherpesvirus 1 (EHV-1) is a DNA virus belonging to the genus *Varicellovirus* in the family *Herpesviridae*, subfamily *Alphaherpesvirinae*. Infection with EHV-1 is included in the World Organization for Animal Health (OIE) List [1] because it causes abortions, respiratory disease and neurological disease, with significant economic losses in the equine industry worldwide. The genome contains 76 open reading frames (ORFs) predicted to encode functional proteins [2]. The *ORF33* gene, encoding the glycoprotein B (gB), possesses a conserved region that is frequently used as target for diagnostic PCR protocols [3]. The *ORF30* gene, which encodes the DNA polymerase gene, is considered a marker of pathogenicity because the potential to cause neuropathogenicity is significantly higher in EHV-1 strains that carry a single nucleotide point (SNP) mutation in this gene [4]. The A to G mutation at nucleotide position 2254 of ORF30 causes a substitution of asparagine (N) to aspartic acid (D) at amino acid position 752 in the catalytic subunit of the viral DNA polymerase. EHV-1 N752 is referred to as a non-neuropathogenic genotype, and D752 as a neuropathogenic genotype. This single amino acid mutation causes replication to a higher level and longer viremia in experimentally infected horses, when compared to animals infected with EHV-1 lacking this particular mutation [5,6]. Comparison between the genomic sequences of the neuropathogenic strain Ab4 and of the non-neuropathogenic strain V592 showed that the major amino acid residue differences were in a membrane-associated virion component encoded by the *ORF68* gene. In particular, a region of about 600 bp in *ORF68* was particularly polymorphic, and therefore it was tentatively adopted as a marker system for efficiently grouping the isolates into six groups [4]. This method was subsequently used for comparing isolates from different geographical regions [7–12]. Recently a Multi-locus analysis approach, based on sequencing heterologous regions in 26 open reading frames, proved a more comprehensive method of strain typing than only *ORF68* sequencing. [12]. An extensive study, covering at least 80% of the genome for each of 78 EHV-1 isolated between 1982 and 2016 mostly in UK, demonstrated that the V32 protein, encoded by the *ORF34* gene, was the most variable in this viral collection [13]. Considering that genotyping studies on EHV-1 circulating in Italy are very limited [12,14] and that a few sequences of Italian EHV-1 are available in public genetic sequence databases (GenBank and European Nucleotide Archive-ENA, last access 1st July 2019), the aim of this study was to investigate the variability of the *ORF30*, *ORF33*, *ORF34* and *ORF68* genes of EHV-1 positive samples collected from Italian horses with a history of abortion, in comparison with sequences available in genetic databases and the bibliography.

2. Materials and Methods

2.1. Samples

DNA archival samples collected from Italian horses in 2008–2010 and in 2017–2019 were available for this study. Samples from mares that aborted, from aborted fetuses and from a recumbent horse had been collected by veterinarians in sterile containers and had been sent to the laboratory within 24 h in an airtight box containing cold accumulators for the diagnosis of beta-hemolytic C-streptococci infection [15,16]. Some horses were regularly vaccinated against EHV-1 (Table 1). DNA had been obtained immediately after samples arrival to the laboratory by a commercial kit (Genomic DNA isolation kit, Norgen Biotek Corp., Thorold, ON, Canada) following the manufacturer's instructions. Two-hundred DNA samples stored at –20 °C were selected for this study and were tested by nested PCR (nPCR) to detect EHV-1 and EHV-4 [17]. Briefly, primers FC2 (5′-CTTGTGAGATCT AACCGCAC-3′) and RC (5′-GGGTATAGAGCTTTCATGGG-3′) targeting a common sequence of EHV-1 and EHV-4 were used in a first PCR. Subsequently, nPCRs were performed to amplify a 188 bp sequence of EHV-1 with primers FC3 (5′-ATACGATCACATCCAATCCC-3′) and R1 (5′-GCGTTATAGCTATCACGTCC-3′) or to amplify a 677 bp sequence of EHV-4 with primers FC3 and R4 (5′-CCTGCATAATGACAGCAGTG-3′). First round PCR mixture included 50 µL 2× Taq PCR Master Mix (Qiagen, Hilden, Germany), 25 pmol each primer (FC2 and RC), 2 µL DNA, and PCR grade water up to 50 µL final volume. Second round PCRs included 50 µL 2× Taq PCR Master Mix (Qiagen), 25 pmol each primer (FC3 and R1 or FC3 and R4), 5 µL of the product of the first round PCR, and PCR grade water up to 50 µL final volume. Amplification conditions of both first and second round PCRs were 94 °C for 5 min, 40 cycles at 94 °C for 1 min, 60 °C for 1 min and 72 °C for 1 min, with a final extension at 72 °C for 7 min followed by refrigeration at 4 °C. PCR products (10 µL each) were visualized in 1.5% agarose gel. Twenty positive samples collected from twelve different stables located in three different geographical regions (Marche, Tuscany and Veneto) or collected from the same stable but in different years were selected for sequencing and analysis of the genes *ORF30*, *ORF33*, *ORF34* and *ORF68* (Table 1).

Table 1. Samples included in the study.

Code	Stable	Source	Vaccination
08m27	A	Organs	no
08m160	B	Organs	yes
09m34T	B	Organs	yes
09m45	C	NS	yes
09m68	D	NS	yes
09m142	E	NS	yes
09m209	E	NS	yes
09m217	F	NS	yes
10m01	G	Organs	no
10m106	J	Organs	yes
17m07	J	Organs	yes
17m13	K	Organs	yes
17m15	K	NS	yes
18m30	F	Organs	yes
19m04	E	Organs	no
19m05	C	Organs	no
19m08	L	CSF	no
19m10	M	Organs	yes
19m13	N	Organs	no
19m14	M	Organs	no

Organs were obtained by mixing lung and spleen samples from aborted fetuses; nasal swab (NS) samples were obtained from maresthat aborted; cerebrospinal fluid (CSF) sample was obtained from the recumbent horse 19m08. The year of collection is indicated by the numbers befor the "m" letter in the sample code (e.g., 09 means 2009, 19 means 2019). Stables from where the samples were collected are reported as alphabetical letters, no further information is provided to respect their privacy.

2.2. PCR and Sequencing

All primers used in PCR and sequencing reactions are reported in Table 2. Nested PCR protocols were used because single PCR (sPCR) showed limited sensitivity. A 256 bp product of the *ORF30* gene including the polymorphic site A2254G, which is suspected to determine the viral pathogenicity [6], was obtained by nPCR [18]. After alignment of all EHV-1 *ORF33*, *ORF34* and *ORF68* sequences available in GenBank and in ENA, the most variable sequences were selected to be amplified by nPCR. A new set of primers (FC2int/RCint) was designed to amplify a 940 bp sequence of the *ORF33* gene by nPCR using the products obtained by first PCR with primers FC2/RC [17] as template.

After preliminary unsuccessful tests with primers used to amplify the *ORF68* gene by sPCR [10,12], new primers were designed (68p1-Fe/68p1-Re and 68p2-Fi/68p2-Ri) to obtain by nPCR a 774 bp product including the sequence used for grouping the isolates [4]. Studies on the *ORF34* gene are limited and primers to amplify this gene are not described. Two pairs of primers were designed to amplify the entire ORF34 gene by nPCR (1058F/1893R and 1090Fi/1784Ri). The primers are complementary to a terminal sequence of the *ORF33* gene and an initial sequence of the *ORF35* gene. All new primers were designed with Primer3Plus [19] and were tested by Nucleotide BLAST to evaluate their similarity with EHV-1 or with other unspecific sequences. The PCR protocols were optimized by using an EHV-1 isolate as positive control and an EHV-1-negative equine sample as negative control. All new nPCR protocols were dedicated only for sequencing and have been not tested for diagnostic purposes.

After the optimization of the amplification protocols, the mixture of the first PCRs included 25 μL 2× Taq PCR Master Mix (Qiagen), 500 nM each primer (F8/R2, FC2/RC, 1058F/1893R, or 68p1-Fe/68p1-Re), 2 μL DNA, and PCR grade water up to 50 μL final volume. PCR conditions were 94 °C for 5 min, 45 cycles of 94 °C for 1 min, appropriate annealing temperature (Table 2) for 1 min, 72 °C for 1 min, and a final extension of 72 °C for 7 min. The second PCRs were carried out with the same amplification conditions but with primers F7/R3, FC2int/RCint, 1090Fi/ 1784Ri or 68p2-Fi/68p2-Ri; 2 μL of the first PCR products were used as template for nPCRs with primers F7/R3 or 1090Fi/ 1784Ri and 5 μL of the first PCR products were used as template for nPCRs with primers FC2int/RCint or 68p2-Fi/68p2-Ri.

PCR products were visualized in 1.0% agarose gel and positive samples were submitted to an external laboratory for sequencing by Sanger method (BMR Genomics, Padua, Italy). Both sense and antisense strands were sequenced. If discordant results were obtained or if new SNPs were observed in the sequences, PCRs and sequencing were repeated. To limit the number of identical sequences included in GenBank, only representative sequences were deposited (Accession numbers MN226968-MN226990, Table S1).

Table 2. Primers used for PCR and sequencing reactions.

Gene	Primer Name	Sequence (5′–3′)	Product Size (bp)	Annealing Temperature (°C)	Reference
ORF30	F8	GTGGACGGTACCCCGGAC	380	60	[18]
	R2	GTGGGGATTCGCGCCCTCACC			
	F7 *	GGGAGCAAAGGTTCTAGACC	256	60	[18]
	R3 *	AGCCAGTCGCGCAGCAAGATG			
ORF33	FC2	CTTGTGAGATCTAACCGCAC	1181	60	[17]
	RC	GGGTATAGAGCTTTCATGGG			
	FC2int *	CCGCACCTACGACCTAAAAA	940	58	This study
	RCint *	CGATCCCCTGCATAATCACT			
ORF34	1058F	GGCCCCAAGGATATTTAAGC	855	58	This study
	1893R	GTTTGAGGCGGTTACGTCAG			
	1090Fi *	CCGAGGTTTCATCCTCATTC	714	58	This study
	1784Ri *	GCGGACATATTCGTGTCTCA			
ORF68	68p1-Fe	AAGCATTGCCAAACAGTTCC	846	55	This study
	68p1-Re	CGAACACTCCCCAGAGTAGG			
	68p2-Fi *	TGAGCCGACAATGTTTCGTA	774	57	This study
	68p2-Ri *	GTTCCATCCACGTCACGTCT			

* Primers used in sequencing reactions.

2.3. Sequence Analysis

Nucleotide sequences were manually checked and edited with the program BioEdit 7.0.5 [20]. Sequences were aligned by MUSCLE [21]. Phylogenetic tree with representative ORF34 sequences were inferred with the program MEGA 7.0.21 [22]. The best-fitting nucleotide substitution model were estimated; Kimura 2-parameter model with gamma-distributed rates among sites was used for *ORF33* analysis and Tamura-3-parameter model was used for *ORF34* analysis, both with bootstrap values based on 1000 repetitions. Phylogeny was estimated by both the Neighbor-Joining algorithm (NJ) and the maximum likelihood (ML) method.

3. Results

A total of 20 sequences of *ORF30* and *ORF34*, 10 sequences of *ORF33* and seven sequences of *ORF68* were suitable for analysis. The sequences were aligned with those available in public databases and SNPs were investigated.

Two out of 20 samples had adenine (A) in position 2254 of the *ORF30* gene and 18 had guanine (G), showing that most animals were infected by the EHV-1 neuropathogenic variant N752. Other SNPs were not present in the 256 bp sequence analyzed in comparison with the reference strain Ab4.

The nucleotide sequence of the EHV-1 Ab4 strain (GenBank Accession number AY665713.1) [2] served as a basis also for the comparison of nucleotide changes in the *ORF33*, *ORF34* and *ORF68* genes.

A limited variability of the *ORF33* sequences was observed (Figure 1). A non-synonymous SNP was found in seven out of 10 samples at position A1526T (Table 3), corresponding to the amino acidic substitution N509I. The same seven samples showed also a synonymous mutation at position G2391A. Samples 08m27 and 10m01 showed a change at A1531G, corresponding to the amino acidic substitution N601D. Furthermore, longer sequences were obtained by first round PCR from samples 09m34, 09m45, 09m142 and 19m10. Sample 19m10 showed three non-synonymous changes at I810V, I838P and A861G, while no further changes were found in the other three samples (Table 3).

Figure 1. Phylogenetic tree showing the evolutionary history of the *ORF33* sequences from nt 1525 to nt 2409 (Neighbor-Joining method with bootstrap test with 1000 replicates). The evolutionary distances were computed using the Kimura 2-parameter method and are in the units of the number of base substitutions per site. The rate variation among sites was modeled with a gamma distribution (shape parameter = 5). Sequences obtained in this study are marked with a diamond (♦).

Table 3. Nucleotide variations in *ORF33* gene of samples and of the reference isolate Ab4.

Code	ORF33 SNPs					
	nt 1526	nt 1531	nt 2391	nt 2429	nt 2513	nt 2583
Ab4	A	A	G	A	A	C
08m27	A	G	G	-	-	-
08m160			unsp.			
09m34	T	A	A	A	A	C
09m45	T	A	A	A	A	C
09m68	T	A	A	-	-	-
09m142	T	A	A	A	A	C
09m209			unsp.			
09m217			unsp.			
10m01	A	G	G	-	-	-
10m106	T	A	A	-	-	-
17m07	-	-	-	-	-	-
17m13	T	A	A	-	-	-
17m15	T	A	A	-	-	-
18m30	-	-	-	-	-	-
19m04	-	-	-	-	-	-
19m05	-	-	-	-	-	-
19m08	-	-	-	-	-	-
19m10	A	A	G	G	C	G
19m13	-	-	-	-	-	-
19m14	-	-	-	-	-	-

Unsp. means that unspecific products have been obtained and sequenced; - means that negative results were obtained or that parts of sequences are missing. Samples 09m34, 09m45, 09m142 and 19m10 show longer sequences because visible products were obtained by first round PCR and good quality sequences were obtained.

The new nPCR protocol to amplify the *ORF34* gene showed a high sensitivity because all samples that resulted positive according to the diagnostic nPCR method of Wang et al. [17] were also positive by this nPCR and all samples provided a high amount of PCR products, which resulted sufficient and suitable for sequencing. Analysis of the 714 bp *ORF34* nucleotide sequences obtained in this study showed a synonymous mutation at T60C in four samples, one of which showed also two non-synonymous changes at C380T and T410C, corresponding to changes at amino acid positions T127I and V137A respectively. Two samples showed a non-synonymous SNP at C149T, while the other 14 samples showed the same *ORF34* sequence as the reference strain Ab4 (Table 4). A total of 114 *ORF34* gene sequences were obtained from GenBank database and were aligned with the sequences obtained in this study. Sequence analysis suggested that all the EHV-1 *ORF34* gene sequences available in GenBank and all sequences obtained in this study could be categorized tentatively into twelve groups (Table 4). A simplified tree including only selected sequences representative of each observed nucleotide variation is reported in Figure 2.

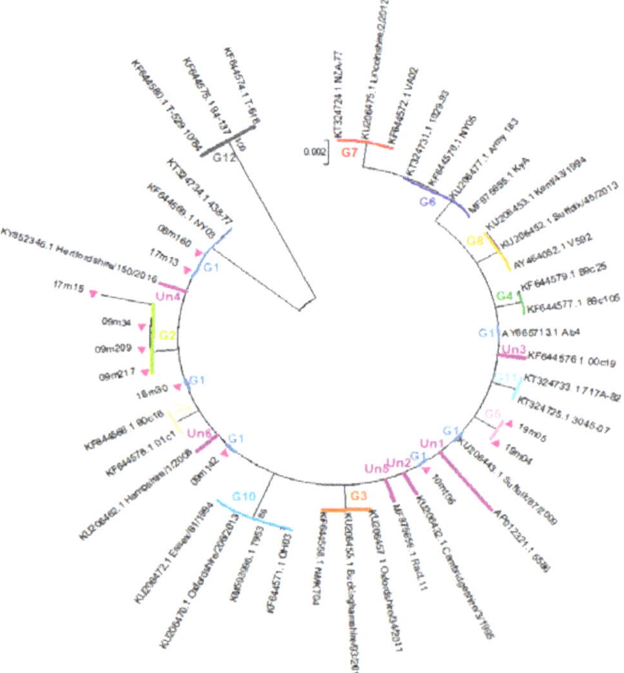

Figure 2. The evolutionary history of the *ORF34* sequences was obtained by Neighbor-Joining method with bootstrap test with 1000 replicates. The evolutionary distances were computed using the Tamura 3-parameter method and are in the units of the number of base substitutions per site. Sequences obtained in this study are marked with a triangle ▲. The letter "G" followed by a number indicates the number of the group where the sequences are located. The letters "Un" followed by a number indicates the sequences not located in any Group.

Table 4. Variations of the 483 bp sequences of the *ORF34* gene are shown.

	Group	33	60	62	71	73	81	95	104	110	115	148	149	156	159	197	216	256	257	282	285	303	304	317	380	391	402	405	410	414	422	428	477	N
AY665713.1_Ab4	1	C	T	C	C	G	G	C	T	G	G	A	C	G	A	A	G	G	C	T	A	C	T	C	C	T	C	A	T	T	C	C	G	70
09m142	1	.	.	.	.	.	.	.	.	.	.	.	.	.	.	.	.	.	.	.	.	.	.	.	.	.	.	.	.	.	.	.	.	
09m209	2	.	C	.	.	.	.	.	.	.	.	.	.	.	.	.	.	.	.	.	.	.	.	.	.	.	.	.	.	.	.	.	.	0
09m34	2	.	C	.	.	.	.	.	.	.	.	.	.	.	.	.	.	.	.	.	.	.	.	.	.	.	.	.	.	.	.	.	.	
17m15	2	.	C	.	.	.	.	.	.	.	.	.	.	.	.	.	.	.	.	.	.	.	.	.	T*	.	.	.	C*	.	.	.	.	
KU206455.1_Buckin.93/2011	3	.	.	T*	.	.	.	.	.	.	.	.	.	.	.	.	.	.	.	.	.	.	.	.	.	.	.	.	.	.	.	.	.	5
KF644568.1_NMKT04	3	.	.	T*	.	.	.	.	.	.	.	.	.	.	.	.	.	.	.	.	.	.	.	.	.	.	.	.	.	.	.	.	.	
KF644579.1_89c25	4	.	.	.	.	A*	.	.	.	.	.	.	.	.	.	.	.	.	.	.	.	.	.	.	.	.	.	.	.	.	.	.	.	2
KF644577.1_89c105	4	.	.	.	.	A*	.	.	.	.	.	.	.	.	.	.	.	.	.	.	.	.	.	.	.	.	.	.	.	.	.	.	.	
19m04	5	.	.	.	.	.	.	.	.	.	.	.	T*	.	.	.	.	.	.	.	.	.	.	.	.	.	.	.	.	.	.	.	.	0
19m05	5	.	.	.	.	.	.	.	.	.	.	.	T*	.	.	.	.	.	.	.	.	.	.	.	.	.	.	.	.	.	.	.	.	
MF975655.1_KyA	6	.	.	.	.	.	.	.	.	.	.	.	.	T*	.	.	.	.	.	.	.	.	.	.	.	.	.	.	.	.	.	.	.	4
KF644570.1_NY05	6	.	.	.	.	.	.	.	.	.	.	.	.	T*	.	.	.	.	.	.	.	.	.	.	.	.	.	.	.	.	.	.	.	
KT324724.1_NZA-77	7	.	.	.	.	.	.	.	.	.	.	.	.	T*	.	.	.	.	.	.	.	A	.	.	.	.	.	.	.	.	.	.	.	6
KU206475.1_Lincoln.2/2012	7	.	.	.	.	.	.	.	.	.	.	.	.	T*	.	.	.	.	.	.	.	A	.	.	.	.	.	.	.	.	.	.	.	
KU206453.1_Kent/43/1994	8	.	.	.	.	.	.	.	.	.	.	.	.	.	.	G*	.	.	.	.	.	.	.	.	.	.	.	.	.	.	.	.	.	4
AY464052.1_V592	8	.	.	.	.	.	.	.	.	.	.	.	.	.	.	G*	.	.	.	.	.	.	.	.	.	.	.	.	.	.	.	.	.	
KF644578.1_01c1	9	.	.	.	.	.	.	.	.	.	.	.	.	.	.	.	.	.	T*	.	.	.	.	.	.	.	.	.	.	.	.	.	.	2
KF644566.1_90c16	9	.	.	.	.	.	.	.	.	.	.	.	.	.	.	.	.	.	T*	.	.	.	.	.	.	.	.	.	.	.	.	.	.	

Table 4. *Cont.*

Sequence	Group	33	60	62	71	73	81	95	104	110	115	148	149	156	159	197	216	256	257	282	285	303	304	317	380	391	402	405	410	414	422	428	477	N
KU206470.1_Oxfor.206/2013	10	.	.	.	.	.	.	.	.	.	.	.	.	.	.	.	.	.	.	.	C	.	.	.	.	.	.	G	.	.	.	.	.	
KM593996.1_T953	10	.	.	.	.	.	.	.	.	.	.	.	.	.	.	.	.	.	.	.	C	.	.	.	.	.	.	G	.	.	.	.	.	10
KT324733.1_717A-82	11	.	.	.	.	.	.	.	.	.	.	.	.	.	.	.	.	.	.	.	.	.	.	.	.	.	.	.	.	C	.	.	.	
KT324725.1_3045-07	11	.	.	.	.	.	.	.	.	.	.	.	.	.	.	.	.	.	.	.	.	.	.	.	.	.	.	.	.	C	.	.	.	2
KF644580.1_T-529_10/84	12	T	.	.	A*	.	A	.	C*	.	C*	G*	.	.	G	.	A	.	.	C	.	.	.	.	.	.	T	.	.	.	.	.	A	
KF644575.1_94-137	12	T	.	.	A*	.	A	.	C*	.	C*	G*	.	.	G	.	A	.	.	C	.	.	.	.	.	.	T	.	.	.	.	.	A	3
KF644574.1_T-616	12	T	.	.	A*	.	A	.	C*	.	C*	G*	.	.	G	.	A	.	.	C	.	.	.	.	.	.	T	.	.	.	.	.	A	
AP012321.1_5586	Un1	.	.	.	.	.	.	T*	.	A	.	.	.	.	.	.	.	.	.	.	.	.	.	.	.	.	.	.	.	.	.	T*	.	
KU206432.1_Cambrid..3/1995	Un2	.	.	.	.	.	.	.	.	.	.	.	.	.	.	.	.	C*	.	.	.	.	.	.	.	.	.	.	.	.	.	.	.	
KF644576.1_00c19	Un3	.	.	.	.	.	.	.	.	.	.	.	.	.	.	.	.	.	.	.	.	.	C*	.	.	.	.	.	.	.	.	.	.	6
KY852346.1_Hertf.150/2016	Un4	.	.	.	.	.	.	.	.	.	.	.	.	.	.	.	.	.	.	.	.	.	.	T*	.	.	.	.	.	.	.	.	.	
MP975656.1_RacL11	Un5	.	.	.	.	.	.	.	.	.	.	.	.	.	.	.	.	.	.	.	.	.	.	.	.	C*	.	.	.	.	.	.	.	
KU206462.1_Hampsh.1/2008	Un6	.	.	.	.	.	.	.	.	.	.	.	.	.	.	.	.	.	.	.	.	.	.	.	.	.	.	.	.	.	T*	.	.	

Variations marked with an asterisk (*) are nonsynonymous point mutations, which changes the single nucleotide into a codon that does not translate into the same amino acid; mutations without an asterisk are synonymous mutations. Dots (.) indicate sequence identity. Only representative samples obtained in this study or representative sequences of the 114 *ORF34* sequences available in GenBank are shown in each group with similar type of mutations; sequences identical to the selected representative sequences were not included in the table. A group was generated when at least two samples or database sequences showed the same variations at the same positions. Single samples with a particular sequence variation were included in the Unassigned group (Un). Six different unassigned groups were found and named from Un1 to Un6. Column N indicates the number of sequences present in GenBank (last access 21st July 2019) and showing nucleotide variations typical of each specific group. The isolate Buckinghamshire/9/93 was tentatively included in Group 1 although it had a frameshift caused by a deletion of nucleotides 231–232 [13].

A group was generated when at least two sequences were identical and showed the same variations at the same positions. Single samples showing a unique sequence variation, which was absent in any other sequence, were included in the Unassigned group (Un). In summary, 15 out of 20 sequences obtained in this study were located in Group 1, three samples were located in group 2 and two samples were located in Group 5. Groups 2 and 5 included only sequences obtained from this study, while sequences available in GenBank were located in the other Groups, mainly in Group 1 ($n = 70$) and in Group 10 ($n = 10$) (Table 4).

Analysis of the *ORF68* according to the grouping criteria previously proposed [4] showed that none of the sequences obtained in this study could be included in the six proposed groups, nor in the additional groups proposed later [7,10]. Indeed, in all samples a 118 bp nucleotide deletion was present at positions 701–818, resulting in a shorter amino acid sequence. The same deletion is present in KyA and Racl11 strains (MF975655 and MF975656), both isolated in the USA. *ORF68* sequences of Italian samples are very similar to sequences of KyA and Racl11, but these latter have a SNP at C236A, and KyA has further changes at T689G and T690C (Table 5). All *ORF68* sequences found in this study were identical and sample 09m142 represents all *ORF68* in Table 5.

Table 5. Nucleotide variations in *ORF68* of samples and of isolates grouped according to Nugent et al., 2006 [4].

	Group	236	336	344	620	626	629	689–690	701	710	713	719	738–739	743	755	783	818	821	825
AY665713.1_Ab4	1	C	C	G	C	T	G	TT	G	T	C	G	GG	C	C	G	G	G	C
DQ172400.1_US85_1_1	1	*	.	.	.	.	.	..	.	.	.	.	GGG	.	.	.	.	.	.
DQ172310.1_AR85_1_1	1	*	.	.	.	.	.	..	.	.	.	.	..	.	.	.	.	.	.
MH976707.1_IRL/559/2009	2	.	.	.	.	.	.	..	.	.	.	.	..	.	.	.	.	.	.
DQ172408.1_US89_1_1	2	*	.	.	.	.	.	..	.	.	.	.	..	T	.	.	.	.	.
DQ172394.1_US79_1_1	2	*	.	.	T	.	.	..	.	.	.	.	..	.	.	.	.	.	G
DQ172309.1_AR79_1_1	2	*	.	.	.	.	.	..	.	C	.	.	..	.	.	.	.	.	.
DQ172384.1_US03_5_2	2	*	.	.	.	.	.	..	.	.	.	.	..	.	.	T	.	.	.
MH976708.1_IRL/471/2008	3	.	.	.	.	.	A	..	.	.	.	T	..	.	.	.	.	.	.
MH976706.1_IRL/307/2015	3	.	.	.	.	.	A	..	.	.	.	T	..	.	.	.	.	A	.
DQ172365.1_GB89_2_1	3	*	.	.	.	.	A	..	.	A	.	T	..	.	.	.	.	.	.
MH976709.1_IRL/268/2001	4	.	.	.	.	.	A	..	.	.	.	.	..	.	.	.	.	.	.
DQ172332.1_GB00_1_1	4	*	.	A	.	.	A	..	.	.	.	.	..	.	.	.	.	.	.
MH976705.1_IRL/837/2007	5	.	.	.	.	C	A	..	.	G	A	.	..	.	.	.	.	.	.
DQ172375.1_US01_1_2	5	*	.	.	.	.	A	..	.	G	A	.	..	.	.	.	.	.	.
AY464052.1_V592	6	.	T	.	.	.	A	..	.	.	.	.	..	.	T	.	.	.	.
MH976703.1_IRL/069/1995	6	.	T	.	.	.	A	..	.	.	.	.	..	.	T	.	.	.	.
DQ172359.1_GB85_1_1	6	*	T	.	.	.	A	..	.	.	.	.	..	.	T	.	.	.	.
09m142		.	.	.	.	.	A	..	Start gap	-	-	-	-	-	-	-	End gap	.	.
MF975656.1_RacL11		A	.	.	.	.	A	..	Start gap	-	-	-	-	-	-	-	End gap	.	.
MF975655.1_KyA		A	.	.	.	.	A	GC	Start gap	-	-	-	-	-	-	-	End gap	.	.

Sequences are aligned with reference to the prototype isolate Ab4 (AY665713.1). Dashes (-) indicate gaps in the alignment, dots (. or ..) indicate sequence identity, asterisks (*) indicate nucleotides not available in GenBank or ENA. Sample 09m142 represents all *ORF68* sequences found in this study, which show identical sequences. EHV-1 RacL11 and KyA and all EHV-1 detected in this study showed a 118 bp deletion between nucleotide positions 701 and 818, resulting in the deletion of a sequence of 40 amino acids.

4. Discussion

This study describes the sequence variations in important EHV-1 genes detected in archival samples of Italian horses and contributes to the knowledge of EHV-1 circulating in Italy. Unfortunately, limited data are available on Italian EHV-1 strains; only two sequences of the *ORF30* gene are deposited in GenBank (HM125711.1 and HM125712.1) and three strains have been recently investigated by a multi-locus sequence analysis approach [12], although sequences are not present in genetic databases. Two of these isolates (ITA/055/2011 and ITA/056/2011 were of the non-neuropathogenic type because they showed an adenine at position 2254 [12]. Three isolates (ITA/944/2003, 314102/BS/2009 and 16656/BS/2010) showed the substitution A2254G, that is considered a marker of the neuropathogenic genotype.

In this study we sequenced and deposited in GenBank sequences of *ORF30*, *ORF33*, *ORF34* and *ORF68* of EHV-1 detected in horses from three Italian Regions. A total of 18 out of 20 samples (90%) showed the mutation A2254G in the *ORF30*, confirming that N752D strains are common in outbreaks involving abortion. Although some samples included in this study were from the same stable, many of them were genetically different (*ORF33*, *ORF34*) and were considered distinct strains. The circulation of the N752D variant in Italy appears much higher than in other countries. Thirty-four out of 269 isolated in Ireland between 1990 and 2017 showed N752D [12], and only two out of 56 EHV-1 isolated from aborted fetuses in India had the neuropathogenic marker [8]. The N752D genotype was not found in EHV-1 isolated in Brazil [23], in Turkey [24] and in the 27 strains isolated in Poland between 1993 and 2017 [11]. Even in other countries the prevalence of the neuropathogenic genotype was rather low, as in Japan (2.7%), the USA (10.8–19.4%), Argentina (7.4%), France (24%), and Germany (10.6%) [25–30]. On the contrary, 90 out of 91 EHV-1 infecting equids in Ethiopia [9] and 12 out of 13 EHV-1 in Uruguay [31] had the variant N752D. Previous investigations in Italy showed that more than 60% of EHV-1 isolated mainly from aborted fetuses since 1980s in Italy possessed the mutation N752D in the *ORF30* [14], demonstrating that the hypervirulent type EHV-1 are spread in Italy since decades. Retrospective studies in the USA demonstrated that viruses with the neuropathogenic genotype increased from 3.3% in the 1960s to 14.4% in the 1990s and to 19.4% in the 2000–2006, suggesting that viruses with the neuropathogenic genotype are continuing to increase in prevalence within the latent reservoir of the virus [28]. Although vaccination remains the main tool for reducing viral spread and clinical disease, there is the evidence that vaccination against EHV-1 sometimes does not prevent abortion and spread of the neuropathogenic genotype [32]. In the present study abortion and spread of N752D strains were observed in vaccinated mares., but also in horses that are still unvaccinated, therefore, it is not surprising that the spread of type N752D in Italy has further increased. For this reason it is important to promote the application of corret vaccination programs in horses together with the research on new efficacious vaccines and new vaccination schedules.

The *ORF33* gene sequences obtained from some samples in this study were identical to the sequence of the reference strain Ab4 and to many other sequences present in GenBank [12,13,33]. Most samples showed the SNPs A1526T (N509I) and G2391A (synonymous), as reported in some EHV-1 from UK. Two samples showed a new SNP at position A1531G, corresponding to the amino acidic substitution N601D. Furthermore, sample 19m10 showed three new and non-synonymous changes at I810V, I838P and A861G. These data confirm that this sequence of the *ORF33* gene is generally highly conserved and it is a good target for diagnostic methods, although some SNPs were observed. Although a few single changes in a sequence usually do not affect the diagnostic sensitivity of standard or real-time PCR protocols, continuous monitoring of changes in this sequence are important to avoid false negative results due to viral mutations.

An extensive study on EHV-1 isolated in the UK demonstrated the highest sequence variability in the *ORF34* gene [13]. In the present study, some samples showed an *ORF34* sequence identical to the sequence of the reference strain Ab4. Other samples showed new SNPs that have not been reported before. A complete analysis of the *ORF34* sequences available in GenBank or reported in bibliography [12,13,33] showed that some SNPs are repeated in groups of strains. In the present study a method based on the comparison of SNPs was tentatively proposed for grouping different *ORF34* sequences. Twelve groups were found and named from 1 to 12. Group 1 includes the reference strain Ab4, Group 12 includes the strains isolated from zebra, onager and Thomson's gazelle [34]. Sequences detected in the present study were located in Groups 1, 2 and 5. Most sequences available in GenBank were located in Group 1. Single strains with unique SNPs were provisionally included in the Unassigned group and will be included in a new group when other strains with the same SNPs will be found. Considering that limited investigations have been carried out so far, we can speculate that further SNPs will be found in the *ORF34* gene and that new groups will be described. Studies carried out on the *ORF34* product of EHV-1 suggest that the *ORF34* protein is required for optimal replication of EHV-1 in cultured cells at early times of infection [35]. The impact of different mutations in the *ORF34* gene on viral replication is not known.

Sequencing and analysis of the *ORF68* gene are widely used for grouping EHV-1 [4,7,10,11]. An extensive study on *ORF68* gene of EHV-1 isolated worldwide has been recently carried out [12]. After that six groups have been originally proposed [4], further SNPs have been described and new groups have been proposed [7,10,11]. All *ORF68* sequences detected in this study were not included in any group because they showed a 118 bp deletion in the nucleotide sequence 701–818 that have been not observed in any existing group. In particular, the sequence between nucleotides 1 and 700 was identical to those of the isolate IRL/268/2001 (MH976709.1), which was located in Group 4. However, members of Group 4 do not have the deletion found in this study, and it seems incorrect to locate our sequences in this group. The 118 bp deletion results in an amino acid sequence shorter than others and with a different sequence of the terminal 10 amino acids, resulting in unknown biological consequences. The same 118 bp deletion in the nucleotide sequence 701–818 is present in isolates RacL11 (MF975656.1) and KyA (MF975655.1). RacL11 has been isolated from an aborted foal and is more pathogenic than the attenuated Kentucky A (KyA), which is a candidate vaccine strain [36]. However, these isolates differ to our samples because they show also the variation C236A, that is not present in our samples.

5. Conclusions

This study describes the genetic characteristics of *ORF30*, *ORF33*, *ORF34* and *ORF68* genes of Italian EHV-1 detected in samples from horses with a history of abortion or recumbency. A very high prevalence of the N752D strains was found. Sequencing of *ORF33* gene confirmed the high conservation of this gene and showed few SNPs, some of which have not been previously reported. In this work a new efficient nPCR protocol to amplify the *ORF34* gene is described. Analysis of *ORF34* sequences obtained in this study and of those available in genetic databases showed new SNPs and suggested the existence of at least 12 different groups. Analysis of the *ORF68* sequences demonstrated an infrequent deletion of 118 bp in all Italian samples.

In conclusion, this study confirms the high variability of the *ORF34* gene and further investigations should assess whether this gene could be a useful marker for epidemiological studies. Furthermore, the presence of the 118 bp deletion in EHV-1 strains from other geographical areas and the pathogenic properties of isolates with this deletion should be evaluated.

Supplementary Materials: The following are available online at http://www.mdpi.com/1999-4915/11/9/851/s1, Table S1: GenBank Accession Numbers of selected sequences obtained in this study.

Author Contributions: V.C., P.M., and M.S. performed clinical examination, collected and provided the samples. S.P. conceived and performed the experiments and analyzed the data. S.P. wrote the article, V.C. Revised the manuscript. All authors read and approved the article.

Funding: This work has been partially supported by the University of Camerino (BVI000071—FAR Preziuso Silvia).

Conflicts of Interest: The authors declare no conflict of interest.

References

1. OIE-Listed Diseases, Infections and Infestations in Force in 2019. Available online: www.oie.int (accessed on 27 July 2019).
2. Telford, E.A.; Watson, M.S.; McBride, K.; Davison, A.J. The DNA sequence of equine herpesvirus-1. *Virology* **1992**, *189*, 304–316. [CrossRef]
3. OIE (World Organization for Animal Health). Chapter 2.5.9: Equine rhinopneumonitis (infection with equid herpesvirus-1 and -4). In *OIE Terrestrial Manual*; OIE: Paris, France, 2017.
4. Nugent, J.; Birch-Machin, I.; Smith, K.C.; Mumford, J.A.; Swann, Z.; Newton, J.R.; Bowden, R.J.; Allen, G.P.; Davis-Poynter, N. Analysis of Equid Herpesvirus 1 Strain Variation Reveals a Point Mutation of the DNA Polymerase Strongly Associated with Neuropathogenic versus Nonneuropathogenic Disease Outbreaks. *J. Virol.* **2006**, *80*, 4047–4060. [CrossRef] [PubMed]
5. Allen, G.P.; Breathnach, C.C. Quantification by real-time PCR of the magnitude and duration of leucocyte-associated viraemia in horses infected with neuropathogenic vs. non-neuropathogenic strains of EHV-1. *Equine Vet. J.* **2006**, *38*, 252–257. [CrossRef] [PubMed]
6. Goodman, L.B.; Loregian, A.; Perkins, G.A.; Nugent, J.; Buckles, E.L.; Mercorelli, B.; Kydd, J.H.; Palù, G.; Smith, K.C.; Osterrieder, N.; et al. A Point Mutation in a Herpesvirus Polymerase Determines Neuropathogenicity. *PLoS Pathog.* **2007**, *3*, e160. [CrossRef] [PubMed]
7. Malik, P.; Balint, A.; Dan, A.; Palfi, V. Molecular characterisation of the *ORF68* region of equine herpesvirus-1 strains isolated from aborted fetuses in Hungary between 1977 and 2008. *Acta Vet. Hung.* **2012**, *60*, 175–187. [CrossRef] [PubMed]
8. Anagha, G.; Gulati, B.R.; Riyesh, T.; Virmani, N. Genetic characterization of equine herpesvirus 1 isolates from abortion outbreaks in India. *Arch. Virol.* **2017**, *162*, 157–163. [CrossRef] [PubMed]
9. Negussie, H.; Gizaw, D.; Tessema, T.S.; Nauwynck, H.J. Equine Herpesvirus-1 Myeloencephalopathy, an Emerging Threat of Working Equids in Ethiopia. *Transbound Emerg. Dis.* **2017**, *64*, 389–397. [CrossRef]
10. Stasiak, K.; Dunowska, M.; Hills, S.F.; Rola, J. Genetic characterization of equid herpesvirus type 1 from cases of abortion in Poland. *Arch. Virol.* **2017**, *162*, 2329–2335. [CrossRef]
11. Matczuk, A.K.; Skarbek, M.; Jackulak, N.A.; Bazanow, B.A. Molecular characterisation of equid alphaherpesvirus 1 strains isolated from aborted fetuses in Poland. *Virol. J.* **2018**, *15*, 186. [CrossRef]
12. Garvey, M.; Lyons, R.; Hector, R.D.; Walsh, C.; Arkins, S.; Cullinane, A. Molecular Characterisation of Equine Herpesvirus 1 Isolates from Cases of Abortion, Respiratory and Neurological Disease in Ireland between 1990 and 2017. *Pathogens* **2019**, *8*, 7. [CrossRef]
13. Bryant, N.A.; Wilkie, G.S.; Russell, C.A.; Compston, L.; Grafham, D.; Clissold, L.; McLay, K.; Medcalf, L.; Newton, R.; Davison, A.J.; et al. Genetic diversity of equine herpesvirus 1 isolated from neurological, abortigenic and respiratory disease outbreaks. *Transbound Emerg. Dis.* **2018**, *65*, 817–832. [CrossRef] [PubMed]
14. Autorino, G.L.; Corradi, V.; Frontoso, R.; Galletti, S.; Manna, G.; Mascioni, A.; Pallone, A.; Ricci, I.; Rosone, F.; Simula, M.; et al. P.3 Gestione di un focolaio neurologico da Equine herpesvirus 1 (EHV-1). In *Workshop Nazionale di Virologia Veterinaria*; Delogu, R., Falcone, E., Monini, M., Ruggeri, F.M., Di Martino, B., Marsilio, F., Monaco, F., Savini, G., Eds.; Istituto Superiore di Sanità: Riassunti, Roma, Italy, 2014; p. 15.
15. Preziuso, S.; Cuteri, V. A Multiplex Polymerase Chain Reaction Assay for Direct Detection and Differentiation of β-Hemolytic Streptococci in Clinical Samples from Horses. *J. Equine Vet. Sci.* **2012**, *32*, 292–296. [CrossRef]
16. Preziuso, S.; Pinho, M.D.; Attili, A.R.; Melo-Cristino, J.; Acke, E.; Midwinter, A.C.; Cuteri, V.; Ramirez, M. PCR based differentiation between Streptococcus dysgalactiae subsp. equisimilis strains isolated from humans and horses. *Comp. Immunol. Microbiol. Infect. Dis.* **2014**, *37*, 169–172. [CrossRef] [PubMed]
17. Wang, L.; Raidal, S.L.; Pizzirani, A.; Wilcox, G.E. Detection of respiratory herpesviruses in foals and adult horses determined by nested multiplex PCR. *Vet. Microbiol.* **2007**, *121*, 18–28. [CrossRef]
18. Allen, G.P. Antemortem detection of latent infection with neuropathogenic strains of equine herpesvirus-1 in horses. *Am. J. Vet. Res.* **2006**, *67*, 1401–1405. [CrossRef]
19. Untergasser, A.; Nijveen, H.; Rao, X.; Bisseling, T.; Geurts, R.; Leunissen, J.A.M. Primer3Plus, an enhanced web interface to Primer3. *Nucleic Acids Res.* **2007**, *35*, W71–W74. [CrossRef]
20. Hall, T.A. BioEdit: A user-friendly biological sequence alignment editor and analysis program for Windows 95/98/NT. *Nucl. Acids Symp. Ser.* **1999**, *41*, 95–98.

21. Edgar, R.C. MUSCLE: Multiple sequence alignment with high accuracy and high throughput. *Nucleic Acids Res.* **2004**, *32*, 1792–1797. [CrossRef]

22. Kumar, S.; Stecher, G.; Tamura, K. MEGA7: Molecular Evolutionary Genetics Analysis Version 7.0 for Bigger Datasets. *Mol. Biol. Evol.* **2016**, *33*, 1870–1874. [CrossRef]

23. Mori, E.; Lara, d.C.C.S.H.; Cunha, E.M.S.; Villalobos, E.M.C.; Mori, C.M.C.; Soares, R.M.; Brandao, P.E.; Fernandes, W.R.; Richtzenhain, L.J. Molecular characterization of Brazilian equid herpesvirus type 1 strains based on neuropathogenicity markers. *Braz. J. Microbiol.* **2015**, *46*, 565–570. [CrossRef]

24. Turan, N.; Yildirim, F.; Altan, E.; Sennazli, G.; Gurel, A.; Diallo, I.; Yilmaz, H. Molecular and pathological investigations of EHV-1 and EHV-4 infections in horses in Turkey. *Res. Vet. Sci.* **2012**, *93*, 1504–1507. [CrossRef]

25. Perkins, G.A.; Goodman, L.B.; Tsujimura, K.; Van de Walle, G.R.; Kim, S.G.; Dubovi, E.J.; Osterrieder, N. Investigation of the prevalence of neurologic equine herpes virus type 1 (EHV-1) in a 23-year retrospective analysis (1984–2007). *Vet. Microbiol.* **2009**, *139*, 375–378. [CrossRef]

26. Vissani, M.A.; Becerra, M.L.; Olguín Perglione, C.; Tordoya, M.S.; Miño, S.; Barrandeguy, M. Neuropathogenic and non-neuropathogenic genotypes of Equid Herpesvirus type 1 in Argentina. *Vet. Microbiol.* **2009**, *139*, 361–364. [CrossRef] [PubMed]

27. Pronost, S.; Leon, A.; Legrand, L.; Fortier, C.; Miszczak, F.; Freymuth, F.; Fortier, G. Neuropathogenic and non-neuropathogenic variants of equine herpesvirus 1 in France. *Vet. Microbiol.* **2010**, *145*, 329–333. [CrossRef] [PubMed]

28. Smith, K.L.; Allen, G.P.; Branscum, A.J.; Frank Cook, R.; Vickers, M.L.; Timoney, P.J.; Balasuriya, U.B.R. The increased prevalence of neuropathogenic strains of EHV-1 in equine abortions. *Vet. Microbiol.* **2010**, *141*, 5–11. [CrossRef]

29. Fritsche, A.K.; Borchers, K. Detection of neuropathogenic strains of Equid Herpesvirus 1 (EHV-1) associated with abortions in Germany. *Vet. Microbiol.* **2011**, *147*, 176–180. [CrossRef]

30. Tsujimura, K.; Oyama, T.; Katayama, Y.; Muranaka, M.; Bannai, H.; Nemoto, M.; Yamanaka, T.; Kondo, T.; Kato, M.; Matsumura, T. Prevalence of equine herpesvirus type 1 strains of neuropathogenic genotype in a major breeding area of Japan. *J. Vet. Med. Sci.* **2011**, *73*, 1663–1667. [CrossRef]

31. Castro, E.R.; Arbiza, J. Detection and genotyping of equid herpesvirus 1 in Uruguay. *Rev. Sci. Tech.* **2017**, *36*, 799–806. [CrossRef]

32. Damiani, A.M.; de Vries, M.; Reimers, G.; Winkler, S.; Osterrieder, N. A severe equine herpesvirus type 1 (EHV-1) abortion outbreak caused by a neuropathogenic strain at a breeding farm in northern Germany. *Vet. Microbiol.* **2014**, *172*, 555–562. [CrossRef] [PubMed]

33. Vaz, P.K.; Horsington, J.; Hartley, C.A.; Browning, G.F.; Ficorilli, N.P.; Studdert, M.J.; Gilkerson, J.R.; Devlin, J.M. Evidence of widespread natural recombination among field isolates of equine herpesvirus 4 but not among field isolates of equine herpesvirus 1. *J. Gen. Virol.* **2016**, *97*, 747–755. [CrossRef] [PubMed]

34. Guo, X.; Izume, S.; Okada, A.; Ohya, K.; Kimura, T.; Fukushi, H. Full genome sequences of zebra-borne equine herpesvirus type 1 isolated from zebra, onager and Thomson's gazelle. *J. Vet. Med. Sci.* **2014**, *76*, 1309–1312. [CrossRef] [PubMed]

35. Said, A.; Damiani, A.; Osterrieder, N. Ubiquitination and degradation of the *ORF34* gene product of equine herpesvirus type 1 (EHV-1) at late times of infection. *Virology* **2014**, *460–461*, 11–22. [CrossRef] [PubMed]

36. Shakya, A.K.; O'Callaghan, D.J.; Kim, S.K. Comparative Genomic Sequencing and Pathogenic Properties of Equine Herpesvirus 1 KyA and RacL11. *Front. Vet. Sci.* **2017**, *4*, 211. [CrossRef] [PubMed]

 *viruses* 

Article

Molecular Surveillance of EHV-1 Strains Circulating in France during and after the Major 2009 Outbreak in Normandy Involving Respiratory Infection, Neurological Disorder, and Abortion

Gabrielle Sutton [1,2,*], Marie Garvey [3], Ann Cullinane [3], Marion Jourdan [4], Christine Fortier [1,2], Peggy Moreau [5], Marc Foursin [5], Annick Gryspeerdt [6], Virginie Maisonnier [4], Christel Marcillaud-Pitel [4], Loïc Legrand [1,2], Romain Paillot [1,2,†] and Stéphane Pronost [1,2,†]

[1]   LABÉO Frank Duncombe, 14280 Saint-Contest, France; christine.fortier@laboratoire-labeo.fr (C.F.); loic.legrand@laboratoire-labeo.fr (L.L.); romain.paillot@laboratoire-labeo.fr (R.P.); stephane.pronost@laboratoire-labeo.fr (S.P.)

[2]   NORMANDIE UNIV, UNICAEN, BIOTARGEN, 14000 Caen, France

[3]   Irish Equine Centre, Johnstown, Naas, County Kildare, Eircode: W91 RH93, Ireland; MGarvey@irishequinecentre.ie (M.G.); ACullinane@irishequinecentre.ie (A.C.);

[4]   RESPE, 14280 Saint-Contest, France; technicien@respe.net (M.J.); contact@respe.net (V.M.); c.marcillaud-pitel@respe.net (C.M.-P.);

[5]   Clinique équine de la Boisrie, 61500 Chailloué, France; peg_moreau@yahoo.fr (P.M.); clin.eq.laboisrie@laposte.net (M.F.)

[6]   Equi Focus Point Belgium, 8900 Ypres, Belgium; annick.gryspeerdt@gmail.com

*   Correspondence: gabrielle.sutton@laboratoire-labeo.fr; Tel.: +33-2-31-41-93-61

†   The authors have contributed equally to the work.

Received: 14 August 2019; Accepted: 1 October 2019; Published: 4 October 2019

Abstract: Equine herpesvirus 1 (EHV-1) is an Alphaherpesvirus infecting not only horses but also other equid and non-equid mammals. It can cause respiratory distress, stillbirth and neonatal death, abortion, and neurological disease. The different forms of disease induced by EHV-1 infection can have dramatic consequences on the equine industry, and thus the virus represents a great challenge for the equine and scientific community. This report describes the progress of a major EHV-1 outbreak that took place in Normandy in 2009, during which the three forms of disease were observed. A collection of EHV-1 strains isolated in France and Belgium from 2012 to 2018 were subsequently genetically analysed in order to characterise EHV-1 strain circulation. The open reading frame 30 (ORF30) non-neuropathogenic associated mutation A_{2254} was the most represented among 148 samples analysed in this study. ORF30 was also sequenced for 14 strains and compared to previously published sequences. Finally, a more global phylogenetic approach was performed based on a recently described Multilocus Sequence Typing (MLST) method. French and Belgian strains were clustered with known strains isolated in United Kingdom and Ireland, with no correlation between the phylogeny and the time of collection or location. This new MLST approach could be a tool to help understand epidemics in stud farms.

Keywords: equine herpesvirus type 1; outbreak; respiratory disease; abortion; neuropathogenic strain; myeloencephalopathy; phylogeny; ORF30; MLST

---

1. Introduction

Equine herpesvirus 1 (EHV-1) is a member of the *Varicellovirus* genus, in the Alphaherpesvirus sub-family [1]. EHV-1 is known to infect horses as principal hosts but some cases have been reported in other equids and non-equid mammals [2–5]. Virus transmission occurs through direct contact between horses, infectious aerosols, fomites, and/or indirectly by humans. More recently, transmission of herpesvirus after survival in water was reported. Experimentally, EHV-1 was shown to be stable and infectious in water for over a week under different conditions of pH, salinity, temperature, and turbidity, and up to three weeks under some of these conditions [6]. EHV-1 infection may induce several clinical forms of disease including respiratory infection (also called rhinopneumonitis) associated with pyrexia, cough, and respiratory distress, abortion in pregnant mares, stillbirth and neonatal death, and neurological disorders with symptoms ranging from mild ataxia to complete paralysis of the animal (EHV-1 inducing neurological disease is usually referred as Equine Herpesvirus Encephalomyelitis (EHM)) [7]. However, numerous factors may affect the nature and the extent of clinical signs of disease, such as age, sex, physical condition, prior infection history, and/or the nature of the EHV-1 strain [8]. After infection, the trigeminal ganglia and leukocytes have been identified as the sites of latency for EHV-1 [9,10]. The virus can be reactivated after environmental stimuli (e.g., stress) or therapeutic treatment (e.g., dexamethasone) and replicate in mucous tissues with subsequent dissemination to other hosts [11]. EHV-1 latency mechanisms are poorly understood, although gene regulation has been shown to play a major role in the process [12].

The linear double-stranded genome of EHV-1 is 150 kb long and consists of a unique long (U_L) and unique short region (U_S). Each of these regions is surrounded by small inverted sequences (Terminal Repeat Long (TRL)/Internal Repeat Long (IRL) and Terminal Repeat Short (TRS)/Internal repeat Short (IRS), respectively). The genome contains 80 open reading frames (ORFs), some of which show more genetic variability than others and contain sufficient variation for phylogenetic analysis [13]. Previous studies performed ORF and/or whole genome sequencing to study genetic polymorphism [14,15]. Whole genome comparison of two characterised strains, Ab4 [16] and V592 [17], identified a polymorphic region within ORF68 (herpes simplex virus type 1 US2 homologue), which was used to examine the genetic heterogeneity of EHV-1 isolates in several different countries [15,18]. Furthermore, a single point mutation of adenine to guanine at nucleotide position 2254 within ORF30 (DNA polymerase catalytic subunit) is often associated with EHV-1 different forms of disease [19–24]. Although it is not exclusive, A_{2254} (non-neuropathogenic) and G_{2254} (neuropathogenic) strains were more frequently associated with abortion and neurological disease, respectively [19,25]. Additionally, genome comparison of Ab4 and V592 identified non-synonymous substitutions that could be used for multi-locus sequence typing (MLST) of EHV-1 strains [19,26].

Due to its impact on animal welfare and performance, EHV-1 represents a major threat to the equine industry and a great interest and challenge for the equine veterinary and scientific community. This report aims to illustrate how challenging EHV-1 infection could be in the field through the description of a multi-syndromic outbreak that lasted over 9 months in 2009 and 2010. This outbreak involved 28 animals and different forms of disease were observed, including two EHM cases and four abortions. Both neuropathogenic and non-neuropathogenic strains were isolated from this major outbreak and were investigated alongside samples collected from 2012 to 2018. Different molecular analytic methods were applied to specimens from these outbreaks (2009, and from 2012 to 2018), including the ORF30 substitution A2254G typing, complete ORF30 sequencing, and Multi Locus Sequence Typing (MLST). These samples included EHV-1 strains isolated from the 2018 EHV-1 epizootic, which is considered to be the most important recorded in France for the last 30 years, with at least 56 outbreaks reported and leading to the cancellation of numerous equestrian events.

2. Materials and Methods

2.1. EHV-1 Outbreak Data Collection

Data related to EHV-1 outbreaks were collected from the diagnostic and equine research Institute LABEO and the French Epidemiological Surveillance Network for Equine Pathologies (Réseau d'Epidémio-Surveillance en Pathologie Equine (RESPE)) and associated equine veterinary practitioners (see Table 1). From August 2009 to June 2010, 242 samples were collected on a stud farm during one single EHV-1 infectious episode (four tissue samples, 115 nasal swabs, one cerebrospinal fluid, 114 blood samples, and eight vaginal swabs). Concerning this outbreak, which happened on the same premise, mares and foals from the stud farm were divided in six groups while yearlings were kept apart. Mares with foals were divided in groups 1, 2, 3, and 4, while mares without foals were divided in groups 5 and 6. From 2012 to 2018, 180 samples were collected from individual EHV-1 outbreaks by the diagnostic and equine research Institute LABEO (France) (85 tissue samples, 82 nasal swabs, one tracheal liquid, four cerebro-spinal fluids, and 19 blood samples). Detailed information is presented in Supplementary Materials (Table S1).

Table 1. Number of individual outbreaks of equine herpesvirus 1 (EHV-1) reported to Réseau d'Epidémio-Surveillance en Pathologie Equine (RESPE) from 2012 to 2018.

Year	Respiratory	Abortion	Neurological	ND [1]	Total
2012	MD [2]	MD	MD	MD	MD
2013	1	1	3	0	5
2014	14	10	3	0	27
2015	11	16	1	0	28
2016	11	8	7	3	32
2017	5	9	6	3	27
2018	15	16	5	20	56

[1] ND = clinical form of disease not defined or reported. [2] MD = missing data.

2.2. Biological Samples And Nucleic acids

Samples collected included body fluids (nasopharyngeal swabs, cerebrospinal fluid, tracheal liquid, vaginal swabs), fetal organs (lungs, liver, and placenta), or whole blood samples (EDTA). Nasal swabs were processed in 4 mL Eagle Minimal Essential medium or phosphate-buffered saline (PBS). For nucleic acid extraction, 140 μL of biological fluids extracts, 30 mg of organs, or 2–3 mL of blood were used. DNA was extracted using different extraction kits according to the type of sample: fluid extracts and organs were treated as previously described [27], blood samples were processed using the DNA Blood Maxi Kit (Qiagen, Germany) prior to May 2015, and the NucleoSpin Blood L (Macherey-Nagel, Germany) post May 2015 according to manufacturer's recommendations. Twenty strains amongst the 180 strains collected were selected for ORF30 sequencing, including three strains collected from outbreaks of EHV-1 in Belgium. Belgian strains were available through a diagnostic collaboration with Equi Focus Point Belgium (equine infectious diseases surveillance network in Belgium) and were incorporated in this study due to the close epidemiological relationship between these two neighbouring countries. Sequenced strains were identified according to their location (French Region or Country) and year of collection, as follows: LOC/ID NUMBER/YEAR.

2.3. EHV-1 Identification And Quantification

Real-time Polymerase Chain Reaction (PCR) was performed on purified nucleic acids using Diallo et al., (2006) [28] primers and probe targeting glycoprotein B of EHV-1 (Forward Primer F1: 5′-CAT GTC AAC GCA CTC CCA-3′; Reverse Primer R1: 5′-GGG TCG GGC GTT TCT GT-3′ and probe FAM-CCC TAC GCT GCT CC-MGB-NFQ). Viral load quantification was performed as previously described [27,29]. A standard curve based on a cloned sequence was used. Results are expressed as a copy number per mL of nasopharyngeal swab extract. A volume of 2.5 μL of each sample was added to the PCR mixture composed of 12.5 μL TaqMan Universal PCR Master Mix (ThermoFisher Scientific, Waltham, MA, USA), 1.25 μL of each primer (at a concentration of 20 μM) and a defined volume of probe depending on the titration. The mixture was completed with nuclease free water to a final volume of 25 μL.

2.4. EHV-1 Strain Typing (ORF30 DNA Polymerase, Position 2254)

Real-time discriminatory allelic PCR was performed on purified nucleic acids using Allen et al. (2007) [22] primers targeting the EHV-1 ORF30 DNA polymerase (Forward Primer 5′-CCA CCC TGG CGC TCG-3′; Reverse Primer 5′-AGC CAG TCG CGC AGC AAG ATG-3′) and probes ("non-neuropathogenic" A2254 Probe VIC-CAT CCG TCA ACT ACT C-MGB; "neuropathogenic" G2254 Probe 6-FAM-TCC GTC GAC TAC TC-MGB). The same mix as described in 2.3. was used for PCR, with 0.6 µL of each primer (at a concentration of 20 µM) and a defined volume of each probe.

2.5. ORF30 Sequencing And Analysis

Twenty EHV-1 samples were chosen for ORF30 sequencing, based on the disease type and the year of sampling (see Table S2 in Supplementary Materials). Due to a lesser number of neurological outbreaks and low viral load of these samples of neurological origin, only one sample (BRE/13/2018) isolated during a neurological episode was selected. The EHV-1 strains NORM/1/2009 and NORM/2/2010 correspond to strains isolated from a foal during the 2009 outbreak (Section 3.1.2.) and the strain isolated from the foetus after Mare E abortion (Section 3.1.3), respectively. Biofidal (France) performed primer design and ORF30 (3.665 kb) sequencing. First, a primer set (EHV-1-ORF30-PCR-F: 5′-GAACGTGCGAGTGCTGTTTT-3′ and EHV-1-ORF30-PCR-RC: 5′TGTGAAGGTCTGTTCGACGG-3′) was designed to amplify a 5kb region of EHV-1 including ORF30. The amplification mixture was composed of 5 µL of 5× PrimeSTAR Buffer (Takara, Japan), 2 µL deoxyribo–nucleotide triphosphates (dNTPs) Mix (2.5 mM each), 1 µL of each primer (10 µM), 0.5 µL PrimeSTAR GXL DNA Polymerase (Takara, Kusatu, Shiga, Japan), 13.5 µL RNase Free Water, and 2 µL of sample. Thermocycling conditions were as follows: initial denaturation at 98 °C for 30 s, 35 cycles of denaturation at 96 °C for 20 s, annealing at 61 °C for 20 s, elongation at 68 °C for 5 min, and final elongation at 68 °C for 5 min. Sequencing PCR was subsequently performed using the two external (PCR) primers and nine internal (sequencing) primers detailed in Supplementary Materials (Table S3). Amplicons were purified using ExoSAP-IT [TM] kit (ThermoFisher Scientific). Sanger sequencing was performed using Big Dye Terminator Sequencing Mix (ThermoFisher Scientific). Samples were purified before electrophoresis using BigDye XTerminator kit (ThermoFisher Scientific). Then, 3730XL DNA Analyzer was then used for sequence electrophoresis. Single nucleotide polymorphisms (SNP) and consensus sequences were analysed using BioEdit version 7.0.5.3 (Tom Hall, Carlsbad, CA, US) [30] and CodonCode Aligner version 8.0.2 (Codon Code Corporation, Centerville, MA, US) [31] software. Sequences were analysed using MEGA 7 software version 7.2.26 (Koichiro Tamura, Glen Stecher, Sudhir Kumar, the Pennsylvania State University, University Park, PA, US) [32] and phylogenetic trees were built using Neighbour-Joining statistical method [33] and Maximum Composite Likelihood model [34]. Finally, all ORF30 sequences used to build the tree were converted into a Nexus format using Seqret EMBOSS (The European Bioinformatics Institute, Hinxton, Cambridgeshire, UK) [35], and a median joining network was built using Population Analysis with Reticulate Trees (PopART) software (Jessica Leigh, David Bryant and Mike Steel, Dunedin, New Zealand) [36].

2.6. Multi-locus Sequence Typing (MLST) Analysis

For Multi Locus Sequence analysis, 37 loci in 26 ORFs were analysed based on non-synonymous changes identified between different published data including Ab4 and V592 protein coding regions as reported by Garvey et al. (2019) [14,19,26]. The MSLT was performed according to methodology described by Garvey et al. (2019) [26]. The method and reactional products used were the same as described by the authors. The MLST based on 37 non-synonymous U_S and U_L amino acid changes (located in 26 ORFs; ORF2, 5, 8, 11, 13, 14, 15, 22, 29, 30, 31, 32, 33, 34, 36, 37, 39, 40, 42, 45, 46, 50, 52, 57, 73, and 76) was performed on eight EHV-1 strains. A reduced MLST based on 14 of the 37 loci (located in 6 ORFs; ORF11, 13, 30, 37, 52 and 76), suspected to be determinant in clade attribution, was performed on seven other EHV-1 strains (see Supplementary Table S2). Amplification products of target ORFs were purified and sequenced using forward primers (except for ORF13, which needed both forward and reverse sequencing). Sequences were analysed using CodonCode Aligner and BioEdit software. Clade attribution is based on U_L sequences distribution and was correlated with MLST segregation by Garvey et al. [14,26]. Sequence analysis was conducted on concatenated amino acids sequences. Both a phylogenetic tree and a network were built using MEGA7 and Splits Tree4 [37], respectively. The phylogenic tree was built using Maximum Likelihood method and Jones Taylor Thornton model, while the network was built using a Neighbour-Net method [38].

2.7. Statistical Analysis

Chi-squared was used to test the null hypothesis that there was no correlation between the A_{2254}/G_{2254} typing and the form of disease (strains isolated from respiratory cases, abortion cases, or neurological cases).

3. Results

3.1. The 2009 Multi-Syndromic EHV-1 Outbreak

The outbreak occurred on a thoroughbred stud farm in Normandy that consisted of 169 horses in total (60 broodmares—44 with and 16 without a foal—and 65 yearlings). The first case was reported at the end of August 2009. Two cases of EHM, 23 cases of respiratory disease, and four abortion cases were reported over a period of nine and a half months (see Figure 1). The mares and their foals were kept on one site but separated into six groups (see Table 2). Mares were vaccinated twice a year against equine influenza (EI) with the EIV-HA coding recombinant canarypoxvirus–based vaccine (ProteqFlu-Te, Merial SAS, Lyon, France) and against rhinopneumonitis on the fifth, seventh, and ninth months of gestation with a whole EHV-1/4 inactivated carbomer-adjuvanted vaccine (Duvaxyn EHV1-4, Fort DodgeAnimalHealth, Overland Park, KS, US). Due to reports of EHV-1 outbreaks in other countries in the weeks and months prior to this case (29 August), a booster dose of a whole inactivated EHV-1 vaccine (Pneumequine, Merial) had been given to mares before their return from abroad (Ireland, the UK, and the US), approximately two and a half months before the commencement of the outbreak on the French stud farm. None of the foals had received a primary vaccination in August. Yearlings were kept on a second site located a few kilometers from the main yard. They had received two immunisations one month apart and a booster immunisation six months later (from April 2009 to June 2009, depending on the foals age).

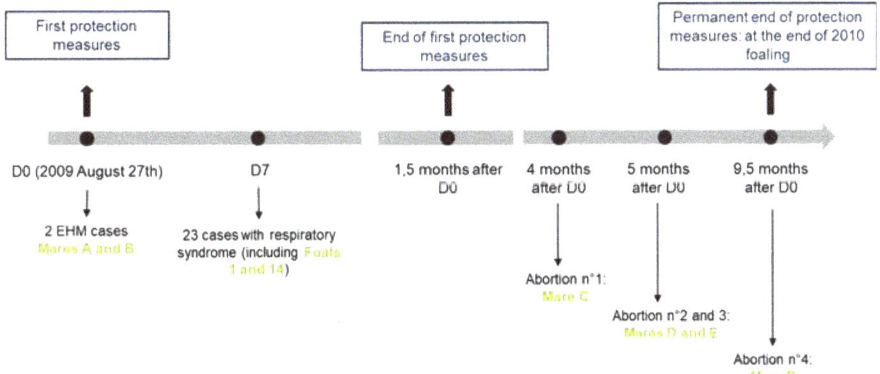

Figure 1. The 2009 EHV-1 outbreak chronology from D0 = day of the first case declaration, to 9.5 months after D0. D7 = Day 7 after the first case was declared. The biosecurity measures that were taken during this outbreak are described in blue boxes.

Table 2. Group composition for mares and foals on the French stud farm.

	Mares				Foals		
Group	Total Number	Respiratory Disease	Abortion [1]	EHM	Total Number	Sex	Respiratory Disease
1	13	0	3 [1]	2	13 [2]	Female	6
2	9	0	1	0	9 [2]	Male	4
3	10	0	0	0	10 [3]	Female	0
4	12	4	2 [1]	0	12 [3]	Male	9
5	6 [2]	0	0	0	0	0	0
6	10 [3]	0	0	0	0	0	0

[1] only four out of six were EHV-1-induced abortion (cf Section 3.1.3.). [2] returned from/born abroad. [3] stay in/born in France.

3.1.1. Neurological Cases

- Index Case: Mare A

On 27 August 2009 (D0), a pregnant 12-year-old broodmare (information not available concerning pregnancy duration) with an accompanying foal was found in lateral decubitus position and was unable to stand (see Figure 1 and clinical and treatment details are presented in Supplementary Materials S4). Despite treatment and due to a persistent nystagmus, a generalised stiffness and apparition of seizures, the mare was euthanized 14 h after being found in the paddock. Cerebrospinal fluid was collected on the atlantooccipital joint just after the death, and post-mortem examination was performed. EHV-1 was detected by PCR in the cerebrospinal fluid, the brain, and spinal cord. The strain was characterised as "G2254, neuropathogenic" in all three biological compartments.

- Second Case: Mare B

Twelve hours after the onset of clinical signs in mare A, another pregnant mare (11 years old and 3.5 months pregnant, with an accompanying foal from the same group, mare B) presented with depression, shivering, and incoordination from the posterior limbs (see Figure 1 and Supplementary Materials S5). EHV-1 was detected on nasopharyngeal swab and whole blood sample, and typed as "G2254, neuropathogenic." Forty-eight hours after the first clinical signs were observed, the mare's condition improved. Mare B made a full recovery one month later.

3.1.2. Respiratory Infections

Between D7 and D10, five foals from group 1 and one foal from group 2 presented with hyperthermia and nasal discharge. Horses were treated depending on their clinical signs. Foals presenting a hyperthermia higher than 39.5 °C received anti-inflammatory drugs (0.3 mg/kg flunixine meglumine intravenously or 17 mg/kg aspirine orally). Foals were also treated with mucolytic drugs (0.5 mg/kg bromhexine orally once a day). Those with a persisting hyperthermia (> 48 h) or more severe clinical signs (coughing, breathing abnormalities) were treated with antibiotics (5 mg/kg trimethoprime-sulfadiazine orally once a day for five days).

Nasopharyngeal swabs were taken from foals and from other animals in each group. Four mares and 13 foals were found positive for EHV-1 in groups 1, 2, and 4, whereas all horses tested in groups 3, 5, and 6 (15 in total) were negative. Further whole blood samples and nasopharyngeal swabs were taken from groups 1, 2, and 4 animals over a six-week period and analysed by qPCR to monitor virus shedding and cell-associated viraemia, respectively. For example, on D7, a foal (foal 1; Table 3) tested positive (high transmission risk) with a high viral load, while foal 14 tested negative. This other foal (foal 14) tested positive on D14, and the viral load significantly increased on D21. It then decreased on D28 and was negative on D42. Overall, virus shedding in foals reached a maximum of $2.14.10^8$ viral particle/mL of nasopharyngeal swab extract between Day 7 and Day 49. Virus loads in blood samples were lower than on swabs extracts and reached a maximum of $2.98.10^3$ viral particle/mL between Day 14 and Day 49 (see Table 3).

Table 3. Monitoring of the viral shedding on 14 horses from groups 1, 2, and 3 over a period of seven weeks (copy/mL).

	Horse	D7 NS	D7 Blood	D14 NS	D14 Blood	D21 NS	D21 Blood	D28 NS	D28 Blood	D35 NS	D35 Blood	D42 NS	D42 Blood	D49 NS	D49 Blood
Group 1 and Group 2	1	2.14E + 08	nd	6.75E + 05	1.05E + 03	7.04E + 02	NEG	POS	NEG	NEG	NEG	NEG	NEG	NEG	NEG
	2	3.02E + 08	nd	4.52E + 04	NEG	7.04E + 02	NEG	6.05E + 03	NEG	NEG	NEG	NEG	NEG	NEG	NEG
	3	9.29E + 07	nd	NEG	NEG	NEG	NEG	8.09E + 02	NEG	2.14E + 03	4.58E + 02	1.11E + 05	NEG	NEG	NEG
	4	1.91E + 06	nd	1.81E + 05	2.98E + 03	7.34E + 04	NEG	NEG	POS	NEG	NEG	NEG	NEG	NEG	NEG
	5	1.66E + 06	nd	NEG	NEG	3.67E + 04	NEG	NEG	NEG	POS	NEG	NEG	NEG	NEG	NEG
	6	1.08E + 07	nd	1.30E + 04	NEG	1.23E + 03	NEG	6.05E + 03	NEG	POS	NEG	NEG	NEG	NEG	NEG
	7	2.23E + 05	nd	NEG	POS	POS	NEG	NEG	POS	NEG	POS	NEG	NEG	NEG	NEG
	8	1.28E + 05	nd	3.93E + 04	POS	POS	NEG	NEG	NEG	NEG	NEG	NEG	NEG	NEG	NEG
	9	4.84E + 04	nd	NEG	NEG	NEG	POS	NEG	NEG	NEG	NEG	NEG	NEG	NEG	NEG
	10	NEG	nd	NEG	POS	NEG	NEG	NEG	NEG	NEG	NEG	NEG	NEG	NEG	NEG
	11	NEG	nd	NEG	NEG	NEG	NEG	NEG	NEG	NEG	NEG	NEG	NEG	NEG	NEG
	12	POS	nd	NEG	NEG	NEG	NEG	NEG	NEG	NEG	NEG	NEG	NEG	NEG	NEG
	13	NEG	nd	NEG	NEG	NEG	NEG	NEG	NEG	NEG	5.64E + 02	NEG	NEG	NEG	NEG
	14	NEG	nd	6.75E + 05	NEG	1.31E + 08	NEG	POS	NEG	6.48E + 03	NEG	NEG	NEG	NEG	NEG
	15	nd	nd	nd	nd	POS	NEG	POS	NEG	NEG	NEG	POS	NEG	NEG	NEG
Group 4	16	2.26E + 04	nd	NEG	POS	NEG	NEG	NEG	NEG	NEG	NEG	NEG	NEG	NEG	NEG
	17	1.18E + 06	nd	4.27E + 03	NEG	NEG	NEG	NEG	NEG	NEG	NEG	NEG	NEG	NEG	NEG
	18	1.97E + 04	nd	POS	NEG	NEG	NEG	NEG	NEG	NEG	NEG	NEG	NEG	NEG	NEG

Viral load in nasal swab (NS) from D0 to D42, and in blood from D7 to D42. POS = positive samples non quantifiable; NEG = negative samples. The boxes in gray indicate the period when the horse became negative (when both the NS and blood samples from horses were negative).

3.1.3. Abortion Cases

Between four and five months after the first neurological case (see Figure 1), a mare (C) from group 1 and two mares (D and E) from group 4 aborted and were confirmed EHV-1 positive. Tests were carried out on fetal tissues (liver, lung, and kidney) and the mares' placentas. Mare C foetus kidney tested negative. The EHV-1 strains were identified as "non-neuropathogenic" (A_{2254}) for the three cases.

Six months after developing neurological signs of disease, mare B (9.5 months pregnant) showed signs of dystocic abortion (liquid loss and contractions without delivery of the foal). Intra-uterine examination revealed an anterior dorsal position of the foal and bended legs and head. The foal was moved in the appropriate position under epidural anaesthesia. It was found dead and icteric. Delivery was immediate and complete (see Supplementary Materials S6). The following day, a uterine wash was performed. The mare was isolated in a stable. and sanitary measures were put in place (Section 3.1.4. Biosafety measures). Post-mortem examination of the foetus revealed a densification of lung parenchyma with interlobular oedema. Placenta and lung/liver samples tested positive for EHV-1. The strains were identified as "neuropathogenic" (G_{2254}), as previously detected for this mare during this neurological episode (Section 3.1.1.).

Uterine swabs were performed eight days and 30 days post abortion on mares B, C, D and E. All samples were positive for EHV-1 eight days post abortion and negative 30 days post abortion. Mare B was moved to Ireland for breeding five weeks post-abortion. Mare B was confirmed pregnant two months post-abortion and subsequently delivered a healthy foal after a normal pregnancy.

Overall, six of the 60 mares from this stud farm aborted during the year. EHV-1 infection was confirmed for mares B, C, D, and E. One of the two other mares aborted due to *Enterobacter amnigenus* infection, while no cause could be confirmed for the last mare (the dead foal could not be recovered).

3.1.4. Biosafety Measures

The first two cases (Mares A and B) were reported to the RESPE. A safety perimeter was put in place around sick horses. Gloves, gowns, and over-boots were used to manipulate the animals. Movement restrictions were put in place, including foot baths with disinfectant installed at the paddock entrance, car wheels cleaning, and movement restriction for horses.

Mares and foals that tested positive for EHV-1 were moved to isolation from the other horses. All mares were sampled (nasal swabs) before leaving or returning to the stud farm. A negative PCR result for EHV-1 was a prerequisite for movement. Booster vaccination against rhinopneumonitis was also administered. The following year, all pregnant mares stayed in France, including mare B.

To conclude, the overall EHV-1-induced morbidity rate reached 16.6% of the herd (28 clinically affected animals out of 169), including 6.7% of the broodmares, 1.2% EHM cases, and 13.6% respiratory cases.

3.2. Surveillance and Phylogeny from 2009 to 2018

3.2.1. Outbreaks, Forms of Disease, and ORF30 A2254G Typing

This epidemiological study involved the 2009 outbreak described in Section 3.1 and EHV-1 outbreaks from 2012 to 2018. Samples collected from 2012 to 2018 are represented in Supplementary Materials (Table S1). From 2012 to 2017, 42 respiratory cases, 44 abortion cases, 20 neurological cases, and six cases with no clinical information were reported. Significantly, in 2018, several outbreaks of EHV-1 occurred. During this year, an increased number of outbreaks (56 in total) were reported to the RESPE, including 15 respiratory outbreaks, 16 cases of abortion, five neurological outbreaks, and 20 outbreaks with no clinical information (see Table 4 and Figure 2). The majority of the outbreaks were reported on the western part of the country, which has the highest concentration of breeding farms. Neurological outbreaks occurred in Normandy and Brittany and lead to the euthanasia of five animals. From these outbreaks, 71 samples were received and analysed (i.e., 15 samples from respiratory cases,

19 samples from abortion cases and 18 samples from neurological cases; clinical information was missing for 19 samples). Of these 71 samples, 12 could not be typed for A2254G substitution (two samples from respiratory cases, one sample from an abortion case, six samples from neurological cases, and three samples missing clinical information).

Table 4. Details of EHV-1 strains, per year, disease type, and strain typing (ORF30 A2254G).

Year	Strains	Respiratory Syndrome		Abortion		Neurological Syndrome		ND [1]		No A/G Typing	Outbreaks (RESPE)
		A_{2254}	G_{2254}	A_{2254}	G_{2254}	A_{2254}	G_{2254}	A_{2254}	G_{2254}		
1 major outbreak 2009	30	24	0	2	2	0	2	0	0	0	/
2012	8	0	0	0	2	0	1	0	0	5 *	Missing data
2013	19	0	0	6	7	2	1	0	1	2 *	4
2014	10	0	0	8	0	0	0	0	0	2 *	26
2015	32	0	0	17	0	1	1	1	0	12 *	27
2016	18	1	0	8	0	2	4	0	1	2 *	32
2017	22	1	0	2	1	0	7	1	2	8 *	23
2018	71	9	4	17	1	10	2	5	11	12 *	56

Abbreviations: ND [1] = clinical form of disease not defined or reported. * Respiratory syndrome was assigned only if corresponding clinical information was submitted with nasal swab samples. / = for 2009, all EHV-1 strains included in this study were linked to the outbreak described in Section 3.1.

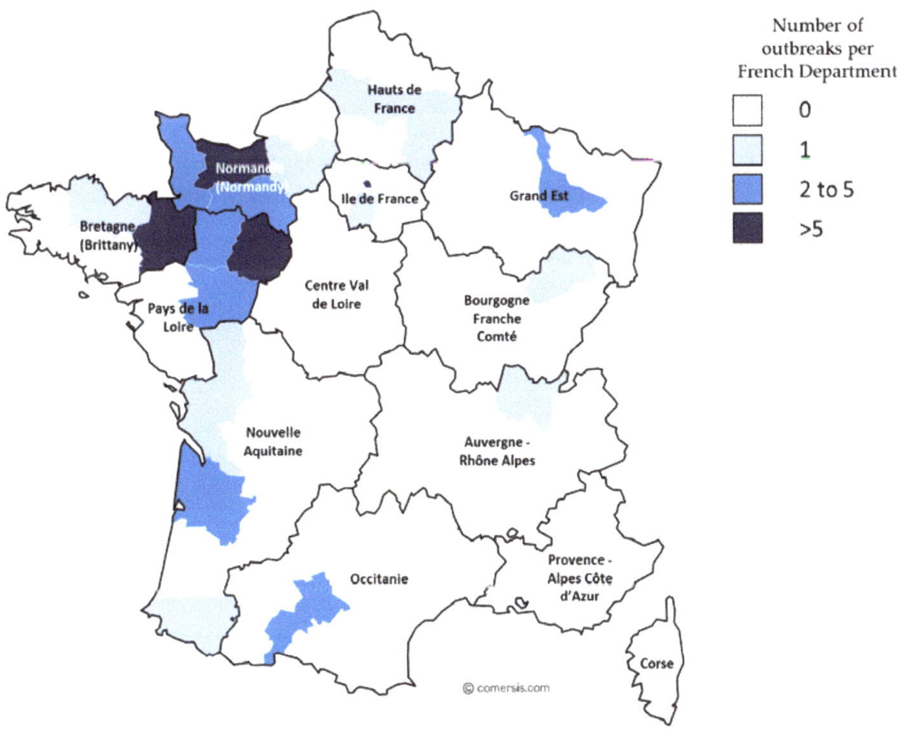

Figure 2. Map of France illustrating EHV-1 outbreaks reported to the RESPE in 2018, with frequency and geographical location.

Among all the strains collected from 2012 to 2018, 137 could be typed for the neuropathogenic/non-neuropathogenic mutation at the ORF30 2254 position. As shown in Table 4, both A_{2254} (non-neuropathogenic type) and G_{2254} (neuropathogenic type) strains have been isolated in respiratory cases, abortion, and neurological outbreaks. The Chi square test indicates a significant statistical association between the ORF30 types and both abortion and neurological disorder (*p*-value = 0.000202, see Table 5). Ninety-one A_{2254} strains have been isolated over the years (66% of all the strains typed), when compared with 46 (34% of all the strains typed) G_{2254} strains.

Table 5. ORF30 2254 mutation according to disease from EHV-1 isolates (2009 to 2018).

	Type	Respiratory	Abortion	Neurological	Information Missing	Total
A_{2254}	Non-neuro.	11 (73%)	58 (84%)	15 (48%)	7 (32%)	91(66%)
G_{2254}	Neuro.	4 (27%)	11 (16%)	16 (52%)	15 (68%)	46 (34%)
		Chi square *p*-value	Resp./Ab.	0.325642		
			Resp./Neu.	0.109608		
			Ab./Neu.	0.000202 *		
			All three	0.000996 *		

Non-neuro. = non neuropathogenic; Neuro. = neuropathogenic. (%: percentage of ORF30 2254 type among the disease category). Chi Square test null hypothesis "There is no correlation between the disease category and the ORF30 2254 type. Resp./Ab. = Chi square test for Respiratory and Abortion categories; Resp./Neu.= Chi square test for Respiratory and Neurological categories; Ab./Neu.= Chi square test for Abortion and Neurological categories; All three= Chi square test for Respiratory, Abortion and Neurological categories. * significant result at *p*-value $p < 0.05$.

3.2.2. ORF30 Sequence Analysis

Probably because of the DNA quantity and purity, only 14 strains could be completely sequenced. BRE/13/2018, the only strain isolated on a neurological outbreak, could not be sequenced. ORF30 sequences from these 14 EHV-1 strains (see Supplementary Table S2) were compared to the neuropathogenic reference strain Ab4 (Genbank accession number AY665713). SNP and amino acid substitutions are reported in Supplementary Table S7.

Twelve of the 14 strains had an adenine residue at position 2254 of the polymerase gene. Seven of these also had a mutation in position 96 (G96A) and five of them in position 2968 (G2968A) compared to reference AY665713 (Ab4). The mutation (A > G) in position 2254 and 2968 led to an amino acid change (N752D and K990E, respectively). Other punctual mutations were observed on the other strains sequenced and five of them induced an amino acid change. One nucleotide on NORM/5/2012 sequence could not be determined during sequencing electrophoresis and was identified as a K (either thymine or guanine). NORM/17/2018 and NORM/18/2018 showed 100% identity at nucleotide and amino acid level. After further investigation about the outbreak location, it appeared that both strains NORM/17/2018 and NORM/18/2018 were collected, respectively, on February 2018 and March 2018 on the same premise after the abortion of two mares present in the stud farm. Although this comparison only concerns ORF30, it is possible that the same strain infected the two pregnant mares and induced their abortion. Both a phylogenetic tree (Neighbour-joining method and Maximum Composite Likelihood model) and a Median Joining Network were constructed (see Figure 3) based on the ORF30 sequences collected since 2009 and sequences obtained in a recent study in UK [14]. Two groups could be observed. The first group (Group 1 in Figure 3) formed a cluster of 32 strains including UK strains and reference strain Ab4. Of these 32, 26 contained the G_{2254} substitution in ORF30 (neuropathogenic type). Both G_{2254} strains collected in France (NORM/4/2012 and ILEDEFR/14/2018) belong to this group. The second group (Group 2 in Figure 3) is divided in three clusters. One of these clusters contains similar strains (NORM/2/2010, NELLEAQU/9/2015, BELG/12/2017, NORM/17/2018, NORM/18/2018) and one UK G_{2254} strain. NORM/5/2012 unidentified nucleotide in position 2876 did not discriminate the strain from ILEDEFR/3/2012 and NORM/8/2015. ORF30 analysis provides a first statement concerning strain phylogeny and potential neuropathogenicity but it only represents a small part of the genome. Multi-locus typing of EHV-1 as described by Garvey et al. 2019 was used to segregate EHV-1 strains into U_L clades (Bryant et al. 2018) [14,26].

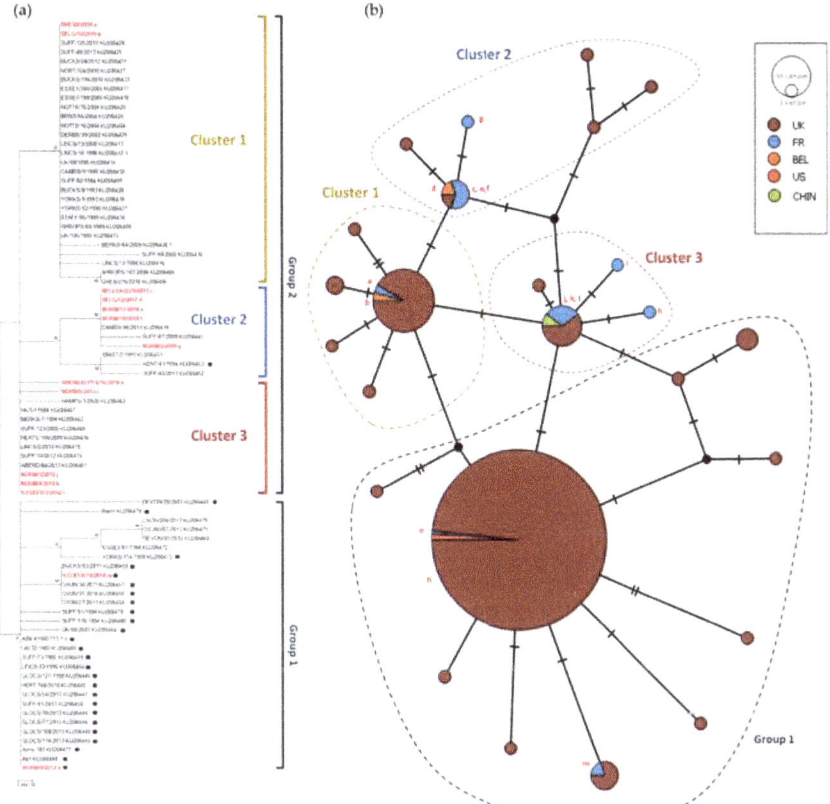

Figure 3. (**a**) Maximum Composite Likelihood method and Tamura-Nei matrix model phylogenetic tree based on ORF30 nucleotide sequences. Eighty-two strains including 14 strains isolated in France and Belgium from 2009 to 2018 (red bold text), 67 strains sequenced by Bryant et al. (2018) [14], and reference strain AY665713 (Ab4) are represented in this tree. Dots represent strains with the neuropathogenic type (G_{2254}). Boostrap values after 1000 replication are indicated at major nodes. (**b**) Median Joining Network based on the same ORF30 nucleotide sequences as for the phylogenetic tree.

3.2.3. Phylogeny and Multi-locus Analysis

MLST analysis for the 15 strains is represented in Table 6 and Figure 4. The analysis grouped five strains as U_L clade 10 (including two strains isolated from the same outbreak: NORM/17/2018 and NORM/18/2018) and three strains as U_L clade 7. Clades 8 and 13 contain two strains each, while clades 1, 6, and 11 contain only one strain. There is no correlation between sites and/or year of collection and clade distribution as also observed with ORF30 sequence comparison. In comparison with ORF30 clusters, strains assigned in group 2 cluster 2 were also assigned in clade 10 according to their MLST profile. The two strains in ORF30 group 1 were assigned to clade 8. No other correlation could be made between ORF30 classification and U_L clade classification.

Table 6. Multi Locus Sequence Typing of 15 EHV-1 strains collected from 2010 to 2018. Determinant loci for clade identification are shaded in grey. Ab4 and V292 are included as reference strains.

ORF	U_L/MLST	2	5	8	11	11	11	13	13	13	13	13	13	14	14	14	15	22	29	30	30	31	32	33	33	34	36	37	39	40	42	45	46	50	52	57	73	76
LOCI	Clades	59	114	114	189	235	250	305	405	460	492	493	499	618/20	628	692	166	430	12	752	990	90	42	15	976	66	47	265	440	196	1275	427	140	367	386	804	122	128
Ab4 AY665713	1	G	G	D	Q	R	A	S	A	A	E	T	A	—	R	S	D	S	T	D	E	N	S	N	N	D	S	A	S	R	K	E	F	P	A	K	A	F
NORM/2/2010	10	D	V	D	Q	R	A	L	A	T	E	T	A	PSR	R	S	N	S	T	N	K	S	S	H	D	D	S	V	S	R	K	G	S	P	V	R	A	S
ILEDEFR/3/2012	7				Q	M	A	S	A	T	E	T	A							N	E							V							V	R		S
NORM/4/2012	8	D	G	D	Q	R	A	S	A	T	E	T	T	PSR	K	N	N	S	T	D	E	N	S	H	D	D	S	V	S	R	K	G	S	P	V	K	A	S
NORM/6/2013	13	D	G	D	Q	R	A	S	A	A	E	T	A	PSR	R	S	D	S	T	N	E	N	S	N	D	D	S	V	S	R	K	G	S	P	A	K	A	S
BELG/7/2014	7	D	G	D	Q	M	A	S	A	T	E	T	A	PSR	K	N	N	S	T	N	E	S	S	H	D	D	S	V	S	R	K	G	S	P	V	K	A	S
NORM/8/2015	6	D	G	D	Q	R	S*	S	T	A	E	T	A	PSR	R	S	N	S	T	N	E	N	S	H	D	D	S	V	S	R	K	G	S	P	V	K	A	S
NELLEAQU/9/2015	10				Q	R	A	L	A	T	E	T	A							N	K							V							A	R		S
BELG/10/2016	1	D	G	D	Q	R	A	S	A	A	E	T	A	—	R	S	D	S	T	N	E	N	S	N	N	D	S	A	S	R	K	E	F	P	A	K	A	F
BELG/12/2017	10	D	V	D	Q	R	A	L	A	T	E	T	A	PSR	R	S	N	S	T	N	K	S	S	H	D	D	S	V	S	R	K	G	S	P	V	R	A	S
ILEDEFR/14/2018	8				Q	R	A	S	A	T	E	T	T							D	E							V							V	K		S
FRANCE/15/2018	13				Q	R	A	S	A	A	E	T	A							N	E							V							A	K		S
RHONEALPES/16/2018	11				Q	R	A	L	A	T	E	T	A							N	E							V							V	R		S
NORM/17/2018	10				Q	R	A	L	A	T	E	T	A							N	K							V							V	R		S
NORM/18/2018	10	D	V	D	Q	R	A	L	A	T	E	T	A	PSR	R	S	N	S	T	N	K	S	S	H	D	D	S	V	S	R	K	G	S	P	V	R	A	S
BRE/20/2018	7				Q	M	A	S	A	T	E	T	A							N	E							V							V	R		S
V592 AY464052	9	D	V	N	K	R	A	L	A	T	E	T	A	PSR	R	S	N	P	K	N	K	S	L	H	D	G	R	V	L	H	R	G	S	S	V	R	V	S

* ORF 11(A250S) was used to identify clade 6 strains in this study.

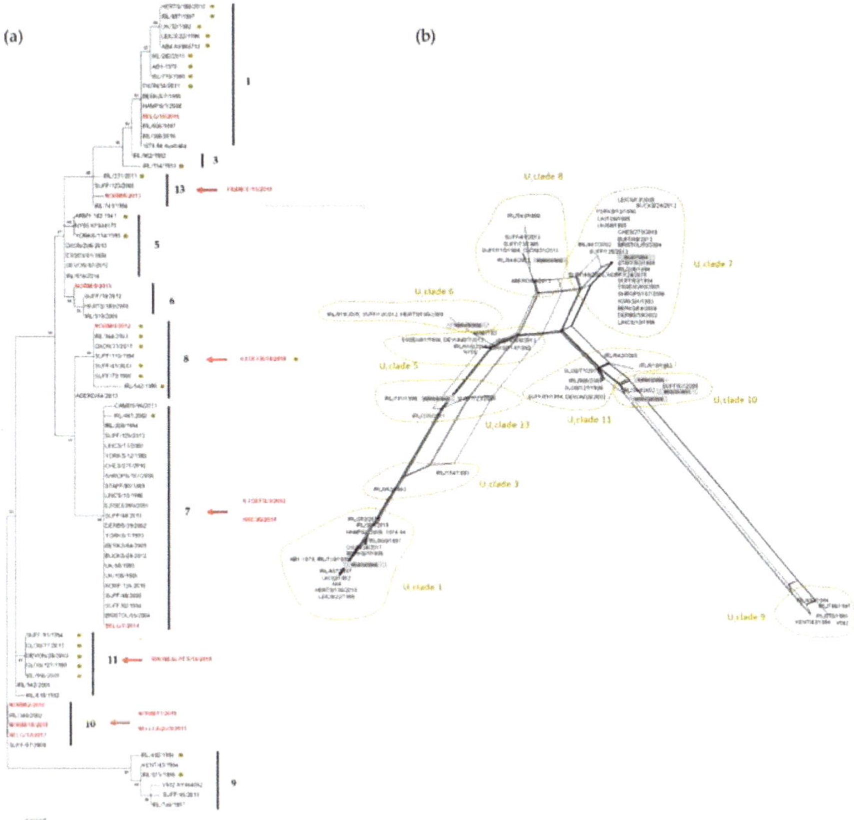

Figure 4. (**a**) Maximum Likelihood phylogenic tree based on Jones Taylor Thornton model built with MLST sequences including 8 French and Belgian strains (in red), 66 EHV-1 UK strains [14], and 22 EHV-1 Irish strains [26]. Dots indicate D752 strains (Neuropathogenic type). Strains with partial concatenated amino acid sequence were not included in the Maximum Likelihood analysis but their supposed position in the tree is indicated with red arrows according to their clade identification (see Table 6). U_L Clades [14] are numbered and represented with vertical lines. Boostrap values after 1000 replication are indicated at major nodes. (**b**) Network built using Neighbour-Net method with the same sequences as the tree. Only 37 amino acid sequences were included in the network, and French and Belgian strains are represented in grey boxes.

MLST sequences from strains collected in Belgium and France were compared to strains sequenced by Bryant et al. [14], and a Maximum Likelihood phylogenic tree was built based on Jones Taylor Thornton model (see Figure 4).

4. Discussion

The 2009 EHV1 outbreak is regretfully a good example of the impact that an EHV-1 infection can have in stud farms and illustrates the diversity of diseases that could be observed and faced by veterinarians. The three clinical forms of disease induced by EHV-1 (respiratory, neurological disorder and abortion) were observed over a nine-and-a-half-month period on a thoroughbred farm (169 horses), which raises questions about the source of EHV-1 infection and its transmission during this outbreak. The significance of this 2009 EHV-1 outbreak was also the observation of the three clinical forms of disease in the same premise over a long period of time. This phenomenon is rarely described in literature [39], and no other outbreak involving all three forms of disease was reported in France since 2009. The number of horses on this premise and the breeding activities may be factors that have contributed to increase the number of cases, which could have been higher in the absence of vaccination. The identification of two different strains, which may be the result from frequent horse movements in France and abroad, might also have influenced the diseases observed during this unusual outbreak. EHV-1 outbreaks are frequent but are usually limited to the report of one or two forms of the disease for the same outbreak (RESPE, personal communication). However, epidemiological links between outbreaks separated in time are not always available or identified. For example, the 2009 outbreak described here lasted nine and a half months with at least two and a half months between the last respiratory infection and the first abortion. Occurrence of multiple forms of disease may be more frequent than currently imagined.

The three clinical forms of disease were reported on a regular basis from 2009 to 2018 on the different outbreaks of this study. In 2018, a large number of cases were reported during a short period (March 2018 to May 2018), with exceptional sanitary measures needed to control contaminations between horses. This EHV-1 crisis is likely to be associated with the fact that a vaccine shortage occurred in 2016, implying a lower vaccination rate.

At the time of the 2009 outbreak, only a few tools were available to conduct EHV-1 molecular investigation. The PCR designed by Diallo et al. (2006, [28]) was used as an EHV-1 detection and quantification test. Viral loads of samples were quantified, and sanitary measures were lifted when undetected. Results obtained at the time indicated that virus titers in total blood samples were lower than in swab extracts, but these two compartments are not always correlated to each other. Monitoring viral loads provided an overview of the virus excretion and risk of transmission. When none of the group 1 foals tested positive after a seven-week isolation period, the day to day management of the stud farm returned to normal. EHV-1 strains were typed (ORF30 SNP A2254G) and both types were identified (neuropathogenic and non-neuropathogenic).

Although those tools gave useful indicative information concerning the different strains isolated from this outbreak, they proved to be limited to establish potential relationship between cases and virus strains. Other tools have been developed (Nugent et al. 2006, Bryant et al. 2018 and Garvey et al. 2019 [14,19,26]) that have motivated our retrospective and molecular analysis of French and Belgian EHV-1 strains isolated from 2009 outbreak to the major 2018 EHV-1 epizootic. Three different molecular analysis were performed: the ORF30 A2254G typing, the complete ORF30 sequencing and the MLST. In 2006, Nugent et al. [19] reported a significant association between the A2254G SNP in the DNA polymerase gene (ORF30), neuropathogenicity. This single point substitution (A2254G) involves an amino acid change (N752D). N752 (A_{2254}) strains were strongly associated to non-neuropathogenic infection cases, while D752 (G_{2254}) strains were strongly associated to neuropathogenic cases [19]. A2254G typing was subsequently used on a regular basis for strain discrimination [40,41]. In our study, 137 strains collected from 2009 to 2018 were typed, and A_{2254} strains were more significantly associated with abortion cases than to neurological cases ($p = 0.0002$), as recently described by Lechmann et al. (2019) [42]. However, no significant correlation between G_{2254} strains and the neurological form of disease was measured. This observation is in agreement with some recent studies showing that A2254G mutation is not exclusively associated to EHM but could be part of a more complex mechanism affecting strains virulence [41]. Both neuropathogenic and non-neuropathogenic strains could be typed among 2018 strains, suggesting that more than one strain was circulating during the crisis. Strain A2254G typing is also interesting as it was reported that the point mutation could change sensitivity to some drugs targeting DNA polymerase activity as it was demonstrated in a study showing that a N752 variant was more sensitive to aphidicolin than the D752 variants. Aphidicolin inhibits some dNTPs binding to a family of DNA polymerase which include herpesvirus DNA polymerase [17,19].

The A2254G typing was completed with ORF30 sequencing on 14 French and Belgian strains isolated between 2012 and 2018 in order to compare sequences and to perform a phylogenetic analysis. Several synonymous and non-synonymous substitutions were identified. Equine herpesvirus DNA polymerase subunit structures and mechanisms are not well known. Despite a low homology between EHV-1 and Human Simplex Virus (HSV) polymerase amino acid sequence (54%), the latter has been described as closest to α polymerase structure [43]. On this basis, SNP found in EHV-1 strains could be attributed to structure domains identified in HSV polymerase [43]. Amino acid substitutions could be localised in the pre-NH2 terminal domain (R59G), in the 3'-5' exonuclease domain (S419L and R429K), in the palm domain (A694V and D752N), in the thumb domain (E990K). Although some of the domain activities have been studied for human herpesviruses [43,44], it is hard to predict the impact of the substitutions observed in EHV-1 strains on the protein activity. A strong homology (99.86% to 100%) was measured among those strains with no obvious evolutionary tendency. Results were in agreement with those published by Bryant et al. in 2018 [14]. According to ORF30 sequencing, EHV-1 evolution is not linked to sampling location or year of collection, with the exception of the strains NORM/17/2018 and NORM/18/2018 that were isolated from the same stud farm, a few weeks apart, and have similar ORF30 sequences. All G_{2254} strains were grouped in the same cluster, suggesting that ORF30 sequencing does not provide further information when compared with the A2254G typing. However, three clusters were identified for A2254 strains, primarily differentiated by one SNP (e.g A2968G). The significance of these different clusters is unknown. MLST analysis provides a more global view concerning EHV-1 strains evolution as it takes into account 37 loci in 26 different ORFs. As observed with ORF30 analysis, there is no obvious correlation between year and location of collection, with the exception of NORM/17/2018 and NORM/18/2018, both located in clade 10, which support a co-circulation of EHV-1 strains from different clades as already described by Bryant et al. and Garvey et al. [14,26]. It is interesting to note that four French strains and one Belgium strain are localised in clade 10. The U_L Clade 12 described by Bryant et al. (2018) [14], which contains a strain from the UK, is not shown here. U_L Clades 2 and 4 were not represented by any of the European strains analysed in this study. It is important to note that the MLST method cannot distinguish U_L Clade 2 and 12 from MLST Clade 1 and 10, respectively [26]. As all the strains compared in this study are from Europe, a broader strain selection would be needed to identify a potential geographical effect on clade differentiation. Finally, two abortion strains isolated from the same premise in a one-month interval after two mare abortions had the exact same MLST profile suggesting that the same strain infected the mares. Obviously, EHV-1 strain surveillance is complicated by the fact that EHV-1 can also establish latency in different sites, implying no viral replication as the viral genome maintains an episomal form blocking transcription and translation of its genes (limited transcription with LATs) [9,11]. This implies that strains circulation and outbreaks are potentially dependent on latency and re-activation.

5. Conclusions

To conclude, this study allowed applying and comparing three different typing approaches to conduct a phylogenetic analysis over a six years period. A significant association was measured between EHV-1 induced abortion and the DNA polymerase A_{2254} genotype of related strains, while no disease association was observed with the G_{2254} genotype. This result suggests that the commonly used "neuropathogenic/non-neuropathogenic" designation is not always appropriate. The ORF30 and MLST analysis highlight the diversity of EHV-1 strains circulating in the French equine population and the difficulty to link strain evolution, time of collection, and location. However, the MLST offers new possibilities for EHV-1 epidemiology.

Supplementary Materials: The following are available online at http://www.mdpi.com/1999-4915/11/10/916/s1, Table S1: Samples type collected from EHV-1 outbreaks from 2012 to 2018, Table S2: Strains identification, location and year of collection. Table S3. ORF30 sequencing Primers. Supplementary Materials S4: Complementary clinical information concerning Mare A. Supplementary Materials S5: Complementary clinical information concerning Mare B. Supplementary Materials S6: Complementary clinical information concerning Mare B abortion. Supplementary Table S7: ORF30 amino acid alignments.

Author Contributions: Here are described the contribution of the authors for this research article: conceptualisation, R.P., S.P., and G.S.; investigation, C.F., M.G., and G.S.; resources, M.J., C.F., P.M., M.F., A.G., V.M., C.M.-P., and L.L.; writing—original draft preparation, R.P., S.P., and G.S.; writing—review and editing,

M.G., A.C., C.F., A.G., and L.L.; supervision, R.P. and S.P.; project administration, R.P. and S.P.; funding acquisition, R.P. and S.P.

Funding: This research was funded by the Fonds Eperon OVERLORD N12-2017, Normandy County Council (17E01598/17EP04324), the IFCE (Institut Français du Cheval et de l'Equitation) grant number 2017-008, CENTAURE European project co-funded by Normandy County Council, European Union in the framework of the ERDF-ESF operationnal programme 2014-2020.

Acknowledgments: We thank Erika Hue (LABÉO) for her help in solving problems and methodology, Farida Mebarki (BIOFIDAL) for her help in sequencing, and Meriadeg Ar Gouilh (NORMANDIE UNIV) for his help with network construction.

Conflicts of Interest: The authors declare no conflict of interest.

References

1. Davison, A.J.; Eberle, R.; Ehlers, B.; Hayward, G.S.; McGeoch, D.J.; Minson, A.C.; Pellett, P.E.; Roizman, B.; Studdert, M.J.; Thiry, E. The Order Herpesvirales. *Arch. Virol.* **2009**, *154*, 171–177. [CrossRef] [PubMed]

2. Abdelgawad, A.; Azab, W.; Damiani, A.M.; Baumgartner, K.; Will, H.; Osterrieder, N.; Greenwood, A.D. Zebra-borne equine herpesvirus type 1 (EHV-1) infection in non-African captive mammals. *Vet. Microbiol.* **2014**, *169*, 102–106. [CrossRef] [PubMed]

3. Wohlsein, P.; Lehmbecker, A.; Spitzbarth, I.; Algermissen, D.; Baumgärtner, W.; Böer, M.; Kummrow, M.; Haas, L.; Grummer, B. Fatal epizootic equine herpesvirus 1 infections in new and unnatural hosts. *Vet. Microbiol.* **2011**, *149*, 456–460. [CrossRef] [PubMed]

4. Chowdhury, S.I.; Ludwig, H. Molecular biological characterization of equine herpesvirus type 1 (EHV-1) isolates from ruminant hosts. *Virus Res.* **1988**, *11*, 127–139. [CrossRef]

5. Ghanem, Y.M.; Fukushi, H.; Ibrahim, E.S.M.; Ohya, K.; Yamaguchi, T.; Kennedy, M. Molecular phylogeny of equine herpesvirus 1 isolates from onager, zebra and Thomson's gazelle. *Arch. Virol.* **2008**, *153*, 2297–2302. [CrossRef]

6. Dayaram, A.; Franz, M.; Schattschneider, A.; Damiani, A.M.; Bischofberger, S.; Osterrieder, N.; Greenwood, A.D. Long term stability and infectivity of herpesviruses in water. *Sci. Rep. UK* **2017**, *7*. [CrossRef]

7. Paillot, R.; Case, R.; Ross, J.; Newton, R.; Nugent, J. Equine Herpes Virus-1: Virus, Immunity and Vaccines. *Open Vet. Sci. J.* **2008**, *2*, 68–91. [CrossRef]

8. Lunn, D.P.; Davis-Poynter, N.; Flaminio, M.J.B.F.; Horohov, D.W.; Osterrieder, K.; Pusterla, N.; Townsend, H.G.G. Equine Herpesvirus-1 Consensus Statement. *J. Vet. Intern. Med.* **2009**, *23*, 450–461. [CrossRef]

9. Chesters, P.M.; Allsop, R.; Purewal, A.; Edington, N. Detection of Latency-Associated Transcripts of Equid Herpesvirus 1 in Equine Leukocytes but Not in Trigeminal Ganglia. *J. Virol.* **1997**, *71*, 7.

10. Aleman, M.; Pickles, K.J.; Simonek, G.; Madigan, J.E. Latent Equine Herpesvirus-1 in Trigeminal Ganglia and Equine Idiopathic Headshaking. *J. Vet. Intern. Med.* **2012**, *26*, 192–194. [CrossRef]

11. Rock, D.L. The molecular basis of latent infections by alphaherpesviruses. *Semin. Virol.* **1993**, *4*, 157–165. [CrossRef]

12. Gulati, B.R.; Sharma, H.; Riyesh, T.; Khurana, S.K.; Kapoor, S. Viral and Host Strategies for Regulation of Latency and Reactivation in Equid Herpesviruses. *Asian J. Anim. Vet. Adv.* **2015**, *10*, 669–689. [CrossRef]

13. Telford, E.A.R.; Watson, M.S.; McBride, K.; Davison, A.J. The DNA sequence of equine herpesvirus-1. *Virology* **1992**, *189*, 304–316. [CrossRef]

14. Bryant, N.A.; Wilkie, G.S.; Russell, C.A.; Compston, L.; Grafham, D.; Clissold, L.; McLay, K.; Medcalf, L.; Newton, R.; Davison, A.J.; et al. Genetic diversity of equine herpesvirus 1 isolated from neurological, abortigenic and respiratory disease outbreaks. *Transbound. Emerg. Dis.* **2018**, *65*, 817–832. [CrossRef] [PubMed]

15. Cuxson, J.L.; Hartley, C.A.; Ficorilli, N.P.; Symes, S.J.; Devlin, J.M.; Gilkerson, J.R. Comparing the genetic diversity of ORF30 of Australian isolates of 3 equid alphaherpesviruses. *Vet. Microbiol.* **2014**, *169*, 50–57. [CrossRef]

16. Crowhurst, F.A.; Dickinson, G.; Burrows, R. An outbreak of paresis in mares and geldings associated with equid herpesvirus 1. *Vet. Rec.* **1981**, *109*, 527–528.

17. Mumford, J.A.; Rossdale, P.D.; Jessett, D.M.; Gann, S.J.; Ousey, J.; Cook, R.F. Serological and virological investigations of an equid herpesvirus 1 (EHV-1) abortion storm on a stud farm in 1985. *J. Reprod. Fertil.* **1987**, *35*, 509–518.

18. Anagha, G.; Gulati, B.R.; Riyesh, T.; Virmani, N. Genetic characterization of equine herpesvirus 1 isolates from abortion outbreaks in India. *Arch. Virol.* **2017**, *162*, 157–163. [CrossRef]

19. Nugent, J.; Birch-Machin, I.; Smith, K.C.; Mumford, J.A.; Swann, Z.; Newton, J.R.; Bowden, R.J.; Allen, G.P.; Davis-Poynter, N. Analysis of Equid Herpesvirus 1 Strain Variation Reveals a Point Mutation of the DNA Polymerase Strongly Associated with Neuropathogenic versus Nonneuropathogenic Disease Outbreaks. *J. Virol.* **2006**, *80*, 4047–4060. [CrossRef]

20. Goodman, L.B.; Loregian, A.; Perkins, G.A.; Nugent, J.; Buckles, E.L.; Mercorelli, B.; Kydd, J.H.; Palù, G.; Smith, K.C.; Osterrieder, N.; et al. A Point Mutation in a Herpesvirus Polymerase Determines Neuropathogenicity. *PLoS Pathog.* **2007**, *3*, e160. [CrossRef]

21. Pronost, S.; Léon, A.; Legrand, L.; Fortier, C.; Miszczak, F.; Freymuth, F.; Fortier, G. Neuropathogenic and non-neuropathogenic variants of equine herpesvirus 1 in France. *Vet. Microbiol.* **2010**, *145*, 329–333. [CrossRef] [PubMed]

22. Allen, G.P. Development of a Real-Time Polymerase Chain Reaction Assay for Rapid Diagnosis of Neuropathogenic Strains of Equine Herpesvirus-1. *J. Vet. Diagn. Investig.* **2007**, *19*, 69–72. [CrossRef] [PubMed]

23. Van de Walle, G.R.; Goupil, R.; Wishon, C.; Damiani, A.; Perkins, G.A.; Osterrieder, N. A Single-Nucleotide Polymorphism in a Herpesvirus DNA Polymerase Is Sufficient to Cause Lethal Neurological Disease. *J. Infect. Dis.* **2009**, *200*, 20–25. [CrossRef] [PubMed]

24. Franz, M.; Goodman, L.; Van de Walle, G.; Osterrieder, N.; Greenwood, A. A Point Mutation in a Herpesvirus Co-Determines Neuropathogenicity and Viral Shedding. *Viruses* **2017**, *9*, 6. [CrossRef] [PubMed]

25. Gryspeerdt, A.; Vandekerckhove, A.; Van Doorsselaere, J.; Van de Walle, G.; Nauwynck, H. Description of an unusually large outbreak of nervous system disorders caused by equine herpesvirus 1 (EHV1) in 2009 in Belgium. *Vlaams Diergeneeskundig Tijdschrif* **2011**, *80*, 147–153.

26. Garvey, M.; Lyons, R.; Hector, R.; Walsh, C.; Arkins, S.; Cullinane, A. Molecular Characterisation of Equine Herpesvirus 1 Isolates from Cases of Abortion, Respiratory and Neurological Disease in Ireland between 1990 and 2017. *Pathogens* **2019**, *8*, 7. [CrossRef]

27. Pronost, S.; Legrand, L.; Pitel, P.-H.; Wegge, B.; Lissens, J.; Freymuth, F.; Richard, E.; Fortier, G. Outbreak of Equine Herpesvirus Myeloencephalopathy in France: A Clinical and Molecular Investigation: Outbreak of EHV-1 myeloencephalopathy. *Transbound. Emerg. Dis.* **2012**, *59*, 256–263. [CrossRef]

28. Diallo, I.S.; Hewitson, G.; Wright, L.; Rodwell, B.J.; Corney, B.G. Detection of equine herpesvirus type 1 using a real-time polymerase chain reaction. *J. Virol. Methods* **2006**, *131*, 92–98. [CrossRef]

29. Thieulent, C.J.; Hue, E.S.; Fortier, C.I.; Dallemagne, P.; Zientara, S.; Munier-Lehmann, H.; Hans, A.; Fortier, G.D.; Pitel, P.-H.; Vidalain, P.-O.; et al. Screening and evaluation of antiviral compounds against Equid alpha-herpesviruses using an impedance-based cellular assay. *Virology* **2019**, *526*, 105–116. [CrossRef]

30. Hall, T. BioEdit: A user-friendly biological sequence alignment editor and analysis program for Windows 95/98/NT. *Nucleic Acids Symp. Ser.* **1999**, *41*, 95–98.

31. CodonCode Aligner. Available online: https://www.codoncode.com (accessed on 27 September 2019).

32. Kumar, S.; Stecher, G.; Tamura, K. MEGA7: Molecular Evolutionary Genetics Analysis Version 7.0 for Bigger Datasets. *Mol. Biol. Evol.* **2016**, *33*, 1870–1874. [CrossRef] [PubMed]

33. Saitou, N.; Nei, M. The neighbor-joining method: A new method for reconstructing phylogenetic trees. *Mol. Biol. Evol.* **1987**, *4*, 406–425. [PubMed]

34. Jones, D.T.; Taylor, W.R.; Thornton, J.M. The rapid generation of mutation data matrices from protein sequences. *Bioinformatics* **1992**, *8*, 275–282. [CrossRef]

35. Madeira, F.; Park, Y.M.; Lee, J.; Buso, N.; Gur, T.; Madhusoodanan, N.; Basutkar, P.; Tivey, A.R.N.; Potter, S.C.; Finn, R.D.; et al. The EMBL-EBI search and sequence analysis tools APIs in 2019. *Nucleic Acids Res.* **2019**, *47*, W636–W641. [CrossRef]

36. PopART. Available online: http://popart.otago.ac.nz (accessed on 27 September 2019).

37. Huson, D.H.; Bryant, D. Application of Phylogenetic Networks in Evolutionary Studies. *Mol. Biol. Evol.* **2006**, *23*, 254–267. [CrossRef]

38. Bryant, D. Neighbor-Net: An Agglomerative Method for the Construction of Phylogenetic Networks. *Mol. Biol. Evol.* **2003**, *21*, 255–265. [CrossRef]

39. Walter, J.; Seeh, C.; Fey, K.; Bleul, U.; Osterrieder, N. Clinical observations and management of a severe equine herpesvirus type 1 outbreak with abortion and encephalomyelitis. *Acta Vet. Scand.* **2013**, *55*, 19. [CrossRef]

40. Vissani, M.A.; Becerra, M.L.; Olguín Perglione, C.; Tordoya, M.S.; Miño, S.; Barrandeguy, M. Neuropathogenic and non-neuropathogenic genotypes of Equid Herpesvirus type 1 in Argentina. *Vet. Microbiol.* **2009**, *139*, 361–364. [CrossRef]

41. Pronost, S.; Cook, R.F.; Fortier, G.; Timoney, P.J.; Balasuriya, U.B.R. Relationship between equine herpesvirus-1 myeloencephalopathy and viral genotype: EHV-1, genotype and EHM. *Equine Vet. J.* **2010**, *42*, 672–674. [CrossRef]

42. Lechmann, J.; Schoster, A.; Ernstberger, M.; Fouché, N.; Fraefel, C.; Bachofen, C. A novel PCR protocol for detection and differentiation of neuropathogenic and non-neuropathogenic equid alphaherpesvirus 1. *J. Vet. Diagn. Investig.* **2019**, *31*, 696–703. [CrossRef]

43. Liu, S.; Knafels, J.D.; Chang, J.S.; Waszak, G.A.; Baldwin, E.T.; Deibel, M.R.; Thomsen, D.R.; Homa, F.L.; Wells, P.A.; Tory, M.C.; et al. Crystal Structure of the Herpes Simplex Virus 1 DNA Polymerase. *J. Biol. Chem.* **2006**, *281*, 18193–18200. [CrossRef] [PubMed]

44. Zarrouk, K.; Piret, J.; Boivin, G. Herpesvirus DNA polymerases: Structures, functions and inhibitors. *Vir. Res.* **2017**, *234*, 177–192. [CrossRef] [PubMed]

 *viruses*

Article

Approach to Strain Selection and the Propagation of Viral Stocks for Venezuelan Equine Encephalitis Virus Vaccine Efficacy Testing under the Animal Rule

Janice M. Rusnak [1,*], Pamela J. Glass [2], Scott C. Weaver [3], Carol L. Sabourin [4], Andrew M. Glenn [1], William Klimstra [5], Christopher S. Badorrek [1], Farooq Nasar [2] and Lucy A. Ward [1]

[1] Joint Program Executive Office for Chemical, Biological, Radiological and Nuclear Defense (JPEO-CBRND), Joint Project Manager-Medical Countermeasure Systems (JMP-MCS), Joint Vaccine Acquisition Program (JVAP), 1564 Freedman Drive, Fort Detrick, MD 21702, USA

[2] Department of Virology, United States Army Medical Research Institute of Infectious Diseases (USAMRIID), 1425 Porter Street, Fort Detrick, MD 21702, USA

[3] Institute for Human Infections and Immunity, World Reference Center for Emerging Viruses and Arboviruses and Department of Microbiology and Immunology, University of Texas Medical Branch, 301 University Boulevard, Galveston, TX 77555, USA

[4] Battelle Biomedical Research Center, 1425 Plain City-Georgesville Road, West Jefferson, OH 43162, USA

[5] Center for Vaccine Research, University of Pittsburgh, 3501 Fifth Avenue, Pittsburgh, PA 15261, USA

* Correspondence: Janice.m.rusnak.ctr@mail.mil; Tel.: +1-301-619-8447

Received: 13 June 2019; Accepted: 30 August 2019; Published: 31 August 2019

Abstract: Licensure of a vaccine to protect against aerosolized Venezuelan equine encephalitis virus (VEEV) requires use of the U.S. Food and Drug Administration (FDA) Animal Rule to assess vaccine efficacy as human studies are not feasible or ethical. An approach to selecting VEEV challenge strains for use under the Animal Rule was developed, taking into account Department of Defense (DOD) vaccine requirements, FDA Animal Rule guidelines, strain availability, and lessons learned from the generation of filovirus challenge agents within the Filovirus Animal Nonclinical Group (FANG). Initial down-selection to VEEV IAB and IC epizootic varieties was based on the DOD objective for vaccine protection in a bioterrorism event. The subsequent down-selection of VEEV IAB and IC isolates was based on isolate availability, origin, virulence, culture and animal passage history, known disease progression in animal models, relevancy to human disease, and ability to generate sufficient challenge material. Methods for the propagation of viral stocks (use of uncloned (wild-type), plaque-cloned, versus cDNA-cloned virus) to minimize variability in the potency of the resulting challenge materials were also reviewed. The presented processes for VEEV strain selection and the propagation of viral stocks may serve as a template for animal model development product testing under the Animal Rule to other viral vaccine programs. This manuscript is based on the culmination of work presented at the "Alphavirus Workshop" organized and hosted by the Joint Vaccine Acquisition Program (JVAP) on 15 December 2014 at Fort Detrick, Maryland, USA.

Keywords: Venezuelan equine encephalitis virus; vaccine; strain selection; Animal Rule; cDNA cloned virus; virus stock propagation

---

1. Introduction

Venezuelan equine encephalitis (VEE) is a mosquito-borne illness, endemic in areas of South America, Central America, Mexico, Florida and Trinidad, that primarily affects equids and humans [1–3]. VEE disease in humans is typically manifested as an acute self-limiting febrile illness of 3 to 5 days duration, with an abrupt onset of fever and chills, severe headache, malaise, myalgia, and nausea.

Unlike equids, encephalitis and mortality in humans from VEEV infection is uncommon (< 1% of cases) and is observed mainly in children and the elderly [4–6].

VEE virus (VEEV) has been designated a Category B biothreat agent as aerosol exposure to as few as 10 to 100 infectious particles results in symptomatic disease in nearly all humans. Aerosol-acquired VEE does not occur naturally. Thereby, as human efficacy studies are not ethical or feasible, licensure in the U.S. of a vaccine to protect against aerosol VEEV exposure must use the Food and Drug Administration (FDA) Animal Rule. The Animal Rule, as set forth in 21 Code of Federal Regulations Part 601.90–95, requires demonstration of vaccine protection in an animal model that is predictive of the response expected in humans. Vaccine efficacy demonstrated in animal models and the immune response to the vaccine observed in humans will be used to predict the likely clinical benefit in humans.

Guidance for virus strain selection for the animal model in the FDA document "Product Development under the Animal Rule- Guidance for Industry" notes selection of the virus strain should consider strain origin, virulence, passage history in cultures and animals, known disease progression in the animal model, and relevancy to human disease [7]. The strain must also result in similar disease and pathophysiology in the selected animal model as occurs in humans by the designated exposure route. Logistical issues (e.g., strain availability, replication characteristics to generate challenge material) and the sponsor's objectives for vaccine protection must also be considered in strain selection.

Selection criteria for VEEV isolates in the Department of Defense (DOD) Joint Vaccine Acquisition Program (JVAP) were reviewed, as well as the strategy in selecting the optimal methodology for the propagation of viral stocks to result in the consistency of the production of VEEV challenge material and reproducibility of animal challenge studies. The criteria for strain selection and strategy for the propagation of viral stocks may serve as a template for other viral vaccine products being developed under the Animal Rule. Lastly, current information and literature on VEEV aerosol challenge studies in animals with the selected strains was reviewed in regards to disease progression and pathophysiology in the animal models and the relevance of the selected strains and animal models to disease in humans.

2. Background

2.1. Alphaviral Committee and Workshop

Based on the recommendation of the FDA Office of Counter-Terrorism and Emergency Coordination Staff in 2012, a committee was organized by the JVAP alphavirus vaccine program to select VEEV strain(s) for animal model development under the Animal Rule, and to obtain concurrence on the strain selection by the FDA Center for Biologics Evaluation and Research (CBER) before developing the master and working viral stocks. The committee was comprised of representatives from JVAP, the U.S. Army Medical Research Institute of Infectious Diseases (USAMRIID), U.S. Army Medical Materiel Development Activity (USAMMDA), and Defense Threat Reduction Agency (DTRA), and involved consultation with multiple VEEV experts and institutes with virus repositories. An "Alphavirus Workshop" was then organized and hosted by the Joint Vaccine Acquisition Program (JVAP) at Fort Detrick, Maryland, on 15 December 2014 to present and discuss the activities of the alphavirus committee members and to address outstanding issues on the generation of VEEV challenge materials.

2.2. Strain Selection

The list of VEEV strains classified under the epizootic varieties (associated with equine-amplified epidemics) and enzootic varieties (maintained in rodent-mosquito cycles and generally not equine amplification-competent) of VEEV subtype I (VEEV IAB, IC, ID, and IE) was comprised after consultation with multiple institutes with known VEEV repositories, and did not include former VEE complex subtypes II–VI and VEEV IF due to their reclassification as distinct species that are now referred to by their prototype strain [8]. Relevant data for each strain included the source of the isolate, passage history, and country/year of initial isolation.

Initial strain selection criteria had to consider the DOD vaccine requirement for vaccine protection against VEEV in an aerosolized bioterrorism event. Further down-selection considered the FDA's guidance for strain selection under the Animal Rule (FDA 2015), which recommended the use of (1) strains isolated from lethal human cases (or associated with human disease/outbreaks), (2) strains with a known and low passage history in animals and cell culture, and (3) strains that mirrored the expected disease state in humans. Strain selection also considered criteria used by the Filovirus Animal Non-clinical Group (FANG) to select virus strains for the filovirus challenge stocks under the Animal Rule—to select isolates with no animal passage (if available) and limited cell passage (Table 1) [9]. Logistical issues considered included selection of strains that were available and accessible to laboratories licensed to work with select agents and strains that could be replicated to produce sufficient challenge material for animal models.

Table 1. Venezuelan equine encephalitis virus (VEEV) strain selection criteria for VEE vaccine animal model [7,8].

Source	Strain Selection Criteria
Department of Defense (DOD)	Strains that will support Food and Drug Administration (FDA) licensure for vaccine protection against at least two VEEV subtype I variants
DOD	Strains relevant to a bioterrorism event
FDA	Strains isolated from lethal human cases or associated with causing human disease
FDA	Strains with known and low passage history
Filovirus Animal Non-clinical Group (FANG)	Strains with no passage history in animals (if available)
	Strains with low passage history in cell culture
FDA	Strains that mirror the expected disease state in humans
Logistics	Strains available and accessible to laboratories licensed to work with select agents
Logistics	Ability to grow strain to yield sufficient challenge material for animal model

2.3. Approach to Selecting a Methodology for the Propagation of Challenge Material

A workshop comprised of alphavirus experts from government, industry, and academia was organized to discuss options for generating viral challenge material that would minimize or avoid variation in the potency of challenge material across multiple sites. Advantages and disadvantages of the propagation of challenge material from uncloned (wild-type), plaque-cloned, versus cDNA-cloned virus were reviewed, and variation in plaque sizes observed in cell cultures of uncloned (wild-type) VEEV was addressed [10–14].

2.4. Comparison of VEE in Humans and Animal Models

Literature and existing studies in mice and nonhuman primates (NHPs) challenged with the two selected VEEV strains were reviewed. Pathophysiological mechanisms of virulence by exposure route and exposure dose in humans and in animal models were compared with regards to time to disease onset, manifestations of disease, morbidity and mortality, and pathophysiology.

3. Summary of Workshop/Committee Proceedings

3.1. VEEV Strain Down-Selection for Animal Model

A total of 12 epizootic and 13 enzootic isolates were identified (Table 2). Down-selection to epidemic VEEV IAB and IC strains was based on the DOD objective to obtain FDA approval for a vaccine that protects against at least two VEEV varieties likely to be used in a bioterrorism event. These two epizootic varieties have been responsible for most epidemics and human cases, in contrast to endemic varieties (ID and IE) that are also associated with human infection but have not produced wide-spread epidemics [4,11,13,15]. In addition, VEEV subtype IAB was the only VEEV variety weaponized in the biological weapons programs of the U.S. and former Soviet Union, and thereby most likely to be involved in a bioterrorism event [16].

3.1.1. VEEV IAB Strains

VEEV IAB strains were responsible for VEE epidemics and outbreaks from 1938 to 1973. Evidence suggested many VEE IAB outbreaks were related to immunization of animals with formalin-inactivated VEEV IAB vaccines containing residual live virus, as further VEE IAB outbreaks were not reported after discontinued use of the inactivated vaccines. However, continuous cryptic circulation as a source of epizootic emergence cannot be excluded [31].

Based on the selection criteria, isolates from two VEEV IAB strains were identified, VEEV Trinidad (TrD) and 69Z1 strains (Table 2). Other strains of VEEV IAB (e.g., human VEEV IAB E123 69 and E541 73 strains isolated in Venezuela in 1969 and 1973, respectively) were not considered as the viral stocks and primary sequences were not readily available. The VEEV 69Z1 human isolate from the 1969 Guatemala outbreak that was available at USAMRIID was unacceptable due to its poorly documented passage history [32–35]. However, a VEEV 69Z1 isolate available at University of Texas Medical Branch (UTMB) had a known and acceptable passage history, consisting of only two passages in suckling mice and one passage on Vero cells.

The VEEV IAB TrD strain was originally isolated from brain tissue of an infected donkey in 1943 during the Trinidad outbreak [2,19,31,36]. As the Trinidad outbreak involved mainly equids, no TrD isolates from humans were available. Manifestation of disease in the 377 officially reported equid cases was most commonly a febrile illness accompanied by loss of appetite and depression, followed by neurological symptoms (e.g., somnolence, incoordination, muscle twitching/spasticity), and death (83% mortality reported but milder cases in rural areas may have been missed). Although not a human isolate, VEEV IAB TrD was extensively studied and weaponized in the U.S. and former Soviet Union weapons programs, and thereby was deemed a likely strain to be involved in a bioterrorism event. The VEEV TrD strain was also commonly used in animal studies to examine pathogenesis and to assess efficacy of vaccines and therapeutics [37–48]. However, most of the available VEEV TrD stocks had multiple passages in animal and culture cell lines since its early isolation in 1943 because adaptation of the virus in cell culture was often required to achieve the high viral titers needed for aerosol challenge of larger animal models [49,50]. In spite of the extensive passage history, several laboratories have demonstrated the continued virulence of these IAB VEEV stocks in animals and in humans, including infection in eight laboratory workers with the VEEV IAB Venezuela 1938 strain that had 52 passages in suckling mice [19,51–53]. Most TrD strains in the USAMRIID repository had a passage history of guinea pig—one, chick embroyo—14, suckling mouse brain—one, and an additional one or two passages in BHK cells (Table 2). However, one VEEV TrD stock had a lower passage history (guinea pig—one, chick embryo—13, duck embryo cells—one) and no passage history in BHK cells [17–19]. Lastly, the consensus sequence of a VEEV TrD stock (L01442.2) was available in the GenBank database and had a documented passage history of once in guinea pigs, six times in Vero cells, and once in BHK cells. This TrD stock was propagated by the CDC using the same original 1943 VEEV TrD donkey brain isolate as the USAMRIID stocks [21,22].

The VEEV TrD strain was selected for animal model development, even though isolated from donkey brain, based on its weaponization history and the extensive study data available. Also, documented cases of aerosol-acquired infection in humans due to VEEV TrD would allow for a comparison of aerosol-acquired disease in humans to the animal models (as required by the Animal Rule) [52]. The USAMRIID TrD stocks with documented passage histories in Table 2 were considered as options for use under the Animal Rule, as well as the TrD (L-01442.2) consensus sequence from the GenBank database.

Table 2. VEEV epizootic (IAB, IC) and enzootic (ID, IE) strains.

Strain Type	Strain	VEEV Subtype [i]	Source	Passage History	Place and Year of Isolation	Reference
Epizootic VEEV Strains	Trinidad Donkey (USAMRIID)	IAB	Donkey brain	GP1, CE14, SMB1, BHK-1 or 2	Trinidad, 1943	[17–20]
	Trinidad Donkey (USAMRIID)	IAB	Donkey brain	GP1, CE13, DE cells 1		
	Trinidad Donkey (CDC)	IAB	Donkey brain	GP1, V6, BHK1	Trinidad, 1943	[21,22]
	69Z1	IAB	Human	SM2, V1	Guatemala, 1969	[20]
	69Z1	IAB	Human	BHK1, unknown	Guatemala, 1969	[20]
	INH-9813	IC	Human serum	V1	Venezuela, 1995	[23]
	INH-6803	IC	Human serum	V1	Venezuela, 1995	[23]
	SH3	IC	Human	V1	Venezuela, 1993	[10,24]
	3908	IC	Human serum	C6/36-1	Venezuela, 1995	[10,15,23,25]
	6119	IC	Human serum	BHK1	Venezuela, 1995	[20,23,26]
	V198	IC	Human serum	SM1, V1, DE1	Colombia, 1962	[26]
	V178	IC	Horse brain	SMB2, V3	Colombia, 1961	[23]
	P676	IC	*Aedes triannulatus*	SM1, V3, BHK1	Venezuela, 1963	[26]
Enzootic VEEV Strains	3880	ID	Human	SM3, V4, BHK1	Panama, 1961	[20,27,28]
	FSL0201	ID	Human serum	V1	Peru, 2000	[29]
	83U434	ID	Hamster	CE cells 1, V1, BHK1	Colombia, 1983	[13,23]
	An9004	ID	Hamster	SM3, V1, BHK1	Colombia, 1969	[20,27]
	66637	ID	Hamster	SM1, V1	Venezuela, 1981	[13,20]
	306425	ID	Hamster	Unknown, BHK1	Colombia, 1972	[20]
	ZPC738	ID	Hamster	Unknown, BHK1	Venezuela, 1997	[13]
	V209A	ID	Mouse	SM2, V2	Colombia, 1960	[20,27]
	68U201	IE	Hamster	SM1, BHK2, CE cells 3	Guatemala, 1968	[30]
	93-42124	IE	Horse (brain)	SM1, CE cells 1	Chiapas, Mexico, 1993	[30]
	CPA-201	IE	Horse	SMB1, RK1, BHK2	Chiapas, Mexico, 1993	[30]
	96-32863	IE	Horse (brain)	SM1, CE1	Oaxaca, Mexico, 1996	[30]

NA = not available. BHK, baby hamster kidney cells; C6/36, mosquito cell line; CE, chick embryo; DE, duck embryo; GP, guinea pig; RK, rabbit kidney cells; SM, suckling mouse; SMB, suckling mouse brain; V, Vero cells; USAMRIID, U.S. Army Medical Research Institute of Infectious Diseases.

3.1.2. VEEV IC Strains

The University of Texas Medical Branch (UTMB) had several available VEEV IC human isolates with a single passage in Vero cells, from which the INH-9813 and INH-6803 strains from the 1995 outbreak in Colombia and Venezuela were selected for further evaluation (Table 2). UTMB also had a VEEV IC SH3 strain (human isolate with a single passage in Vero cells) from a less extensive Venezuelan outbreak in 1993. Other VEEV IC strains listed in Table 2 were not considered as they were not human isolates, or because they had less favorable passage histories than the INH-9813 and INH-6803 human isolates [54].

The VEEV IC INH-9813 and INH-6803 strains from the more extensive 1995 VEE outbreak were selected for further evaluation. The evaluation included preparation and characterization of a master stock (passage 2) and working stock (passage 3) for each strain in American Type Culture Collection (ATCC) Vero E6 cells (CRL-1586). Deep sequence analysis demonstrated that the sequence of each isolate was consistent with a VEEV IC strain. The virus stocks were determined to be pure of contaminating agents, and endotoxin levels were <0.23 EU/mL. The replication kinetic experiments of each strain were essentially equivalent and were comparable to the VEEV TrD strain. Based on the slightly higher titers in the prepared master and working virus stocks, the VEEV INH-9813 strain was ultimately selected.

3.2. Preparation of Challenge Material

The objective of the characterization of the VEEV challenge material from the virus stocks was to maintain purity, identity, potency, and uniformity of lots. The alphavirus workshop reviewed the pros and cons of three methodologies in regards to the generation of virus stocks and challenge material that would result in the reproducibility of potency (Table 3). The use of plaque-purified virus stocks carried the risk of not being able to replicate the performance of wild-type (uncloned) viral stocks if the selected plaque had suboptimal animal model fitness. The majority of experts felt that the cDNA clone-derived stock offered the most controlled pathway, as it would result in an unlimited supply of the virus derived from a well-characterized consensus sequence (expected to produce similar-sized plaques) and would avoid the potential for phenotypic changes associated with additional passages needed for the propagation of new master and working challenge material banks. Potential benefits from using uncloned (wild-type) VEEV to generate virus stocks were outweighed by identified disadvantages that included the lack of control of cell culture-adaptive mutations due to various passages, in vivo virus attenuation, and the potential need for additional culture passages to replenish the master and working challenge banks.

Table 3. Comparison of pros and cons for three methods of the propagation of VEEV stocks.

Population Type	Pros	Cons
Uncloned (wild-type)	Maintenance of wild type diversity	- Cell culture adaptation often results in artificial amino acid substitutions and in vivo attenuation - Risk that multiple major variants present in the population can confound identification of genetic determinants of important phenotypes - Limited supply without additional passages
Plaque-cloned	- Clear consensus sequence (verified by sequencing original population) and lack of multiple variants	- Possible reduced single nucleotide polymorphism diversity - Risk of selection of a suboptimal fitness mutant (change in the consensus and master sequence) - Limited supply without additional passages
cDNA-cloned	- Clear consensus sequence (verified by sequencing the first-generation population) and lack of multiple major variants - Unlimited supply of the same master virus without passages	- Possible reduction in single nucleotide polymorphism diversity

A strategy was developed during the workshop to manufacture virus stocks derived from a cDNA clone of the selected VEEV strain, and to then compare the clone-derived virus to the wild-type VEEV isolate in regards to plaque size, in vitro replication kinetics in cell culture, and in vivo virulence (mouse LD$_{50}$) (Figures 1 and 2). The prepared virus stocks (master and working virus banks) would be prepared in ATCC Vero E6 cells and subjected to deep sequence analysis to characterize the virus population and consensus sequences and purity of the preparations. The use of cDNA clones would depend on the characteristics and virulence of the rescued virus being similar to the wild-type VEEV strain. If the cDNA clone characteristics were noted to be similar to the wild-type VEEV strain, a comparison of ID$_{50}$ and targeted pathology (e.g., brain) of the cDNA clone to the wild-type VEEV isolate in NHPs would be considered.

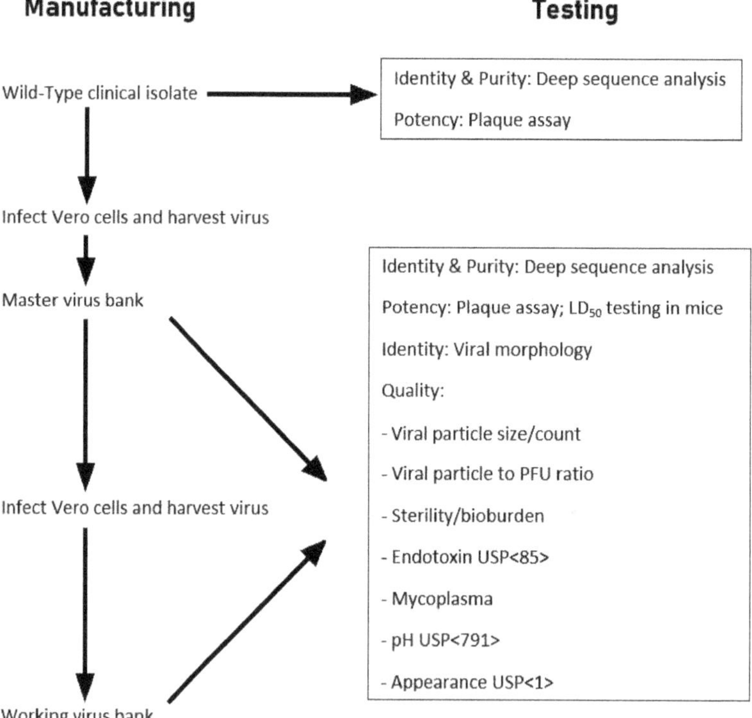

Figure 1. Process flow diagram for initial assessment and manufacturing uncloned (wild-type) VEEV TrD and INH 9813 strains for challenge material.

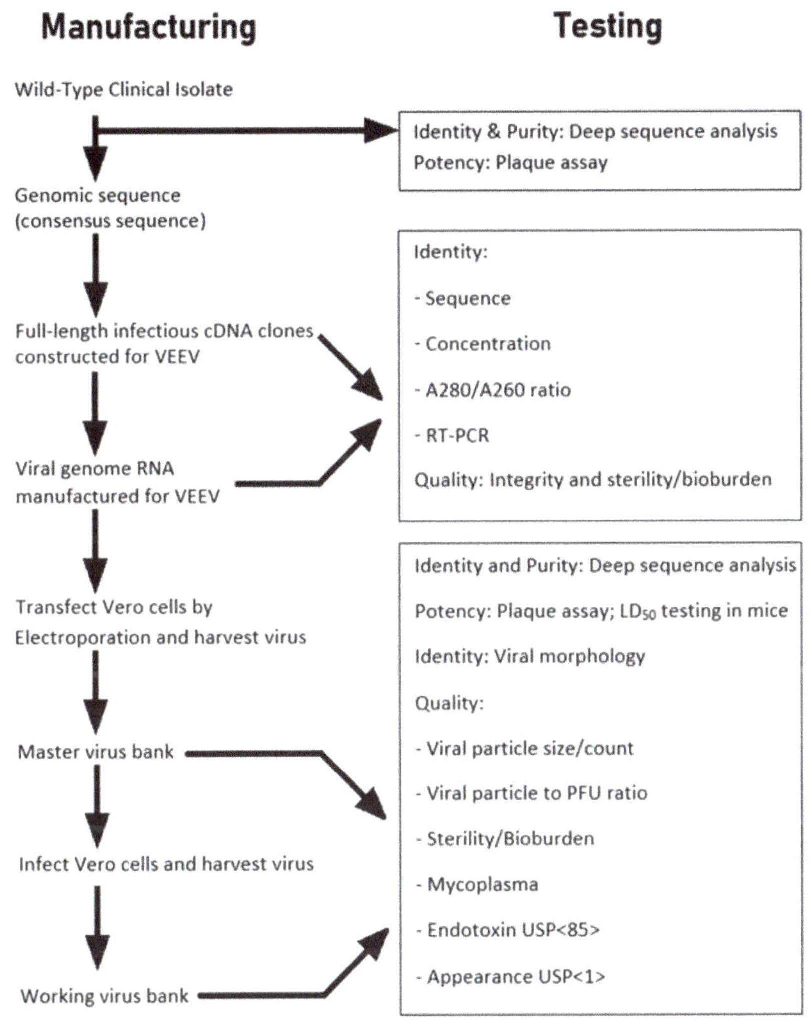

Figure 2. Process flow diagram for initial assessment and manufacturing cloned (cDNA) VEEV TrD and INH 9813 strains for challenge material.

3.3. VEE Disease and Pathophysiology in Humans

3.3.1. Background of VEE Disease

The literature has reported clinical manifestations of VEE to be similar in humans infected with VEEV varieties IAB, IC, ID, and IE, regardless of exposure route [51,55–60]. A recently published retrospective review of detailed medical records of laboratory- and vaccine-acquired VEEV IAB (TrD strain) infections during the U.S. Biowarfare (BW) program noted both aerosol and percutaneous routes of VEEV exposure to result in a self-limited, acute febrile illness that presented initially with an abrupt onset of fever, chills, severe headache, malaise, fatigue, weakness, back pain, myalgia (often in the lower back, thigh, or calf muscles), sore throat, anorexia, and nausea [52]. Incubation periods by both routes were similar (generally 1 to 4 days; range 2 h to 8 days), with most signs and symptoms occurring on the initial day of illness. Common physical examination findings included fever, pharyngeal erythema, conjunctival injection, and lymphadenopathy. Unlike mice and NHPs, in which aerosol-acquired infection resulted in an increased severity of central nervous system (CNS) disease compared to percutaneous-acquired infection [48,52,61–66], CNS signs and symptoms in humans were infrequent regardless of the exposure route, and severe encephalitis (e.g., seizures, paralysis, or coma) was not observed in this adult population (age 21–41 years). The only significant difference noted in aerosol-acquired infection was an association of increased upper respiratory tract-related symptoms and signs that included sore throat (with or without erythema), cervical lymphadenopathy, and neck pain. There were no deaths, and symptoms generally resolved within a week, with asthenia persisting in some cases for an additional 1 or 2 weeks. Fever, viremia, and lymphopenia were common markers of disease in humans by both exposure routes. Fever was observed in nearly all cases, viremia was common on days 1–4 of illness (range day 0–7 of illness), and lymphopenia (defined as <1500 cells/mm^3) was common early in infection (onset generally by day 1–3 of illness). The observed increase in upper respiratory tract-related findings associated with aerosol-acquired VEE in humans is supported by the mouse model in which aerosol and intranasal (IN) challenge were associated with nasal mucosa necrosis, necrotizing rhinitis, and increased viral burden in the upper respiratory tract, and by the IN challenge NHP model that detected VEEV in the cervical lymph nodes within 18 h post-challenge.

Rusnak et al. also compared the aerosol-acquired VEEV cohort from the U.S. BW Program (VEEV IAB strain TrD) to the initial 14 aerosol-acquired VEE cases reported in laboratory workers (mainly VEEV IAB Venezuela 1938 strain), to the 24 aerosol-acquired VEE cases in Russia (subtype unknown) from a single-source laboratory exposure, and to mosquito-borne VEEV IC and ID cohorts [36,52,53]. Regardless of the exposure route or VEEV subtype, infection in adults generally presented with a self-limited febrile illness that uncommonly resulted in severe encephalitis (e.g., seizures, coma, cranial nerve abnormalities, or paralysis) at exposure doses encountered from laboratory accidents, early VEEV vaccine candidates, and mosquito bites. While aerosol-acquired VEE in humans from VEEV IC strains has not been reported, the similar clinical presentation of percutaneous-acquired VEE due to VEEV IC and VEEV IAB strains supports aerosol-acquired VEE from VEEV IAB and IC strains to also have a similar clinical presentation.

Of note, the VEEV exposure doses in humans, whether laboratory-, vaccine-, or mosquito-acquired, were likely lower than aerosol challenge doses of NHPs which often were $\geq 1 \times 10^6$ pfu. The infrequent occurrence of CNS disease in adult humans by aerosol exposure may be species-related, with younger age and elderly being the major risk factors for CNS disease in humans. It is unknown if higher aerosol challenge doses in humans would result in increased CNS disease.

3.3.2. VEEV IAB (TrD) Infection

While the VEEV TrD outbreak (1943–44) mainly affected equids, two deaths in humans were attributed to a VEEV TrD strain, diagnosed by either viral isolation or cross-immunity testing in brain tissue. In addition, three laboratory technicians and three entomology workers developed nonlethal VEE infection due to occupational exposure [2,11,19,36,67]. Subsequent to the Trinidad outbreak, VEEV infection in humans from the TrD strain has been reported in laboratory workers after exposure to contaminated needles or infectious aerosols, and also after VEEV vaccination due to incomplete inactivation of the initial formalin-inactivated VEEV (TrD strain) vaccine administered during the U.S. BW Program [19,52,68], with exposure by both routes resulting in a similar self-limited febrile illness [52].

3.3.3. VEEV IC (INH-9813) Infection

The VEEV IC INH-9813 strain was isolated from human serum during the 1995 outbreak in Venezuela and Colombia that resulted in 75,000 to 100,000 human cases [6,23,69]. The clinical presentation most commonly manifested as a febrile illness of 3 to 4 days duration, presenting with an acute onset of fever, chills, severe headache, myalgia, prostration, vomiting, and sometimes diarrhea. The case-fatality rate during the 1995 outbreak was estimated to be 0.7% based on random surveys of residents in the Manaure municipality of La Guajira state, Colombia, with a conservative estimate of 300 VEE-associated deaths in La Guajira state alone during the outbreak. Neurological manifestations of encephalitis included disorientation, drowsiness, mental depression, and seizures. Thirteen VEEV human isolates from this epidemic characterized antigenically and/or genetically were determined to be closely related to VEEV IC isolates from the 1962–1964 Venezuela outbreak and a 1983 mosquito Panaquire isolate from north central Venezuela [23]. No outbreaks of VEEV IC have been reported since 1995.

Aerosol-acquired VEE due to subtype IC has not been reported in humans, and CNS histopathology in humans is limited mainly to mosquito-borne VEEV IC cases (strain unknown) (Tables 4 and 5) [23,70]. Based on the similar clinical presentation of mosquito-borne disease from VEEV IAB and IC strains, the symptomatology of aerosol-acquired disease and CNS histopathology are also likely to be similar between the two VEEV epidemic varieties [5,52].

Table 4. Summary of the selected VEEV IAB TrD strain in mice, NHPs, and humans.

	BALB/c Mouse	Cynomolgus Macaque	Humans
Route of Exposure	SC (scruff of neck, footpad) Aerosol (whole body)	Aerosol	Aerosol (laboratory exposure) Parenteral (vector-borne, laboratory exposure, vaccine-related)
Disease and Mortality	Lethal infection after SC and aerosol exposure due to encephalitis even at low challenge doses, with death 5–7 days post-challenge [48,64]	Generally non-lethal infection after SC and aerosol exposure at high challenge doses (100% infection after aerosol challenge 1×10^8 pfu); death uncommon. Increased severity of CNS disease after aerosol challenge [52,64].	Generally non-lethal infection after SC or aerosol exposure [52] Infection common after exposure to low doses. Encephalitis uncommon (<0.5%); mainly in children/elderly
Exposure Dose	LD_{50} SC ~10 pfu; LD_{50} aerosol <1 pfu [a]	ID_{50} unknown	Low infective dose (exposure dose unknown in humans)
Disease Onset	24–72 h	24–48 h	24–72 h (range 24 h–8 days)
Signs and Symptoms; Progression of Disease	Nearly 100% mice infected develop disease. Mice infected by the SC route demonstrated initial signs of decreased grooming and ruffled fur 2–3 days after challenge; followed by lethargy, hunched posture, and hind-limb paralysis; death or euthanasia 5–7 days after challenge. Mice infected by aerosol route demonstrated onset of similar signs 2–3 days after challenge, and death or euthanasia 6–7 days after challenge.	Aerosol VEEV TrD challenge results in fever and lethargy within 24–48 h after challenge. Fever resolved by D9. May have mild tremors. Generally NHPs have full recovery.	Self-limiting febrile illness, usually <1 week duration (asthenia may persist 1–2 weeks). Symptoms of high fever, chills, severe headache, back pain, malaise, myalgia, anorexia, nausea, sore throat, fatigue, photophobia, and/or vomiting. Encephalitis uncommon (<1% cases); manifested by decreased sensorium, confusion, gait abnormalities; severe cases with seizures, paralysis, or coma. Most cases recover without neurological sequelae.
Clinical Laboratory	Not done.	Viremia observed initially on D1–D2 that resolved by D3–D4 Lymphopenia observed early in illness. Increase in WBC observed later in some NHPs [37]	Viremia common D1–D4 (range D1–D7 illness) after aerosol and SC challenge. VEEV isolation from pharynx common D1–D4 (range D1–D7) in VEE IAB TrD and VEE IAB Co1938 strains [52]. Lymphopenia common D1–D3 of illness, with improvement after D3 and recovery by D6. Leukopenia may occur D3–D5; resolves by D5–D8 illness [52].
Pathology	Death due to encephalitis. CNS entry occurs via the olfactory system. Histological lesions present in both neural and extraneural tissues; CNS lesions characterized by necrotizing panencephalitis and myelitis; congestion and minor hemorrhage, damaged endothelial cells, perivascular edema, minimal necrosis and infiltration of a few neutrophils and mononuclear cells (no vasculitis) [48,64]	Limited pathology studies in cynomolgus macaques with TrD strain. Intraperitoneal challenge of Rhesus macaques showed histopathology findings initially in lymphoid tissues; lesions in olfactory cortex and thalamus by D6, and in hypothalamus and throughout brain by D8. Predominant lesions of gliosis and multifocal perivascular cuffing composed of lymphocytes. CNS lesions most severe between D14 and D21 post-challenge (observed in 18/20 NHPs). Aerosol challenge rhesus macaques (VEEV-subtype unknown) noted VEEV in nasal mucosa, lungs, cervical/hilar lymph nodes by 18 h, olfactory bulb by 48 h; more severe CNS infection than intraperitoneal challenge (higher CNS viral titers, increased neuronal damage, neuronophagia, and neutrophil infiltration); CNS infection preceded onset of viremia [52,64–66].	CNS pathology available in humans with VEEV IC (see Table 5).

D = day; SC = subcutaneous. [a] Starting concentration and all-glass impinger samples for aerosol exposure quantitated by plaque assay to determine titer (pfu/mL); Guyton and Bide formulas used to calculate the inhaled exposure dose per animal.

Table 5. Summary of the selected VEEV IC INH-9813 strain in mice, nonhuman primates (NHPs), and humans.

	BALB/c Mouse	Cynomolgus Macaque	Humans (VEEV IC Strains)
Route of Exposure	SC (scruff of neck); Aerosol (whole body)	Aerosol (head only)	Parenteral (mosquito-borne). No aerosol-acquired VEEV IC cases reported
Disease and mortality	Lethal infection after SC and aerosol exposure due to encephalitis even at low challenge doses (100% mortality after 100 pfu challenge dose), with death or euthanasia 6–8 days after challenge [71]	Non-lethal infection (100%) in 11 NHPs after aerosol challenge (dose range 5×10^4 to 5.94×10^8 pfu; 30 pfu in a single NHP) [71]	1995 Colombia/Venezuela outbreak with VEE IC INH-9813 strain (mosquito-borne disease) [6,23]. Generally non-lethal infection after SC exposure. No reported cases of aerosol-acquired VEEV IC. Infection common after exposure to low doses. Encephalitis uncommon; mainly in children/elderly
Exposure Dose	LD_{50} SC =5 pfu; LD_{50} aerosol = 53 pfu [a]	ID_{50} aerosol $\leq$ 30 pfu ($n = 1$)	Low infective dose (human infective dose unknown)
Disease Onset	48 h to 6 days	24 to 48 h	27.5 h to 4 days in 11 mosquito-borne VEE IC cases (strain unknown) [5]
Clinical Manifestation	VEEV INH-9813 resulted in infection after SC and aerosol exposure (100% infection). Mice with infection demonstrated initial signs of decreased grooming and ruffled fur 6 days after SC challenge or 48 h after aerosol challenge, followed by lethargy, hunched posture, and hind-limb paralysis (likely due to encephalitis), with death or euthanasia 6–8 days after SC or aerosol challenge	VEEV INH-9813 resulted in 100% infection after aerosol challenge. NHPs demonstrated initial signs of fever and lethargy 24–48 h after challenge, with fever resolving by D7. Biphasic fever noted only at highest dose tested. Tremors of variable duration (4–8 days; maximum 26 days) in 50% of NHPs. All NHPs survived	Mosquito-borne VEEV IC infection (strain unknown) similar to VEE IA/B infection in well-characterized outbreak in Texas (n-88) [5]. Self-limiting febrile illness of 1-week duration (asthenia may persist for 1–2 weeks). Symptoms included high fever, chills, severe headache, myalgias, malaise, and anorexia. Also, nausea, vomiting, sore throat, photophobia, ocular pain, arthralgia, somnolence and drowsiness reported. Encephalitis in 2% adults (mild) and 7.6% children. Encephalitis manifested by decreased sensorium, disorientation, delirium, nuchal rigidity, ataxia, seizures; coma and paralysis in severe cases
Clinical Laboratory	Unknown	Viremia detected starting on D1–D2 post-challenge and resolved by D3–D4. Lymphopenia typically noted early in infection. Elevated WBC count detected later in some NHPs ($n = 6$)	VEE IC outbreak (strain unknown) [5]: Viremia documented in 40 cases from D0–D8 of illness (most common on D3 of illness). Lymphopenia ($<$1490 cells/mm^3) in 80% cases on D1–D4 of illness; increase generally on D4 with recovery by D9. Leukopenia ($<$4500 cells/mm^3) in 75% cases at D1–D2, (recovery of total WBC by D3/neutropenia by D6–D8)
Pathology	Pathology studies not performed	Mild to moderate lymphocytic perivascular cuffing with gliosis in CNS at day 28 post-challenge. No viral antigen detected in CNS	Autopsy of 21 mosquito-borne VEEV IC cases (strain unknown): Cerebrovascular congestion ($n = 14$), edema with inflammatory infiltrates in brain/spinal cord ($n = 17$), intracerebral hemorrhage ($n = 7$), vasculitis ($n = 4$), meningitis ($n = 13$), encephalitis ($n = 7$), cerebritis ($n = 5$). Vasculitis, fibrin thrombi, perivascular hemorrhage and edema, occasional necrosis of blood vessel walls. Inflammatory infiltrates with lymphocytic and mononuclear cells, neutrophils, histiocytes. Lymph nodes and spleen with marked lymphoid depletion/follicular necrosis; hepatocellular degeneration and congestion (11/18 cases); interstitial pneumonia (19/21 cases) and pulmonary edema (11/21 cases) [70]

D = day; SC = subcutaneous. [a] Starting concentration and all-glass impinger samples for aerosol exposure quantitated by plaque assay to determine titer (pfu/mL); Guyton and Bide formulas used to calculate the inhaled exposure dose per animal.

3.4. Animal Models

3.4.1. VEEV Mouse Model

VEEV aerosol and IN challenge of several mouse strains (CD-1, BALB/c, outbred ICR, and C3H/HeN mice) in the literature were demonstrated to be lethal models, with death mainly due to encephalitis. Mice were challenged most commonly with wild-type VEE TrD strain or V3000 strain (VEEV derived from a cDNA clone of the VEE TrD strain with a passage history of once in guinea pig brain and 14 times in chick embryonated eggs) [48,52,61,63,72–77]. These studies demonstrated that, regardless of the exposure route, VEEV in mice entered the CNS mainly via the olfactory system due to the increased susceptibility of olfactory neurons to VEEV infection. Unlike subcutaneous (SC) challenge of VEEV that required a viremia before infection of the olfactory system, aerosol and IN challenge resulted in direct infection of the nasal mucosa and olfactory system with early neuroinvasion that occurred before the onset of viremia.

Aerosol and IN VEEV challenge were associated with increased histopathological findings and viral burden in the upper respiratory tract, nasal mucosa, and CNS compared to parenteral challenge. Aerosol and IN challenge resulted in necrotizing rhinitis, massive infection of the olfactory epithelium, and bilateral infection of the olfactory nerves, bulbs and tracts, with CNS infection noted between 16 and 48 h post-challenge. Viral levels were observed to be three times higher in the olfactory bulb than the brain at 16 to 24 h post-aerosol challenge but were similar to viral levels in the brain at 60 h, supporting virus entry into the brain via the olfactory system. Aerosol challenge also resulted in detectable virus in the lungs within 12 h post-challenge, with subsequent viremia and viral spread to lymphoid tissues [48,52,61–63,78].

3.4.2. VEEV NHP Model

Rhesus (*Macaca mulatta*) and cynomolgus (*Macaca fascicularis*) macaques in the literature have been assessed to be nonlethal models of VEEV infection with VEEV varieties IAB, IC, and IE. NHPs generally had onset of fever, viremia, and lymphopenia within 1 to 3 days following VEEV aerosol or parenteral challenge [52,64,79]. While some NHPs exhibited signs of encephalitis a few days later, nearly all NHPs (similar to humans) survived infection. Also, similar to disease in humans, fever, viremia, and lymphopenia were identified as markers of infection. CNS histopathology of infected NHPs noted multifocal perivascular cuffs composed mainly of lymphocytes, gliosis, satellitosis, neuronal death, and a few microhemorrhages.

Earlier NHP studies comparing aerosol/IN to parenteral VEEV challenge were often limited due to the absence of immunohistochemistry staining, electronmicrography, and VEEV strain characterization. Nevertheless, similar to mice, these NHP studies demonstrated earlier onset and more severe CNS disease after aerosol and IN challenge as compared to parenteral challenge. Unlike the mouse model, VEEV neuroinvasion and neurovirulence were more limited, resulting in a nonlethal infection. Similar to mice studies, studies in NHPs supported aerosol and IN challenge routes to be associated with early and direct CNS infection via the olfactory pathway, including studies that detected VEEV in the olfactory bulb within 48 h post-IN challenge (compared to 6 days after intraperitoneal challenge) and before the onset of viremia. Studies demonstrated that intratracheal challenge (bypassing the upper respiratory tract) or aerosol challenge of NHPs after surgical interruption of the olfactory tracts resulted in delayed and less severe CNS infection similar to parenteral VEEV challenge [52,64–66]. Although clinical and histopathological findings support both the rhesus and cynomolgus macaques as potential nonlethal animal models under the Animal Rule, most recent vaccine trials have used the cynomolgus macaque model to demonstrate vaccine efficacy against aerosol VEEV challenge [72,79–81].

3.4.3. VEEV IAB TrD Animal Models

The pathogenesis and virulence of the VEEV TrD strain and the V3000 cDNA clone-derived virus from the TrD strain were studied in mice, hamsters, and NHPs [64,77,82]. The passage history of

VEEV TrD virus stocks for the majority of animal studies conducted at USAMRIID was recorded as guinea pig—one, chick embryo—14, suckling mouse brain—one, and BHK—one or two. A summary of the host susceptibility and the clinical and pathological responses to the TrD strain is provided in Table 4. Of note, aerosol and IN challenge of mice with wild-type VEEV TrD and V3000 cDNA-cloned strains resulted in nasal mucosal necrosis, necrotizing rhinitis, an increase in viral burden in the upper respiratory tract, and an earlier and more severe CNS infection compared to percutaneous challenge [48,62,63]. NHP studies also noted an earlier onset and more severe CNS infection with aerosol and IN challenge compared to parenteral challenge routes.

3.4.4. VEEV IC INH-9813 Animal Models

Animal model development with wild-type and cDNA clone versions of the VEE IC INH-9813 strain is still in the early stages [71]. Aerosol challenge of BALB/c mice with the VEE IC INH-9813 strain (10 mice/group) resulted in lethal disease at low challenge doses, with 100% lethality at 2150 pfu, 80% lethality at 195 pfu, and 20% lethality at 16-pfu challenge doses (Table 5). Histopathology studies have not yet been performed. Aerosol challenge of cynomolgus macaques (challenge doses ranging from 30 to 5.94×10^8 pfu) with VEE IC INH-9813 resulted in nonlethal disease in all 11 NHPs, with a febrile illness observed even at the lowest 30-pfu challenge dose (one NHP only) (Table 5). Fever, viremia, and lymphopenia were observed as potential markers of infection.

4. Discussion

Selection of the viral strain and methodology for propagating viral stocks for challenge material for vaccine approval under the FDA Animal Rule is critical. The propagation of the virus must result in the uniformity of virus stocks and challenge material, and challenge of the selected animal model with the selected viral strain should result in the reproducibility of disease with a comparable morbidity/mortality as humans.

The VEEV strains selected for the aerosol-challenge animal model should "ideally" be isolates from human cases that have a low passage history in animals and cell cultures. However, the DOD requirement for a vaccine to protect against VEEV strains likely to be involved in a bioterrorism event justified the selection of a VEEV IAB TrD strain, even though a donkey brain isolate with multiple passages, as the TrD strain was the only VEEV strain weaponized. Even after multiple cell culture passages, the TrD strain still demonstrates virulence in animal models and exposed laboratory workers [52]. Furthermore, the availability of documented aerosol-acquired VEE cases in humans due to VEEV IAB (mainly TrD strain) would allow for comparison of aerosol-acquired VEEV in humans to the animal model.

However, down-selection to the VEE IC INH-9813 strain was based mainly on the strain being a human isolate with a limited passage history. While aerosol-acquired VEEV IC infection in humans has not been documented, the similarity of VEEV IAB and IC mosquito-borne disease supports the likely similarity to aerosol-acquired VEEV IAB disease and use of aerosol-acquired VEEV IAB human disease for comparison to the VEEV IC animal model. Initial VEEV INH-9813 aerosol and SC challenge studies in mice and NHPs demonstrated a similar morbidity and mortality as observed with the TrD strain. Based on the available data, the FDA concurred that the animal study data supported proceeding with the evaluation of both the TrD and INH-9813 strains in BALB/c mouse and cynomolgus macaque animal models.

A concern in the propagation of VEEV stocks and challenge material was a potential variation in virulence associated with the different sized plaques observed in cell cultures of wild-type VEEV (both large- and small-sized plaques). Variability in plaque size was also addressed by the FANG in the generation of Ebola virus challenge material under the Animal Rule, as the larger-sized plaques of 8U Ebola (attributed to virus adaptation in cell culture) were less virulent than the smaller-sized 7U Ebola virus plaques [9]. To address and avoid variations in the virulence of viral stocks due to 7U and 8U Ebola virus plaques, the Filovirus group developed a characterization, release, and stability test plan that would be used to evaluate master and working virus stocks generated using uncloned (wild-type) methodology and the cDNA-cloned methodology. However, the propagation of virus stocks using the cDNA-cloned methodology for Ebola virus was demonstrated not to be a viable option due to non-reproducibility of lots (different consensus sequences in each lot produced) and to the lower virulence of the cDNA clone-derived virus compared to the parent clone.

Unlike the experience of the Filovirus group, the use of the cDNA-cloned methodology to generate VEEV stocks was supported in the literature. VEEV cDNA clones derived from wild-type VEEV IC SH3 and VEEV ID ZPC728 strains exhibited in vitro and in vivo characteristics indistinguishable from their parent viruses and demonstrated similar replication kinetics as their parent virus in Vero76 and L929 cell lines [10]. SC VEEV challenge (1000 pfu) of adult National Institute of Health (NIH) Swiss mice with the cDNA-cloned strains resulted in comparable serum viral titers and time to death as their respective parent virus, and in high viremia titers within 24 h that were nearly identical to their parent strain. Disease from all four strains was manifested by decreased activity and hunching by day 3 to 4 post-infection; followed by onset of anorexia, lethargy, and hind limb paralysis by day 4 to 7 that progressed to stupor and coma; and then death due to encephalitis on day 7 or 8. A VEEV ID ZPC738 cDNA clone was also developed and demonstrated to be phenotypically indistinguishable from its enzootic parent strain [11]. Also, a VEEV IAB V3000 cDNA mutant clone (a single I239N mutation with a passage history in guinea pigs once and 14 times in embryonated eggs) derived from wild-type TrD demonstrated a similar LD_{50} as its parent wild-type TrD strain in aerosol-challenge studies in 4-week-old female CD-1 mice, as did a recently produced VEEV IAB V3000 cDNA clone (using similar methodology) without the I239N mutation [74,77,83]. The original V3000 cDNA clone resulted in a similar lethal illness in mice and nonlethal febrile illness with viremia in NHPs as reported with wild-type VEEV TrD and was also used to evaluate the mechanism of neuroinvasion in mice and to assess vaccine efficacy in NHPs [61,63,72,82]. Lastly, initial comparison of the virulence of a newly developed cDNA clone of VEEV TrD GenBank reference sequence L01442.2 to wild-type VEEV TrD (passage history in guinea pigs once, chick embryos 13 times, and duck embryo cells once) showed similar lethality in mice following aerosol challenge, with 100% mortality after challenge with 7 pfu of cDNA-cloned VEEV TrD or 28 pfu of wild-type VEEV TrD [84]. Only the 4-pfu aerosol challenge dose with wild-type VEEV TrD resulted in survival of some mice (estimated LD_{50} of 4.6 pfu).

Based on the experience with cDNA-cloned stocks and the identified risks in generating virus stocks using wild-type or plaque-cloned viruses, the cDNA-cloned methodology was chosen as the path forward to propose to the FDA for the generation of virus stocks and challenge material. The cDNA-cloned methodology would provide a more controllable path with use of a clear consensus sequence (verified by sequencing of the original population), avoid multiple major variants, and result in an unlimited supply of the virus without additional passages. The use of cDNA clones hypothetically has the potential to result in a reduction in single nucleotide polymorphism (SNP) diversity as compared to the wild-type parenteral stock. And though reduction in SNP diversity has been associated with high viral fidelity variants that are also attenuated, animal studies indicate that clone-derived VEEV stocks are as virulent if not more virulent than their originating parent (uncloned) stock [10,11,83–86].

Subsequent communication with the FDA resulted in their concurrence on the usage of the cDNA-cloned methodology to generate viral stocks, provided the virulence of the cDNA-cloned VEEV stocks was comparable or noninferior to the uncloned (wild-type) VEEV stocks as outlined in Figures 1 and 2. Comparison of ID_{50} and targeted pathology (e.g., brain) of the cDNA clone to the wild-type VEEV strain in NHPs would be considered if the cDNA clone characteristics were similar to the wild-type VEEV isolate. After the generation of sufficient data, the overall characterization plan would be reviewed and modified to a more efficient approach as assays are refined and the master and working viral stocks are characterized.

Vaccine licensure under the Animal Rule for the DOD VEEV vaccine initially will be requested only for protection against aerosol challenge with VEEV TrD strain. However, VEEV vaccine candidates demonstrating protection against the TrD strain will likely protect against other IAB and IC epizootic strains, based on phylogenetic/antigenic similarities among the epizootic strains [31,39,46,78,87]. VEEV IAB strains showed only a 96.1 to 99.2% genetic distance between strains on full sequence analysis [31,88]. Only a 15 amino acid difference was observed in the E2 glycoprotein between a VEEV IC epizootic strain derived from a VEEV ID enzootic strain following site-directed mutations in the E2 glycoprotein [11,88]. While phylogenetic differences may be greater with VEEV IE and the former VEEV complex subtypes VEE IF and VEE complex subtypes II to IV, the highly conserved E1 protein among alphaviruses may potentially result in cross-immunogenicity of epizootic-based vaccines against these former VEE complex subtypes via vaccine-induced antibodies to E1 and E2 envelope proteins [88–91].

Vaccine cross-protection from VEEV IAB strain-derived vaccines is supported by serological testing and/or animal challenge studies that demonstrated immunogenic responses and/or cross-protection against other VEEV IAB strains, other subtype I varieties, and/or former VEE complex subtype II and IIIA viruses. The VEEV TC-83 investigational vaccine, a live, attenuated vaccine derived from the VEEV IAB TrD strain, resulted in long-term (9 years) persistence of plaque reduction neutralization antibodies in humans against VEEV IC strain V-198 but not against enzootic VEE IE [80,85]. However, a more immunogenic, live-attenuated V3526 vaccine candidate (also derived from VEEV TrD strain), protected NHPs against VEEV IE (68U201) aerosol challenge even though it failed to elicit measurable neutralizing antibodies to VEEV IE in six of eight NHPs [80,90]. A virus-like replicon particle (VRP) VEEV vaccine candidate demonstrated long-term (12 months) cross-protection in mice (100% survival) against aerosol challenge to VEEV IAB, VEEV IE, and Mucambo virus (former VEEV subtype IIIA), that was associated with a strong VEEV-specific IgG-specific ELISA response but a minimal neutralizing antibody response [39]. VRP vaccination in all NHPs was associated with a measurable neutralizing antibody response to VEE IAB (TrD), VEEV IC (p676 strain), VEEV ID (3880 strain), and VEEV IE (68U201 strain). Lastly, cross-protection against IE and former VEEV subtypes was demonstrated with a humanized monoclonal antibody (Hu1A3B-7) derived from bone marrow donors immunized with TC-83, that protected mice against aerosol challenge to VEEV TrD, VEEV IE, Mucambo virus (former VEE complex subtype IIIA), and Everglades virus (former VEE complex subtype II) [42].

Clinical presentation of VEEV IAB-induced disease in humans by both aerosol and percutaneous exposure routes has been well characterized [52]. Unlike mice and NHPs, an increased severity of encephalitis was not observed with aerosol-acquired VEE in humans. Nevertheless, a greater disease severity in mice and NHPs does not preclude their use as animal models, if the models are stricter than human disease.

5. Conclusions

Systematic approaches were developed for virus strain selection and selection of the optimal methodology for the propagation of viral stocks and challenge material for the animal models, as required for VEEV vaccine approval under the FDA Animal Rule. The VEEV IAB TrD and IC INH-9813 strains were selected for vaccine development challenge strains in the BALB/c mouse and cynomolgus macaque animal models. The selection of cDNA-clone methodology to generate virus stocks and challenge material was based on the use of a stable consensus sequence being a more controlled path, providing an increased likelihood of uniformity of lots and an unlimited virus supply without additional passages. These methodologies may serve as a template for strain selection for other vaccine programs using the Animal Rule.

Author Contributions: All authors attended and contributed to discussions at the alphavirus workshop, and also contributed to the writing and editing of the report. J.M.R. was the primary author of this manuscript. A.M.G. organized the workshop and committee meetings. P.J.G., S.C.W., C.L.S., and W.K. gave presentations at the alphaviral workshop on propagation methodologies for viral stocks, which were used to select the propagation methodology for the VEEV stocks. VEEV strains listed in the manuscript were based on literature review and isolates available at institutions with known VEEV repositories (P.J.G., A.M.G.). Comparison of animal challenge studies and human disease due to the selected VEEV isolates was based on review of the literature (J.M.R., P.J.G.)

and from animal studies pending publication (P.J.G., C.L.S., S.C.W., W.K.). L.A.W., C.S.B., C.L.S., and F.N. provided technical oversight and guidance for challenge material preparation and testing strategies.

Funding: This work was supported by the U.S. DOD Joint Program Executive Office for Chemical, Biological, Radiological, and Nuclear Defense (JPEO-CBRND) Joint Project Manager Medical Countermeasure Systems (JPM-MCS)—Joint Vaccine Acquisition Program (JVAP). Opinions, interpretations, conclusions, and recommendations are those of the authors and are not necessarily endorsed by the U.S. Army. This article was co-written by an officer or employees of the U.S. Government as part of their official duties and therefore is not subject to U.S. copyrights. Funding for work in the presentation by Scott C. Weaver on the use of VEEV cDNA clone methodology for propagation of viral stocks and challenge material in animal models was funded by NIH grant AI120942.

Acknowledgments: We want to thank other participants from the following institutions for contributions to the alphaviral workshop: Medical Countermeasure Systems Joint Vaccine Acquisition Program (MCS-JVAP), MCS Biotherapeutics Program (MCS-BDTx); U.S. Army Medical Research Institute of Infectious Diseases (USAMRIID); Defense Threat Reduction Agency (DTRA); Center for Vaccine Research at the University of Pittsburgh; Battelle Biomedical Research Center; University of Texas Medical Branch (UTMB); Texas Biomedical Research Institute; and U.S. Army Medical Materiel Development Activity (USAMMDA). We also want to thank Sarah Cooper of JPEO-CBRND for her editorial assistance on the manuscript.

Conflicts of Interest: This work was co-written by employees of the U.S. Government (JR, AG, CB, LW, PG, and FN) as part of their official duties. Opinions, interpretations, conclusions, and recommendations are those of the authors and are not necessarily endorsed by the U.S. Army. The authors declare no conflict of interest.

References

1. Weaver, S.C.; Ferro, C.; Barrera, R.; Boshell, J.; Navarro, J.C. Venezuelan equine encephalitis. *Annu. Rev. Entomol.* **2004**, *49*, 141–174. [CrossRef] [PubMed]
2. Randall, R.; Mills, J.W. Fatal encephalitis in man due to the Venezuelan virus of equine encephalomyelitis in Trinidad. *Science* **1944**, *99*, 225–256. [CrossRef] [PubMed]
3. Ehrenkranz, N.J.; Ventura, A.K. Venezuelan equine encephalitis virus infection in man. *Annu. Rev. Med.* **1974**, *25*, 9–14. [CrossRef]
4. Aguilar, P.V.; Estrada-Franco, J.G.; Navarro-Lopez, R.; Ferro, C.; Haddow, A.D.; Weaver, S.C. Endemic Venezuelan equine encephalitis in the Americas: Hidden under the dengue umbrella. *Future Virol.* **2011**, *6*, 721–740. [CrossRef] [PubMed]
5. Bowen, G.S.; Fashinell, T.R.; Dean, P.B.; Gregg, M.B. Clinical aspects of human Venezuelan equine encephalitis in Texas. *Bull. Pan Am. Health Organ.* **1976**, *10*, 46–57. [PubMed]
6. Rivas, F.; Diaz, L.A.; Cardenas, V.M.; Daza, D.E.; Bruzon, L.; Alcala, A.; De la Hoz, O.; Caceres, F.M.; Aristizabal, G.; Martinez, J.W.; et al. Epidemic Venezuelan equine encephalitis in La Guajira, Colombia, 1995. *J. Infect. Dis.* **1997**, *175*, 828–832. [CrossRef]
7. U.S. Food and Drug Administration. Product Development under the Animal Rule. Guidance for Industry. 2015. Available online: https://www.fda.gov/downloads/drugs/guidances/ucm399217.pdf (accessed on 21 October 2018).
8. Powers, A.M.; Huang, H.V.; Roehrig, J.T.; Strauss, E.G.; Weaver, S.C. Togaviridae. In *Virus Taxonomy: Classification and Nomenclature of Viruses: Ninth Report of the International Committee on Taxonomy of Viruses*; King, A.M.Q., Adams, M.J., Carstens, E.B., Lefkowitz, E.J., Eds.; Elsevier Academic Press: San Diego, CA, USA, 2011; pp. 1103–1110.
9. Hirschberg, R.; Ward, L.A.; Kilgore, N.; Kurnat, R.; Schiltz, H.; Albrecht, M.T.; Christopher, G.W.; Nuzum, E. Challenges, progress, and opportunities: Proceedings of the filovirus medical countermeasures workshop. *Viruses* **2014**, *6*, 2673–2697. [CrossRef] [PubMed]
10. Anishchenko, M.; Paessler, S.; Greene, I.P.; Aguilar, P.V.; Carrara, A.; Weaver, S.C. Generation and characterization of closely related epizootic and enzootic infectious cDNA clones for studying interferon sensitivity and emergence mechanisms of Venezuelan equine encephalitis virus. *J. Virol.* **2004**, *78*, 1–8. [CrossRef] [PubMed]
11. Anishchenko, M.; Bowen, R.A.; Paessler, S.; Austgen, L.; Greene, I.P.; Weaver, S.C. Venezuelan encephalitis emergence mediated by a phylogenetically predicted viral mutation. *Proc. Natl. Acad. Sci. USA* **2006**, *103*, 4994–4999. [CrossRef]
12. Martin, D.H.; Dietz, W.H.; Alvaerez, O., Jr.; Johnson, K.M. Epidemiological significance of Venezuelan equine encephalomyelitis virus in vitro markers. *Am. J. Trop. Med. Hyg.* **1982**, *31*, 561–568. [CrossRef]

13. Wang, E.; Barrera, R.; Boshell, J.; Ferro, C.; Freier, J.E.; Navarro, J.C.; Salas, R.; Vasquez, C.; Weaver, S.C. Genetic and phenotypic changes accompanying the emergence of epizootic subtype IC Venezuelan equine encephalitis viruses from an enzootic subtype ID progenitor. *J. Virol.* **1999**, *73*, 4366–4371.

14. Powers, A.M.; Brault, A.C.; Kinney, R.M.; Weaver, S.C. The use of chimeric Venezuelan equine encephalitis viruses as an approach for the molecular identification of natural virulence determinants. *J. Virol.* **2000**, *74*, 4258–4263. [CrossRef] [PubMed]

15. Greene, I.P.; Wang, E.; Deardorff, E.R.; Milleron, R.; Domingo, E.; Weaver, S.C. Effect of alternating passage on adaptation of Sindbis virus to vertebrate and invertebrate cells. *J. Virol.* **2005**, *79*, 14253–14260. [CrossRef] [PubMed]

16. Bronze, M.S.; Huycke, M.M.; Machado, L.J.; Voskuhl, G.W.; Greenfield, R.A. Viral agents as biological weapons and agents of bioterrorism. *Am. J. Med. Sci.* **2002**, *323*, 316–325. [CrossRef] [PubMed]

17. Berge, T.O.; Gleiser, C.A.; Gochenour, W.S., Jr.; Miesse, M.L.; Tigertt, W.D. Studies on the virus of Venezuelan equine encephalomyelitis. II. Modification by specific immune serum of response of central nervous system of mice. *J. Immunol.* **1961**, *87*, 509–517. [PubMed]

18. Jahrling, P.B.; DePaoli, A.; Powanda, M.C. Pathogenesis of a Venezuelan encephalitis virus strain lethal for adult white rats. *J. Med. Virol.* **1978**, *2*, 109–116. [CrossRef]

19. Tigertt, W.D.; Downs, W.G. Studies on the virus of Venezuelan equine encephalomyelitis in Trinidad, W.I. I. The 1943–1944 epizootic. *Am. J. Trop. Med. Hyg.* **1962**, *11*, 822–834. [CrossRef]

20. Powers, A.M.; Oberste, M.S.; Brault, A.C.; Rico-Hesse, R.; Schmura, S.M.; Smith, J.F.; Kang, W.; Sweeney, W.P.; Weaver, S.C. Repeated emergence of epidemic/epizootic Venezuelan equine encephalitis from a single genotype of enzootic subtype ID virus. *J. Virol.* **1997**, *71*, 6697–6705.

21. France, J.K.; Wyrick, B.C.; Trent, D.W. Biochemical and antigenic comparison of the envelope glycoproteins of Venezuelan equine encephalomyelitis virus strains. *J. Gen. Virol.* **1979**, *44*, 725–740. [CrossRef]

22. Kinney, R.M.; Johnson, B.J.; Brown, V.L.; Trent, D.W. Nucleotide sequence of the 26 S mRNA of the virulent Trinidad donkey strain of Venezuelan equine encephalitis virus and deduced sequence of the encoded structural proteins. *Virology* **1986**, *152*, 400–413. [CrossRef]

23. Weaver, S.C.; Salas, R.; Rico-Hesse, R.; Ludwig, G.V.; Oberste, M.S.; Boshell, J.; Tesh, R.B. Re-emergence of epidemic Venezuelan equine encephalomyelitis in South America. *Lancet* **1996**, *348*, 436–440. [CrossRef]

24. Wang, E.; Bowen, R.A.; Medina, G.; Powers, A.M.; Kang, W.; Chandler, L.M.; Shope, R.E.; Weaver, S.C. Virulence and viremia characteristics of 1992 epizootic subtype IC Venezuelan equine encephalitis viruses and closely related enzootic subtype ID strains. *Am. J. Trop. Med. Hyg.* **2001**, *65*, 64–69. [CrossRef] [PubMed]

25. Brault, A.C.; Powers, A.M.; Weaver, S.C. Vector infection determinants of Venezuelan equine encephalitis virus reside within the E2 envelope glycoprotein. *J. Virol.* **2002**, *76*, 6387–6392. [CrossRef] [PubMed]

26. Brault, A.C.; Powers, A.M.; Medina, G.; Wang, E.; Kang, W.; Salas, R.A.; de Siger, J.; Weaver, S.G. Potential sources of the 1995 Venezuelan equine encephalitis subtype IC epidemic. *J. Virol.* **2001**, *75*, 5823–5832. [CrossRef] [PubMed]

27. Rico-Hesse, R.; Roehrig, J.T.; Trent, D.W.; Dickerman, R.W. Genetic variation of Venezuelan equine encephalitis virus strains of the ID variety in Colombia. *Am. J. Trop. Med. Hyg.* **1988**, *38*, 195–204. [CrossRef] [PubMed]

28. Rico-Hesse, R.; Weaver, S.C.; de Siger, J.; Medina, G.; Salas, R.A. Emergence of a new epidemic/epizootic Venezuelan equine encephalitis virus in South America. *Proc. Natl. Acad. Sci. USA* **1995**, *92*, 5278–5281. [CrossRef] [PubMed]

29. Vilcarromero, S.; Aguilar, P.V.; Halsey, E.S.; Laguna-Torres, V.A.; Razuri, H.; Perez, J.; Valderrama, Y.; Gotuzzo, E.; Suarez, L.; Cespedes, M.; et al. Venezuelan equine encephalitis and 2 human deaths, Peru. *Emerg. Infect. Dis.* **2010**, *16*, 553–556. [CrossRef]

30. Moncayo, A.C.; Lanzaro, G.; Kang, W.; Orozco, A.; Ulloa, A.; Arredondo-Jimenez, J.; Weaver, S.C. Vector competence of eastern and western forms of *Psorophora columbiae* (Diptera: Culicidae) mosquitoes for enzootic and epizootic Venezuelan equine encephalitis virus. *Am. J. Trop. Med. Hyg.* **2008**, *78*, 413–421. [CrossRef]

31. Weaver, S.C.; Pfeffer, M.; Marriott, K.; Kang, W.; Kinney, R.M. Genetic evidence of the origins of Venezuelan equine encephalitis virus subtype IAB outbreaks. *Am. J. Trop. Med. Hyg.* **1999**, *60*, 441–448. [CrossRef]

32. Davis, N.L.; Brown, K.W.; Greenwald, G.F.; Zajac, A.J.; Zacny, V.L.; Smith, J.F.; Johnston, R.E. Attenuated mutants of Venezuelan equine encephalitis virus containing lethal mutations in the PE2 cleavage signal combined with a second-site suppressor mutation in E1. *Virology* **1995**, *212*, 102–110. [CrossRef]

33. Johnston, R.E.; Smith, J.F. Selection for accelerated penetration in cell culture coselects for attenuated mutants of Venezuelan equine encephalitis virus. *Virology* **1988**, *162*, 437–443. [CrossRef]
34. Bernard, K.A.; Klimstra, W.B.; Johnston, R.E. Mutations in the E2 glycoprotein of Venezuelan equine encephalitis virus confer heparin sulfate interaction, low morbidity, and rapid clearance from blood of mice. *Virology* **2000**, *276*, 93–103. [CrossRef] [PubMed]
35. Coffey, L.L.; Vasilakis, N.; Brault, A.C.; Powers, A.M.; Tripet, F.; Weaver, S.C. Arbovirus evolution in vivo is constrained by host alteration. *Proc. Natl. Acad. Sci. USA* **2008**, *105*, 6970–6975. [CrossRef] [PubMed]
36. Gilyard, R.T. A clinical study of Venezuelan virus equine encephalomyelitis in Trinidad, B.W.I. *JAMA* **1945**, *818*, 267–277.
37. Dupuy, L.C.; Richards, M.J.; Ellefsen, B.; Chau, L.; Luxembourg, A.; Hannaman, D.; Livingston, B.D.; Schmaljohn, C.S. A DNA vaccine for Venezuelan equine encephalitis virus delivered by intramuscular electroporation elicits high levels of neutralizing antibodies in multiple animal models and provides protective immunity to mice and nonhuman primates. *Clin. Vaccine Immunol.* **2011**, *18*, 707–716. [CrossRef] [PubMed]
38. Parker, M.D.; Buckley, M.J.; Melanson, V.R.; Glass, P.J.; Norwood, D.; Hart, M.K. Antibody to the E3 glycoprotein protects mice against lethal Venezuelan equine encephalitis virus infection. *J. Virol.* **2010**, *84*, 12683–12690. [CrossRef]
39. Reed, D.S.; Glass, P.F.; Bakken, R.R.; Barth, J.F.; Lind, C.M.; da Silva, L.; Hart, M.K.; Rayner, J.; Alterson, K.; Custer, M.; et al. Combined alphavirus replicon particle vaccine induces durable and cross-protective immune responses against equine encephalitis viruses. *J. Virol.* **2014**, *88*, 12077–12086. [CrossRef] [PubMed]
40. Hart, M.K.; Caswell-Stephan, K.; Bakken, R.; Tammariello, R.; Pratt, W.; Davis, N.; Johnston, R.E.; Smith, J.; Steele, K. Improved mucosal protection against Venezuelan equine encephalitis virus is induced by the molecularly defined, live-attenuated V3526 vaccine candidate. *Vaccine* **2000**, *18*, 3067–3075. [CrossRef]
41. Hunt, A.R.; Frederickson, S.; Hinkel, C.; Bowdish, K.S.; Roehrig, J.T. A humanized murine monoclonal antibody protects mice either before or after challenge with virulent Venezuelan equine encephalomyelitis virus. *J. Gen. Virol.* **2006**, *87*, 2467–2476. [CrossRef]
42. O'Brien, L.M.; Goodchild, S.A.; Phillpotts, R.J.; Perkins, S.D. A humanized murine monoclonal antibody protects mice from Venezuelan equine encephalitis virus, Everglades virus and Mucambo virus when administered up to 48 h after airborne challenge. *Virology* **2012**, *426*, 100–105. [CrossRef]
43. Phillpotts, R.J.; O'Brien, L.; Appleton, R.E.; Carr, S.; Bennett, A. Intranasal immunisation with defective adenovirus serotype 5 expressing the Venezuelan equine encephalitis virus E2 glycoprotein protects against airborne challenge with virulent virus. *Vaccine* **2005**, *23*, 1615–1623. [CrossRef] [PubMed]
44. Bennett, A.M.; Lescott, T.; Phillpotts, R.J. Improved protection against Venezuelan equine encephalitis by genetic engineering of a recombinant vaccinia virus. *Viral Immunol.* **1998**, *11*, 109–117. [CrossRef] [PubMed]
45. Bennett, A.M.; Elvin, S.J.; Wright, A.J.; Jones, S.M.; Phillpotts, R.J. An immunological profile of Balb/c mice protected from airborne challenge following vaccination with a live attenuated Venezuelan equine encephalitis virus vaccine. *Vaccine* **2000**, *19*, 337–347. [CrossRef]
46. Martin, S.S.; Bakken, R.R.; Lind, C.M.; Garcia, P.; Jenkins, E.; Glass, P.J.; Parker, M.D.; Hart, M.K.; Fine, D.L. Evaluation of formalin inactivated V3526 virus with adjuvant as a next generation vaccine candidate for Venezuelan equine encephalitis virus. *Vaccine* **2010**, *28*, 3143–3151. [CrossRef]
47. Martin, S.S.; Bakken, R.R.; Lind, C.M.; Garcia, P.; Jenkins, E.; Glass, P.G.; Parker, M.D.; Hart, M.K.; Fine, D.L. Comparison of the immunological responses and efficacy of gamma-irradiated V3526 vaccine formulations against subcutaneous and aerosol challenge with Venezuelan equine encephalitis virus subtype IAB. *Vaccine* **2010**, *28*, 1031–1040. [CrossRef]
48. Steele, K.E.; Davis, K.J.; Stephan, K.; Kell, W.; Vogel, P.; Hart, M.K. Comparative neurovirulence and tissue tropism of wild-type and attenuated strains of Venezuelan equine encephalitis virus administered by aerosol in C3H/HeN and BALB/c mice. *Vet. Pathol.* **1998**, *35*, 386–397. [CrossRef]
49. Klimstra, W.B.; Ryman, K.D.; Johnston, R.E. Adaptation of Sindbis virus to BHK cells selects for use of heparin sulfate as an attachment receptor. *J. Virol.* **1998**, *72*, 7357–7366.
50. Paessler, S.; Ni, H.; Petrakova, O.; Fayzulin, R.Z.; Yun, N.; Anishchenko, M.; Weaver, S.C.; Frolov, I. Replication and clearance of Venezuelan equine encephalitis virus from the brains of animals vaccinated with chimeric SIN/VEE viruses. *J. Virol.* **2006**, *80*, 2784–2796. [CrossRef]

51. Lennette, E.H.; Koprowski, H. Human infection with Venezuelan equine encephalomyelitis virus: A report on eight cases of infection acquired in the laboratory. *JAMA* **1943**, *123*, 1088–1095. [CrossRef]

52. Rusnak, J.M.; Dupuy, L.C.; Niemuth, N.A.; Glenn, A.M.; Ward, L.A. Comparison of aerosol- and percutaneous-acquired Venezuelan equine encephalitis in humans and nonhuman primates for suitability in predicting clinical efficacy under the animal rule. *Comp. Med.* **2018**, *68*, 380–395. [CrossRef]

53. Beck, C.E.; Wyckoff, R.W.G. Venezuelan equine encephalomyelitis. *Science* **1938**, *88*, 530. [CrossRef] [PubMed]

54. McCurdy, K.; Joyce, J.; Hamilton, S.; Nevins, C.; Sosna, W.; Puricelli, K.; Rayner, J.O. Differential accumulation of genetic and phenotypic changes in Venezuelan equine encephalitis virus and Japanese encephalitis virus following passage in vitro and in vivo. *Virology* **2011**, *415*, 20–29. [CrossRef] [PubMed]

55. Peters, C.J.; Dalrymple, J.M. Alphaviruses. In *Fields Virology*, 3rd ed.; Fields, B.N., Knipe, D.M., Hawley, P.M., Eds.; Lippincott-Raven: Philadelphia, PA, USA, 1990; pp. 713–759.

56. Casals, J.; Curnen, E.C.; Thomas, L. Venezuelan equine encephalomyelitis in man. *J. Exp. Med.* **1943**, *77*, 521–530. [CrossRef] [PubMed]

57. Koprowski, H.; Cox, H.R. Human laboratory infection with Venezuelan equine encephalomyelitis virus; report of four cases. *N. Engl. J. Med.* **1947**, *236*, 647–654. [CrossRef] [PubMed]

58. Slepushkin, A.N. An epidemiological study of laboratory infections with Venezuelan equine encephalomyelitis. *Vopr. Virusol.* **1959**, *3*, 311–314.

59. Sutton, L.S.; Brooke, C.C. Venezuelan equine encephalomyelitis due to vaccination in man. *JAMA* **1954**, *155*, 1473–1476. [CrossRef] [PubMed]

60. Franck, P.T.; Johnson, K.M. An outbreak of Venezuelan encephalitis in man in the Panama Canal Zone. *Am. J. Trop. Med. Hyg.* **1970**, *19*, 860–865. [CrossRef] [PubMed]

61. Charles, P.C.; Walters, E.; Margolis, F.; Johnston, R.E. Mechanism of neuroinvasion of Venezuelan equine encephalitis virus in the mouse. *Virology* **1995**, *208*, 662–671. [CrossRef] [PubMed]

62. Ryzhikov, A.B.; Tkacheva, N.V.; Sergeev, A.N.; Ryabchikova, E.I. Venezuelan equine encephalitis virus propagation in the olfactory tract of normal and immunized mice. *Biomed. Sci.* **1991**, *2*, 607–614.

63. Vogel, P.; Abplanlp, D.; Kell, W.; Ibrahim, M.S.; Downs, M.B.; Pratt, W.D.; Davis, K.J. Venezuelan equine encephalitis in BALB/c mice. Kinetic analysis of central nervous system infection following aerosol or subcutaneous inoculation. *Arch. Pathol. Lab. Med.* **1996**, *120*, 164–172.

64. Steele, K.E.; Twenhafel, N.A. Review Paper: Pathology of animal models of alphavirus encephalitis. *Vet. Pathol.* **2010**, *47*, 790–805. [CrossRef] [PubMed]

65. Danes, L.; Kufner, J.; Hruskova, J.; Rychterova, V. The role of the olfactory route on infection of the respiratory tract with Venezuelan equine encephalomyelitis virus in normal and operated *Macaca rhesus* monkeys. I. Results of virological examination. *Acta. Virol.* **1973**, *17*, 50–56. [PubMed]

66. Danes, L.; Rychterova, V.; Kufner, J.; Hruskova, J. The role of the olfactory route of the respiratory tract with Venezuelan equine encephalitis virus in normal and operated *Macaca rhesus* monkeys. II. Results of histological examination. *Acta Virol.* **1973**, *17*, 57–60.

67. Tasker, J.B.; Miesse, M.L.; Berge, T.O. Studies on the virus of Venezuelan equine encephalomyelitis. III. Distribution in tissues of experimentally infected mice. *Am. J. Trop. Med. Hyg.* **1962**, *11*, 844–850. [CrossRef] [PubMed]

68. Smith, D.G.; Mamay, H.K.; Marshall, R.G.; Wagner, J.C. Venezuelan equine encephalomyelitis. Laboratory aspects of fourteen human cases following vaccination and attempts to isolate the virus from the vaccine. *Am. J. Hyg.* **1956**, *63*, 150–164. [CrossRef]

69. Guo-Yun, Y.; Wiley, M.R.; Kugelman, J.R.; Ladner, J.T.; Beitzel, B.F.; Eccleston, L.T.; Morazzani, E.M.; Glass, P.J.; Palacios, G.F. Complete coding sequences of eastern equine encephalitis virus and Venezuelan equine encephalitis virus strains isolated from human cases. *Genome Announc.* **2015**, *3*, e00243-15. [CrossRef]

70. Suzanne, M.; Castro, F.; Bonilla, N.J.; de Urdaneta, A.G.; Hutchins, G.M. The systemic pathology of Venezuelan equine encephalitis virus infection in humans. *Am. J. Trop. Med. Hyg.* **1985**, *34*, 194–202. [CrossRef]

71. Food and Drug Administration-Center for Biologics Evaluation and Research (FDA-CBER). *Development of a Western, Eastern, and Venezuelan Equine Encephalitis (WEVEEV) Vaccine, Animal Model Strain Selection, Version 1.0*; Written Communication with US Army Medical and Materiel Development Authority (USAMMDA); Food and Drug Administration-Center for Biologics Evaluation and Research (FDA-CBER): Fort Detrick, MD, USA, 20 April 2015.

72. Pratt, W.D.; Gibbs, P.; Pitt, M.L.; Schmaljohn, A.L. Use of telemetry to assess vaccine-induced protection against parenteral and aerosol infections of Venezuelan equine encephalitis virus in nonhuman primates. *Vaccine* **1998**, *16*, 1056–1064. [CrossRef]

73. Aronson, J.F.; Grieder, F.B.; Davis, N.L.; Charles, P.C.; Knott, T.; Brown, K.; Johnston, R.E. A single-site mutant and revertants arising in vivo define early steps in the pathogenesis of Venezuelan equine encephalitis virus. *Virology* **2000**, *270*, 111–123. [CrossRef]

74. Grieder, F.B.; Davis, N.L.; Aronson, J.F.; Charles, P.C.; Sellon, D.C.; Suzuki, K.; Johnston, R.E. Specific restrictions in the progression of Venezuelan equine encephalitis virus-induced disease resulting from single amino acid changes in the glycoproteins. *Virology* **1995**, *206*, 994–1006. [CrossRef]

75. Davis, N.L.; Willis, L.V.; Smith, J.F.; Johnston, R.E. In vitro synthesis of infectious Venezuelan equine encephalitis virus from a cDNA clone: Analysis of a viable deletion mutant. *Virology* **1989**, *171*, 189–204. [CrossRef]

76. Davis, N.L.; Powell, N.; Greenwald, G.F.; Willis, L.V.; Johnson, B.J.; Smith, J.F.; Johnston, R.E. Attenuating mutations in the E2 glycoprotein gene of Venezuelan equine encephalitis virus: Construction of single and multiple mutants in a full-length cDNA clone. *Virology* **1991**, *183*, 20–31. [CrossRef]

77. Davis, N.L.; Grieder, F.B.; Smith, J.F.; Greenwald, G.F.; Valenski, M.L.; Sellon, D.C.; Charles, P.C.; Johnston, R.E. A molecular genetic approach to the study of Venezuelan equine encephalitis virus pathogenesis. *Arch. Virol. Suppl.* **1994**, *9*, 99–109. [CrossRef] [PubMed]

78. Ryzhikov, A.B.; Ryabchikova, E.I.; Sergeev, A.N.; Tkacheva, N.V. Spread of Venezuelan equine encephalitis virus in mice olfactory tract. *Arch. Virol.* **1995**, *140*, 2243–2254. [CrossRef] [PubMed]

79. Rossi, S.L.; Russell-Lodrigue, K.E.; Killeen, S.Z.; Wang, E.; Leal, G.; Bergren, N.A.; Vinet-Oliphant, H.; Weaver, S.C.; Roy, C.J. IRES-containing VEEV vaccine protects cynomolgus macaques from IE Venezuelan equine encephalitis virus aerosol challenge. *PLoS Negl. Trop. Dis.* **2015**, *9*, e0003797. [CrossRef] [PubMed]

80. Reed, D.S.; Lind, C.M.; Lackemeyer, M.G.; Sullivan, L.J.; Pratt, W.D.; Parker, M.D. Genetically engineered, live, attenuated vaccines protect nonhuman primates against aerosol challenge with a virulent IE strain of Venezuelan equine encephalitis virus. *Vaccine* **2005**, *23*, 3139–3147. [CrossRef]

81. Dupuy, L.C.; Richards, M.J.; Reed, D.S.; Schmaljohn, C.S. Immunogenicity and protective efficacy of a DNA vaccine against Venezuelan equine encephalitis virus aerosol challenge in nonhuman primates. *Vaccine* **2010**, *28*, 7345–7350. [CrossRef] [PubMed]

82. Pratt, W.D.; Davis, N.L.; Johnston, R.E.; Smith, J.F. Genetically engineered, live attenuated vaccines for Venezuelan equine encephalitis: Testing in animal models. *Vaccine* **2003**, *21*, 3854–3862. [CrossRef]

83. Weaver, S.C.; University of Texas Medical Branch, Galveston, TX, USA; Klimstra, W.B.; University of Pittsburgh, Pittsburgh, PA, USA. Personal communication, 2019.

84. Brittingham, K.; Garver, J.N.; Cauthon, A.G.; Sabourin, C.L.; Burback, B.L. Evaluation of the Median Lethal Dose (LD$_{50}$) of Venezuelan Equine Encephalitis Virus in BALB/c Mice. Unpublished work.

85. Coffey, L.L.; Beeharry, T.; Borderia, A.V.; Blanc, H.; Vignuzzi, M. Arbovirus high fidelity variant loses fitness in mosquitoes and mice. *Proc. Natl. Acad. Sci. USA* **2011**, *108*, 16038–16043. [CrossRef]

86. Warmbrod, K.L.; Patterson, E.I.; Kautz, T.F.; Stanton, A.; Rockx-Brouwer, D.; Kalveram, B.K.; Khanipov, K.; Thangamani, S.; Fofanov, Y.; Forrester, N.L. Viral RNA-dependent RNA polymerase mutants display an altered mutation spectrum resulting in attenuation in both mosquito and vertebrate hosts. *PLoS Pathog.* **2019**, *15*, e1007610. [CrossRef]

87. Burke, D.S.; Ramsburg, H.H.; Edelman, R. Persistence in humans of antibody to subtypes of Venezuelan equine encephalomyelitis (VEE) virus after immunization with attenuated (TC-83) VEE virus vaccine. *J. Infect. Dis.* **1977**, *136*, 354–359. [CrossRef] [PubMed]

88. Forrester, N.L.; Wertheim, J.O.; Dugan, V.G.; Auguste, A.J.; Lin, D.; Adams, P.; Chen, R.; Gorchakov, R.; Leal, G.; Estrada-Franco, J.G.; et al. Evolution and spread of Venezuelan equine encephalitis complex alphavirus in the Americas. *PLoS Negl. Trop. Dis.* **2017**, *11*, e0005693. [CrossRef] [PubMed]

89. Rico, A.B.; Phillips, A.T.; Schountz, T.; Jarvis, D.L.; Tjalkens, R.B.; Powers, A.M.; Olson, K.E. Venezuelan and western equine encephalitis virus E1 liposome antigen nucleic acid complexes protect mice from lethal challenge with multiple alphaviruses. *Virology* **2016**, *499*, 30–39. [CrossRef] [PubMed]

90. Hart, M.K.; Lind, C.; Bakken, R.; Robertson, M.; Tammariello, R.; Ludwig, G.V. Onset and duration of protective immunity to IA/IB and IE strains of Venezuelan equine encephalitis virus in vaccinated mice. *Vaccine* **2002**, *20*, 612–622. [CrossRef]
91. Schmaljohn, A.L.; Johnson, E.D.; Dalrymple, J.M.; Cole, G.A. Non-neutralizing monoclonal antibodies prevent lethal alphavirus encephalitis. *Nature* **1982**, *297*, 70–72. [CrossRef] [PubMed]

Article

Viruses in Horses with Neurologic and Respiratory Diseases

Eda Altan [1,2], Yanpeng Li [1,2], Gilberto Sabino-Santos Jr [1,2], Vorthon Sawaswong [1,2], Samantha Barnum [3], Nicola Pusterla [3], Xutao Deng [1,2] and Eric Delwart [1,2,*]

[1] Vitalant Research Institute, San Francisco, CA 94118, USA; EAltan@vitalant.org (E.A.); YLi@vitalant.org (Y.L.); GSabino-SantosJr@vitalant.org (G.S.-S.J.); vorthon007.giftedcru@gmail.com (V.S.); XDeng@vitalant.org (X.D.)
[2] Department of Laboratory Medicine, University of California, San Francisco, CA 94118, USA
[3] Department of Medicine and Epidemiology, School of Veterinary Medicine, University of California, Davis, CA 95616, USA; smmapes@ucdavis.edu (S.B.); npusterla@ucdavis.edu (N.P.)
* Correspondence: EDelwart@vitalant.org or Eric.Delwart@ucsf.edu; Tel.: +1-(415)-531-0763

Received: 29 August 2019; Accepted: 7 October 2019; Published: 14 October 2019

Abstract: Metagenomics was used to identify viral sequences in the plasma and CSF (cerobrospinal fluid) of 13 horses with unexplained neurological signs and in the plasma and respiratory swabs of 14 horses with unexplained respiratory signs. Equine hepacivirus and two copiparvoviruses (horse parvovirus-CSF and a novel parvovirus) were detected in plasma from neurological cases. Plasma from horses with respiratory signs contained the same two copiparvoviruses plus equine pegivirus D and respiratory swabs contained equine herpes virus 2 and 5. Based on genetic distances the novel copiparvovirus qualified as a member of a new parvovirus species we named *Eqcopivirus*. These samples plus another 41 plasma samples from healthy horses were tested by real-time PCRs for multiple equine parvoviruses and hepacivirus. Over half the samples tested were positive for one to three viruses with eqcopivirus DNA detected in 20.5%, equine hepacivirus RNA and equine parvovirus-H DNA in 16% each, and horse parvovirus-CSF DNA in 12% of horses. Comparing viral prevalence in plasma none of the now three genetically characterized equine parvoviruses (all in the copiparvovirus genus) was significantly associated with neurological and respiratory signs in this limited sampling.

Keywords: *Parvoviridae*; *Eqcopivirus*; horse parvovirus-CSF; equine hepacivirus; equine parvovirus H; bosavirus; virome

---

1. Introduction

The United States has the largest horse population with 9.5 million horses, followed by those of China, Mexico, Brazil, and Argentina with an estimated 2006 world population of 68 million [1].

Equine viral pathogens belonging to diverse viral families have been described including, but not limited to, equid herpesviruses, equine arteritis virus, African horse sickness virus, equine infectious anemia virus, equine coronavirus, Hendra virus, vesicular stomatitis virus, equine influenza virus, West Nile virus, Eastern Equine Encephalitis, Venezuelan Equine Encephalitis, Western Equine Encephalitis, and most recently equine parvovirus-H causing hepatitis [2–9].

A limited number of viral metagenomics studies have analyzed horse samples. A 2013 study identified a novel flavivirus that was named Theiler's disease-associated virus (TDAV) from an outbreak of this liver disease transmitted through equine-origin tetanus anti-toxin [10]. Equine parvovirus-H in the copiparvovirus genera (EqPV-H) was first described in 2018 [2] and was recently shown through epidemiological studies and inoculation studies to be a more likely cause of Theiler's disease caused either by transfusion of contaminated horse serum or through naturally acquired infections (although asymptomatic infections are also frequent) [11–13]. A hepacivirus closely related to HCV (hepatitis C virus), initially characterized by metagenomics from a dog [14], but subsequently found to more frequently infect horses was named equine hepacivirus (EqHV) but was not associated with serum hepatitis [15–17]. Equine pegivirus (EPgV), most closely related to a bat pegiviruses [18], was first identified using a degenerate PCR approach [19] and shown to be a common equine infection but was also not associated with equine liver disease [15].

Beside the above studies focused on Theiler's disease viral metagenomics study of CSF from a horse with neurological signs identified another parvovirus named horse parvovirus-CSF [20]. Metagenomics studies of equine serum pools recently identified the four equine viruses listed above plus unexpectedly the porcine Suid betaherpesvirus 2 [21].

To further characterize the eukaryotic virome of horses and identify possible equine pathogens we used here a combination of viral metagenomics and real-time PCR to analyze plasma and cerebrospinal fluid (CSF) from 13 horses with unexplained neurological signs, plasma and respiratory swabs from 14 horses with unexplained respiratory signs, and plasma from 41 healthy horses, when necessary.

2. Materials and Methods

2.1. Study Samples

The use of horse samples adhered to the animal use guidelines set by UC Davis' Institutional Animal Care and Use Committee (AICUC Protocol # 19988, approval date 5/31/2018).

In this study 68 horses were analyzed using one or two of three types of biological samples (plasma, CSF, and respiratory swabs). Both plasma and CSF were collected from 13 horses that showed acute neurological signs (behavioral changes, spinal ataxia, proprioceptive deficits and/or cranial nerve deficits). Plasma and respiratory swabs were collected from another 14 horses with fever (>38.6 °C) and nasal discharge/cough (the respiratory swab from animal 5 is missing). Plasma was also collected from 41 healthy horses.

The 13 horses with neurological signs each had CSF with evidence of lymphocytic pleocytosis, negative antibody titters to *Sarcocysits neurona* and *Neospora hughesi* by immunodiagnostics (indirect fluorescence antibody tests) and no detection of *S. neurona*, *N. hughesi*, and *Borrelia burgdorferi* by PCR. The 13 horses with respiratory signs each had respiratory swabs that tested non-reactive for EHV-1/-4, EIV, ERAV/ERBV, and *Streptococcus equi* subspecies *equi*.

All 53 samples from these 27 sick horses were analyzed in 12 pools of 4–5 samples using viral metagenomics (Table 1).

Table 1. Distribution of translated sequence reads with similarity E score of $<10^{-10}$ to known mammalian viral proteins.

Disease and Sample Types	Total Reads	Equine Hepacivirus	Equine Pegivirus D	Equid Gamma Herpesvirus2	Equid Gammaherpes Virus5	Horse Parvovirus-CSF	Novel Parvovirus: Eqcopivirus	Bosavirus
Neuro Pool1 Plasma	2,575,048					1023	17,058	
Neuro Pool4 CSF	1,106,824							
Neuro Pool2 Plasma	1,612,796	7						
Neuro Pool5 CSF	1,028,398							
Neuro Pool3 Plasma	2,080,966							
Neuro Pool6 CSF	1,057,992							

Table 1. *Cont.*

Disease and Sample Types	Total Reads	Equine Hepacivirus	Equine Pegivirus D	Equid Gamma Herpesvirus2	Equid Gammaherpes Virus5	Horse Parvovirus-CSF	Novel Parvovirus: Eqcopivirus	Bosavirus
Respiratory Pool7 Plasma	1,509,792		317				36,069	
Respiratory Pool10 Swab	622,376							1719
Respiratory Pool8 Plasma	603,470					1087	16,355	
Respiratory Pool11 Swab	949,260			120	25			880
Respiratory Pool9 Plasma	1,520,736							
Respiratory Pool12 Swab	629,274							395

2.2. Viral Metagenomics

For viral metagenomics plasma, CSF, and respiratory swab samples were clarified by 14,000 rpm centrifugation for five minutes, and filtered using a 0.45-μm filter (Merck Millipore Ltd., Cork, Ireland). Free nucleic acids in the 400 μL filtrates were digested using DNAse and RNAse enzymes to enrich for viral nucleic acids protected within viral capsids. Nucleic acids were then extracted (MagMAX Viral RNA Isolation Kit, Ambion, Inc, Austin, TX, USA) [22] and amplified by random RT-PCR followed by use of the Nextera™ XT Sample Preparation Kit (Illumina) to generate a library for Illumina MiSeq (2 × 250 bases) with dual barcoding as previously described [23].

An in-house analysis pipeline was used to analyze sequence data. Before analyzing, raw data were pre-processed by subtracting human and bacterial sequences, duplicate sequences, and low quality reads. Following de novo assembly using the Ensemble program [24], both contigs and singlets viral sequences were then analyzed using translated protein sequence similarity search (BLASTx v.2.2.7) to all annotated viral proteins available in GenBank. Candidate viral hits were then compared to an in-house non-virus non-redundant (nr) protein database to remove false positive viral hits. To align reads and contigs to reference viral genomes from GenBank and generate complete or partial genome sequences the Geneious R10 program was used.

2.3. Generation of Full Genomes of Novel Horse Parvovirus

Following mini-pool metagenomics sequencing three individual samples were re-sequenced using the same method to generate 3 near full genomes (complete ORFs) of eqcopiviruses (EqCoPV) (deposited in GenBank with accession numbers MN181466-8 for eqcopivirus 8, 9, and 11 derived from respiratory symptoms plasma samples 4, 5, and 7 respectively). Genome gaps left by high throughput sequencing (HTS) were filled by PCR and products were Sanger sequenced.

2.4. Real-Time PCR

Real-time PCR assays were developed to the three parvoviruses detected here using HTS (horse parvovirus-CSF, novel eqcopivirus, and bovine serum-associated bosavirus). Real-time PCR assays were also designed for equine hepaciviruses detected in one plasma pool and for the recently identified hepatotropic equine parvovirus-H (EqPV-H) pathogen using previously described PCR conditions [12,17]. Real-time PCR amplification was performed in a Roche 480 thermocycler. Primers and probes were used in multiplex real-time PCR (eqcopivirus, horse parvovirus CSF, and bosavirus) and a separate multiplex real-time PCR for equine hepacivirus and EqPV-H. Primer, probes, oligonucleotide sequences, and length of amplicons are summarized in Supplementary Table S1. The reaction mix for each sample consisted of 12.5 μL QuantiFast Probe PCR Master Mix (Qiagen, Hilden, Germany), 1M each primer, 0.5 M each probe, and 3 μL DNA or cDNA. Real-time PCR sensitivity was first determined using known concentrations of oligonucleotides representing the PCR target regions. Oligonucleotides were serially diluted and real-time PCR results endpoint detection were estimated at 100 to 125 input genome equivalents.

2.5. Phylogenetic Analysis

The NS1 and VP1 protein sequences of parvoviruses were aligned using Clustal W in Geneious v10.1.3. and phylogenetic trees constructed using the Maximum likelihood method with two substitution models: Le_Gascule_2008 model (LG) with Freqs and gamma distributed, invariant sites (G + I) MEGA software ver. X [25]. The substitution models were selected based on the results of the Best Model search of MEGA X. The percentage of trees in which the associated taxa clustered together is shown next to the branch points. Initial tree(s) for the heuristic search were obtained by applying the neighbor-joining method to a matrix of pairwise distances estimated using the maximum composite likelihood (MCL) approach. The tree is drawn to scale, with branch lengths measured in the number of substitutions per site. The phylogenetic analysis used sequences from nineteen different copiparvoviruses and human parvovirus B19 (AY386330) was used to root the tree.

3. Results

3.1. Viral Metagenomics

Plasma and CSF from 13 horses with neurological signs and plasma and respiratory swabs from another 14 horses with respiratory signs tested negative for a panel of known equine pathogens (see Study Samples in Materials and Methods).

These samples were analyzed by viral metagenomics in 12 pools of 4–5 samples each. A total of 15.2 million reads were generated for an average number of reads of ~1.1 million per pool. The raw sequence data for each pool are available at NCBI's Short Reads Archive under GenBank accession number SRP120619.

Viruses detected with BLASTx translated protein matches (E score $<10^{-10}$) were: a novel copiparvovirus in three pools, bovine serum-associated parvovirus (bosavirus) in three pools, horse parvovirus-CSF in two pools, and equine hepacivirus, equine pegivirus D, and equid gammaherpesvirus 2 and 5 in one pool each (Table 1).

The most commonly detected eukaryotic viral reads belonged to the *Copiparvovirus* genus of *Parvoviridae* family with a frequency ranging from 0.66% to 3.68% of total reads. Novel (divergent) copiparvovirus reads from three pools generated contigs ranging in size from 1417 to 5158 nt that showed closest translated aa identity of 42.5 to 43.9% to NS1 and 43.5 to 44.9% to VP1 encoded by the horse parvovirus-CSF genome (GenBank KR902500). This virus was named equine copiparvovirus (eqcopivirus or EqCoPV).

Reads matching the previously described horse parvovirus-CSF genome were also detected. This parvovirus was reported in the CSF of a horse with neurological signs [20].. A partial NS1 sequence of this virus was also recently reported from a thoroughbred in China sampled in 2018 (QCF41227.1). Two of the three pools containing the novel eqcopivirus also showed the presence of horse parvovirus-CSF although with fewer reads. Both 37.5% and 60.5% of the horse parvovirus-CSF genomes could be assembled using reads from these plasma pools showing 98.8% and 97% nucleotide similarity with the original horse parvovirus-CSF derived genome in GenBank (KR902500).

Equine hepacivirus contigs of 522 and 657 nucleotides were also generated from seven reads in a pool of plasma from horses with neurological signs. Contigs showed 97.9–98.7% nt identities to equine hepacivirus (JQ434008) in GenBank.

A total of 13 different equid gammaherpesvirus 2 and 5 contigs, ranging in size from 446 to 1049 nucleotides, were generated from 120 and 25 reads in a pool of swab from horses with respiratory symptom. Contigs showed 92–100% aa identities to equid gammaherpesvirus 2 (NC_001650.2) and 5 (NC_026421.1) in GenBank.

Two partial equine pegivirus contigs of 446 and 903 nucleotides were generated from 317 reads in a pool of plasma from horses with respiratory signs. Contigs showed 92–93% nt identities to equine pegivirus 1 (KC410872) in GenBank.

Another copiparvovirus (bosavirus) was also detected but likely originated from fetal bovine Plasma spiked into the respiratory swab viral transport medium. A near complete genome from a respiratory swab pool showed 100% aa identity to that of bosavirus in GenBank (NC_031959) [26].

3.2. Generation of Near-Full Length Genomes of Novel Equine Copiparvovirus (Eqcopivirus)

The length of the longest eqcopivirus contig was 5159 nucleotides (nt) with typical copiparvovirus genome organization of two complete major ORFs (NS and VP), and 5′ and 3′ UTRs missing the genome's terminal hairpin sequences. The study strain had a 43.8% G+C content and it has a nt distribution of 39.3% A, 16.8% T, 22.8% G, and 20.9% C.

The expected NS1 ATP- or GTP-binding Walker A loop motif (GxxxxGKT/S; GPPSVGKS) and Walker B motif (EE) were all found [27,28]. The phospholipase A2 (PLA2) catalytic residues (HDLGY) and its highly conserved calcium-binding site (YTGPG) were also found at the *N*-terminus of the capsid protein [29].

The predicted capsid proteins (VP1) of eqcopivirus showed closest aa identity of 43.4% (coverage: 77%) and 37.7% (coverage: 53%) to the corresponding proteins of its two closest relative the horse parvovirus-CSF (KR902500) and equine parvovirus-H (MG136722), respectively. The nonstructural protein (NS1) proteins had 43.4% (coverage: 74%) and 31.3% (coverage: 69%) aa identity to the corresponding proteins of horse parvovirus-CSF (KR902500) and equine parvovirus-H (MG136722), respectively. The ORF structure of eqcopivirus is shown in Figure 1A. Phylogenetic analyses of the eqcopiviruses and bosavirus NS and VP proteins acquired in this study are shown in Figure 1B.

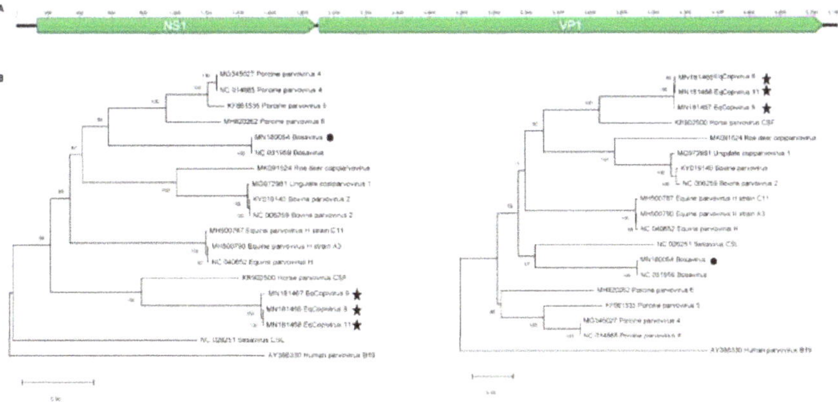

Figure 1. (**A**) Genome ORF (Open Reading Frame) structure of eqcopivirus genome. (**B**) Phylogenetic analysis of NS1 (left) and VP1 (right) proteins in the *Copiparvovirus* genus. Bootstrap values from a hundred replicate runs are shown. Symbols are used to highlight the new genomes described here.

3.3. Real-Time PCR Results

Real-time PCR assays were developed to the three parvoviruses detected here using HTS (horse parvovirus-CSF, novel eqcopivirus, and bosavirus). Real-time PCR assays using previously described conditions were used for equine hepaciviruses and the recently identified hepatotropic equine parvovirus-H pathogen [12,19]. These PCR assays were then used on all 94 available equine plasma, CSF, and respiratory swabs (from 68 horses) described above. The positive detections are shown with Ct values that are inversely proportional to their nucleic acid target concentration (Table 2).

Table 2. Real-time PCR results and Ct values. Within each of the 2 clinical groups different samples from the same animals are labeled with the same number. Lowest to highest Ct labeled red to white.

Clinical Group	Sample Type	Pool	Equine Hepacivirus	EqPV-H	Eqcopivirus	Horse Parvovirus-CSF	Bosavirus
Neurological Signs	Plasma 1	Pool1			30.32		
	Plasma 2						
	Plasma 3						
	Plasma 4		27.7			20	
	Plasma 5	Pool2					
	Plasma 6		34.65	36.73			
	Plasma 7		23.6				
	Plasma 8						
	Plasma 9	Pool3					
	Plasma 10			33.82			
	Plasma 11						
	Plasma 12						
	Plasma 13						
	CSF 1	Pool4					
	CSF 2						
	CSF 3						
	CSF 4						
	CSF 5	Pool5					
	CSF 6						
	CSF 7						
	CSF 8						
	CSF 9	Pool6				30	
	CSF 10						
	CSF 11						
	CSF 12				36.5		
	CSF 13						
Respiratory Signs	Plasma 1	Pool7		37.78			
	Plasma 2						
	Plasma 3						
	Plasma 4				34		
	Plasma 5				27.53		
	Plasma 6	Pool8					
	Plasma 7				30.16		
	Plasma 8						
	Plasma 9			34.6	34.01	21	
	Plasma 10	Pool9					
	Plasma 11						
	Plasma 12						
	Plasma 13						
	Plasma 14						
	Swab 1	Pool10					26.85
	Swab 2						25.65
	Swab 3		25.28				26.4
	Swab 4				36.7		26.82
	Swab 6	Pool11					25.02
	Swab 7				32.62		27.75
	Swab 8						25.57
	Swab 9						26.24
	Swab 10	Pool12				21.2	24.92
	Swab 11					23.1	23.37
	Swab 12						26.63
	Swab 13				32.41		23.29
	Swab 14		37.2			26	25.3

Table 2. *Cont.*

	Sample Type	Pool	Equine Hepacivirus	EqPV-H	Eqcopivirus	Horse Parvovirus-CSF	Bosavirus
Healthy Control Group	Plasma 1						
	Plasma 2					16.8	
	Plasma 3		21.8	37.3			
	Plasma 4		26.5				
	Plasma 5				36.5		
	Plasma 6						
	Plasma 7						
	Plasma 8						
	Plasma 9						
	Plasma 10						
	Plasma 11				28.08		
	Plasma 12						
	Plasma 13						
	Plasma 14				30.21		
	Plasma 15						
	Plasma 16						
	Plasma 17		26.08				
	Plasma 18						
	Plasma 19		23.02	36.9			
	Plasma 20						
	Plasma 21						
	Plasma 22						
	Plasma 23		23.53				
	Plasma 24						
	Plasma 25		25.11				
	Plasma 26						
	Plasma 27			33.83			
	Plasma 28						
	Plasma 29						
	Plasma 30						
	Plasma 31			33.99	30.09		
	Plasma 32			37.93	26.34	30.1	
	Plasma 33						
	Plasma 34						
	Plasma 35						
	Plasma 36			33.22			
	Plasma 37						
	Plasma 38						
	Plasma 39						
	Plasma 40			34.69	28.63		
	Plasma 41				30.12		

Bosavirus (originally reported in bovine serum pools) [26] was detected in every one of the 13 respiratory swab samples with nearly identical C_T value. These respiratory swabs had been preserved in a transport media spiked with fetal bovine serum from the same commercial product. None of the other equine samples (plasma and CSF) were preserved using fetal bovine serum and none showed the presence of bosavirus parvovirus DNA by HTS or real-time PCR. Because bosavirus was originally described as a contaminant of fetal bovine serum [26] we ascribe its detection to the use of contaminated fetal bovine serum spiked into every respiratory swab sample.

The most commonly detected virus was the new eqcopivirus detected in 16/94 samples (14/68 horses). Its highest detection rate was in 4/14 plasmas from respiratory cases. Three of the four plasma positive samples also had matching respiratory swabs, two of which were also positive indicating the presence of eqcopivirus genomes in both plasma and respiratory swabs and revealing a possible mode of transmission through respiratory fluids (Table 2). When the eqcopivirus prevalence in plasma samples from respiratory cases (4/14 or 28.6%) was compared to that in plasma from healthy animals (7/41 or 17%) the difference (measured using Fisher's exact test) yielded a non-significant p value of 0.443 (Table 3). The next two most prevalent viruses were the hepatotropic parvovirus-H and equine hepacivirus both found in 11/89 samples including 3 co-infections. All detections were in plasma samples except for two hepacivirus positive respiratory swabs (Table 2). No evidence of higher virus prevalence was detected between cases and health controls (Table 3). Lastly the horse parvovirus-CSF was detected in eight samples from eight horses in all sample types (plasma, CSF, respiratory swabs) also with no obvious association with either neurological or respiratory disease when comparing prevalence in plasma samples (Table 3). When the real-time PCR Ct values were compared (using unpaired two tailed T-test with Welsh's correction) no statistically significant differences were found between cases and controls.

Table 3. Rate of real-time PCR detection for equine viruses and association with unexplained neurological or respiratory diseases using Fisher's exact test.

	Neuro Plasma $n = 13$	Neuro CSF $n = 13$	Respiratory Plasma $n = 14$	Respiratory Swab $n = 13$	Healthy Control Plasma $n = 41$	Total Samples $n = 94$	P Value = Plasma Neurological Versus Healthy	P Value = Plasma Respiratory Versus Healthy
Equine flavivirus	3 (23%)	0	0	2 (15%)	6 (15%)	11 (12%)	0.32	0.67
EqPV-H	2 (15%)	0	2 (15%)	0	7 (17%)	11 (12%)	1	1
Eqcopivirus	1 (7%)	1 (7%)	4 (30%)	3 (23%)	7 (17%)	15 (16%)	0.443	0.663
Horse parvovirus-CSF	1 (7%)	1 (7%)	1 (7%)	3 (23%)	2 (5%)	8 (8.5%)	1	1

4. Discussion

Virus metagenomics analysis of plasma, CSF and respiratory swabs from horses with unexplained neurological and respiratory signs showed the presence of three parvoviruses, one of which (bosavirus) was likely introduced by the addition of fetal bovine serum to the respiratory swabs transport medium. Additionally detected by metagenomics were equine hepacivirus, equine pegivirus [19], and equine herpesvirus 2 and 5. Another copiparvovirus (equine parvovirus-H) was also detected using real-time PCR (but not by HTS) likely due to its lower concentration as reflected by higher real-time PCR C_T values.

The presence of viral nucleic acids detected by PCR or HTS does not prove the presence of infectious viruses. Nonetheless, the repeated detection of viral genomes in nuclease-treated, presumably sterile samples such as plasma and CSF, and the typically rapid clearance of viral particles from the circulation by the liver support the possibility of active replication [30–32].

When the prevalence of the plasma-associated viruses was determined in individual samples using real-time PCR none showed a statistically higher rate of detection or higher viral loads as determined by Ct when comparing plasma from 13 neurological or 14 respiratory cases to plasma from 41 healthy horses. Only the new eqcopivirus showed a trend toward a higher rate of detection in plasma from unexplained respiratory cases (30% versus 17%) ($p = 0.443$). CSF and respiratory swabs where available only from neurological or respiratory cases respectively. The rate of virus detection could therefore not be measured in these anatomical compartments in healthy animals. Horse parvovirus-CSF and eqcopivirus were also detected in 1/13 (different) CSF samples and in 3/13 (different) respiratory swabs each. The initial characterization of horse parvovirus-CSF genome was also from a CSF sample from a different unexplained neurological case [20].

PCR previously showed EqPV-H DNA to be present in the plasma of 13% healthy horses in the US [2] and 12% of healthy race horses in China [13], rates of viremia similar to that detected here in 6/41 (15%) of healthy horses.

The *Parvoviridae* family consists of non-enveloped, icosahedral, viruses with single stranded DNA genomes of 4 to 6 Kb [27,28]. Eight ICTV approved genera, including *Copiparvovirus*, are currently included in the *Parvoviridae* family [33]. The eqcopivirus genome described here represents the third parvovirus species confirmed to infect horses by virtue of detection in multiple equine plasma, CSF, and respiratory swab samples. That all parvoviruses identified to date in horses belong only to the copiparvovirus genus is somewhat unexpected and future studies are likely to expand the list of known equine parvoviruses. While a higher rate of eqcopivirus DNA detection was found in the plasma of horses with unexplained respiratory signs (30%) versus plasma from healthy horses (17%) that difference, based on the limited number of samples tested, was not statistically significant. Because CSF and respiratory swabs from healthy animals were not available only PCR results from plasma could be compared for disease association. Viruses resulting in neurological and respiratory disease may be expected to be present in CSF or respiratory fluids respectively as were both horse parvovirus-CSF and eqcopivirus. The collection of CSF and respiratory swabs samples from healthy animals may be necessary to better test for possible associations between these copiparvoviruses and unexplained neurological and respiratory signs. The now repeated detection of horse parvovirus-CSF [20] ($n = 2$) and eqcopivirus ($n = 1$) in CSF of horses indicates a possible role for these viruses in these neurological signs which will require further studies. The availability of eqcopivirus genome sequences will now also allow the design of hybridization probes to determine whether infected cells can be identified in fixed brain or lung tissues from animals with unexplained neurological or respiratory signs.

Supplementary Materials: The following are available online at http://www.mdpi.com/1999-4915/11/10/942/s1, Table S1: Real time PCR conditions.

Author Contributions: Conceptualization: E.A., Y.L, G.S.-S.J., S.B., N.P., and E.D; data curation: E.A., Y.L., and X.D.; formal analysis: E.A., Y.L., G.S.-S.J., X.D., and V.S.; methodology: E.A.; supervision, E.D.; writing—original draft: E.A.; writing—review and editing: G.S.-S.J., N.P. and E.D. All authors contributed for the final version of the manuscript.

Funding: Vitalant Research Institute.

Acknowledgments: The authors would like to acknowledge Claire Quiner for statistical assistance and the Vitalant Research Institute for funding.

References

1. Global Horse Population Report. Food and Agriculture Organization (FAO) of the United Nations. Available online: http://faostat.fao.org/fastat/servlet (accessed on 28 August 2019).
2. Divers, T.J.; Tennant, B.C.; Kumar, A.; McDonough, S.; Cullen, J.; Bhuva, N.; Jain, K.; Chauhan, L.S.; Scheel, T.K.H.; Lipkin, W.I. , et al. New parvovirus associated with serum hepatitis in horses after inoculation of common biological product. *Emerg. Infect. Dis* **2018**, *24*, 303–310. [CrossRef] [PubMed]
3. Gilkerson, J.R.; Bailey, K.E.; Diaz-Mendez, A.; Hartley, C.A. Update on viral diseases of the equine respiratory tract. *Vet. Clin. North. Am. Equine Pract* **2015**, *31*, 91–104. [CrossRef] [PubMed]
4. Kumar, B.; Manuja, A.; Gulati, B.R.; Virmani, N.; Tripathi, B.N. Zoonotic viral diseases of equines and their impact on human and animal health. *Open Virol. J.* **2018**, *12*, 80–98. [CrossRef] [PubMed]
5. Middleton, D. Hendra virus. *Vet. Clin. North. Am. Equine Pract.* **2014**, *30*, 579–589. [CrossRef] [PubMed]
6. Pusterla, N.; Vin, R.; Leutenegger, C.M.; Mittel, L.D.; Divers, T.J. Enteric coronavirus infection in adult horses. *Vet. J.* **2018**, *231*, 13–18. [CrossRef] [PubMed]
7. Timoney, P.J.; McCollum, W.H. Equine viral arteritis. *Vet. Clin. North. Am. Equine Pract* **1993**, *9*, 295–309. [CrossRef]
8. Nowotny, N.; Kolodziejek, J.; Jehle, C.O.; Suchy, A.; Staeheli, P.; Schwemmle, M. Isolation and characterization of a new subtype of borna disease virus. *J. Virol.* **2000**, *74*, 5655–5658. [CrossRef] [PubMed]
9. Chapman, G.E.; Baylis, M.; Archer, D.; Daly, J.M. The challenges posed by equine arboviruses. *Equine Vet. J.* **2018**, *50*, 436–445. [CrossRef] [PubMed]
10. Chandriani, S.; Skewes-Cox, P.; Zhong, W.; Ganem, D.E.; Divers, T.J.; Van Blaricum, A.J.; Tennant, B.C.; Kistler, A.L. Identification of a previously undescribed divergent virus from the flaviviridae family in an outbreak of equine serum hepatitis. *Proc. Nat. Acad. Sci. USA* **2013**, *110*, E1407–E1415. [CrossRef]
11. Tomlinson, J.E.; Tennant, B.C.; Struzyna, A.; Mrad, D.; Browne, N.; Whelchel, D.; Johnson, P.J.; Jamieson, C.; Lohr, C.V.; Bildfell, R.; et al. Viral testing of 10 cases of theiler's disease and 37 in-contact horses in the absence of equine biologic product administration: A prospective study (2014-2018). *J. Vet. Intern. Med.* **2019**, *33*, 258–265. [CrossRef]

12. Tomlinson, J.E.; Kapoor, A.; Kumar, A.; Tennant, B.C.; Laverack, M.A.; Beard, L.; Delph, K.; Davis, E.; Schott Ii, H.; Lascola, K.; et al. Viral testing of 18 consecutive cases of equine serum hepatitis: A prospective study (2014-2018). *J. Vet. Intern. Med.* **2019**, *33*, 251–257. [CrossRef]

13. Lu, G.; Sun, L.; Ou, J.; Xu, H.; Wu, L.; Li, S. Identification and genetic characterization of a novel parvovirus associated with serum hepatitis in horses in china. *Emerg. Microbes. Infect.* **2018**, *7*, 170. [CrossRef] [PubMed]

14. Kapoor, A.; Simmonds, P.; Gerold, G.; Qaisar, N.; Jain, K.; Henriquez, J.A.; Firth, C.; Hirschberg, D.L.; Rice, C.M.; Shields, S.; et al. Characterization of a canine homolog of hepatitis c virus. *Proc. Nat. Acad. Sci. USA* **2011**, *108*, 11608–11613. [CrossRef] [PubMed]

15. Lyons, S.; Kapoor, A.; Schneider, B.S.; Wolfe, N.D.; Culshaw, G.; Corcoran, B.; Durham, A.E.; Burden, F.; McGorum, B.C.; Simmonds, P. Viraemic frequencies and seroprevalence of non-primate hepacivirus and equine pegiviruses in horses and other mammalian species. *J. Gen. Virol.* **2014**, *95*, 1701–1711. [CrossRef] [PubMed]

16. Lyons, S.; Kapoor, A.; Sharp, C.; Schneider, B.S.; Wolfe, N.D.; Culshaw, G.; Corcoran, B.; McGorum, B.C.; Simmonds, P. Nonprimate hepaciviruses in domestic horses, united kingdom. *Emerg. Infect. Dis.* **2012**, *18*, 1976–1982. [CrossRef] [PubMed]

17. Burbelo, P.D.; Dubovi, E.J.; Simmonds, P.; Medina, J.L.; Henriquez, J.A.; Mishra, N.; Wagner, J.; Tokarz, R.; Cullen, J.M.; Iadarola, M.J.; et al. Serology-enabled discovery of genetically diverse hepaciviruses in a new host. *J. Virol.* **2012**, *86*, 6171–6178. [CrossRef] [PubMed]

18. Epstein, J.H.; Quan, P.L.; Briese, T.; Street, C.; Jabado, O.; Conlan, S.; Ali Khan, S.; Verdugo, D.; Hossain, M.J.; Hutchison, S.K.; et al. Identification of gbv-d, a novel gb-like flavivirus from old world frugivorous bats (pteropus giganteus) in bangladesh. *PLoS Path.* **2010**, *6*, e1000972. [CrossRef]

19. Kapoor, A.; Simmonds, P.; Cullen, J.M.; Scheel, T.K.; Medina, J.L.; Giannitti, F.; Nishiuchi, E.; Brock, K.V.; Burbelo, P.D.; Rice, C.M.; et al. Identification of a pegivirus (gb virus-like virus) that infects horses. *J. Virol.* **2013**, *87*, 7185–7190. [CrossRef] [PubMed]

20. Li, L.; Giannitti, F.; Low, J.; Keyes, C.; Ullmann, L.S.; Deng, X.; Aleman, M.; Pesavento, P.A.; Pusterla, N.; Delwart, E. Exploring the virome of diseased horses. *J. Gen. Virol.* **2015**. [CrossRef] [PubMed]

21. Paim, W.P.; Weber, M.N.; Cibulski, S.P.; da Silva, M.S.; Puhl, D.E.; Budaszewski, R.F.; Varela, A.P.M.; Mayer, F.Q.; Canal, C.W. Characterization of the viral genomes present in commercial batches of horse serum obtained by high-throughput sequencing. *Biologicals* **2019**. [CrossRef]

22. Victoria, J.G.; Kapoor, A.; Li, L.; Blinkova, O.; Slikas, B.; Wang, C.; Naeem, A.; Zaidi, S.; Delwart, E. Metagenomic analyses of viruses in stool samples from children with acute flaccid paralysis. *J. Virol.* **2009**, *83*, 4642–4651. [CrossRef] [PubMed]

23. Li, L.; Deng, X.; Mee, E.T.; Collot-Teixeira, S.; Anderson, R.; Schepelmann, S.; Minor, P.D.; Delwart, E. Comparing viral metagenomics methods using a highly multiplexed human viral pathogens reagent. *J. Virol. Methods* **2015**, *213*, 139–146. [CrossRef] [PubMed]

24. Deng, X.; Naccache, S.N.; Ng, T.; Federman, S.; Li, L.; Chiu, C.Y.; Delwart, E.L. An ensemble strategy that significantly improves de novo assembly of microbial genomes from metagenomic next-generation sequencing data. *Nucl. Acids Res.* **2015**. [CrossRef] [PubMed]

25. Le, S.Q.; Gascuel, O. An improved general amino acid replacement matrix. *Mol. Biol. Evol.* **2008**, *25*, 1307–1320. [CrossRef] [PubMed]

26. Sadeghi, M.; Kapusinszky, B.; Yugo, D.M.; Phan, T.G.; Deng, X.; Kanevsky, I.; Opriessnig, T.; Woolums, A.R.; Hurley, D.J.; Meng, X.J.; et al. Virome of us bovine calf serum. *Biologicals* **2017**, *46*, 64–67. [CrossRef] [PubMed]

27. Cotmore, S.F.; Tattersall, P. Parvoviruses: Small does not mean simple. *An. Rev. Virol* **2014**, *1*, 517–537. [CrossRef]

28. Mietzsch, M.; Penzes, J.J.; Agbandje-McKenna, M. Twenty-five years of structural parvovirology. *Viruses* **2019**, *11*, 362. [CrossRef]

29. Zadori, Z.; Szelei, J.; Lacoste, M.C.; Li, Y.; Gariepy, S.; Raymond, P.; Allaire, M.; Nabi, I.R.; Tijssen, P. A viral phospholipase a2 is required for parvovirus infectivity. *Develop. Cell* **2001**, *1*, 291–302. [CrossRef]

30. Zhang, L.; Dailey, P.J.; He, T.; Gettie, A.; Bonhoeffer, S.; Perelson, A.S.; Ho, D.D. Rapid clearance of simian immunodeficiency virus particles from plasma of rhesus macaques. *J. Virol* **1999**, *73*, 855–860.

31. Ramratnam, B.; Bonhoeffer, S.; Binley, J.; Hurley, A.; Zhang, L.; Mittler, J.E.; Markowitz, M.; Moore, J.P.; Perelson, A.S.; Ho, D.D. Rapid production and clearance of hiv-1 and hepatitis c virus assessed by large volume plasma apheresis. *Lancet* **1999**, *354*, 1782–1785. [CrossRef]

32. Zhang, L.; Dailey, P.J.; Gettie, A.; Blanchard, J.; Ho, D.D. The liver is a major organ for clearing simian immunodeficiency virus in rhesus monkeys. *J. Virol.* **2002**, *76*, 5271–5273. [CrossRef] [PubMed]
33. Cotmore, S.F.; Agbandje-McKenna, M.; Canuti, M.; Chiorini, J.A.; Eis-Hubinger, A.M.; Hughes, J.; Mietzsch, M.; Modha, S.; Ogliastro, M.; Penzes, J.J.; et al. Ictv virus taxonomy profile: Parvoviridae. *J. Gen. Virol.* **2019**, *100*, 367–368. [CrossRef] [PubMed]

Article

No Evidence of Mosquito Involvement in the Transmission of Equine Hepacivirus (Flaviviridae) in an Epidemiological Survey of Austrian Horses

Marcha Badenhorst [1], Phebe de Heus [1], Angelika Auer [2], Till Rümenapf [2], Birthe Tegtmeyer [3], Jolanta Kolodziejek [2], Norbert Nowotny [2,4], Eike Steinmann [5] and Jessika-M.V. Cavalleri [1,*]

[1] University Equine Clinic - Internal Medicine, Department for Companion Animals and Horses, University of Veterinary Medicine Vienna, Veterinärplatz 1, 1210 Vienna, Austria; marcha.badenhorst@vetmeduni.ac.at (M.B.); phebe.de-heus@vetmeduni.ac.at (P.d.H.)

[2] Institute of Virology, University of Veterinary Medicine Vienna, Veterinärplatz 1, 1210 Vienna, Austria; angelika.auer@vetmeduni.ac.at (A.A.); till.ruemenapf@vetmeduni.ac.at (T.R.); jolanta.kolodziejek@vetmeduni.ac.at (J.K.); norbert.nowotny@vetmeduni.ac.at (N.N.)

[3] Institute for Experimental Virology, TWINCORE Centre for Experimental and Clinical Infection Research, Medical School Hannover (MHH) – Helmholtz Centre for Infection Research (HZI), Feodor-Lynen-Strasse 7, 30625 Hannover, Germany; birthe.tegtmeyer@twincore.de

[4] Department of Basic Medical Sciences, College of Medicine, Mohammed Bin Rashid University of Medicine and Health Sciences, Building 14, Dubai Healthcare City, Dubai, UAE

[5] Department of Molecular and Medical Virology, Ruhr-University Bochum, 44801 Bochum, Germany; eike.steinmann@rub.de

* Correspondence: jessika.cavalleri@vetmeduni.ac.at; Tel.: +43-1-250775510

Received: 28 September 2019; Accepted: 29 October 2019; Published: 1 November 2019

Abstract: Prevalence studies have demonstrated a global distribution of equine hepacivirus (EqHV), a member of the family Flaviviridae. However, apart from a single case of vertical transmission, natural routes of EqHV transmission remain elusive. Many known flaviviruses are horizontally transmitted between hematophagous arthropods and vertebrate hosts. This study represents the first investigation of potential EqHV transmission by mosquitoes. More than 5000 mosquitoes were collected across Austria and analyzed for EqHV ribonucleic acid (RNA) by reverse transcription quantitative polymerase chain reaction (RT-qPCR). Concurrently, 386 serum samples from horses in eastern Austria were analyzed for EqHV-specific antibodies by luciferase immunoprecipitation system (LIPS) and for EqHV RNA by RT-qPCR. Additionally, liver-specific biochemistry parameters were compared between EqHV RNA-positive horses and EqHV RNA-negative horses. Phylogenetic analysis was conducted in comparison to previously published sequences from various origins. No EqHV RNA was detected in mosquito pools. Serum samples yielded an EqHV antibody prevalence of 45.9% (177/386) and RNA prevalence of 4.15% (16/386). EqHV RNA-positive horses had significantly higher glutamate dehydrogenase (GLDH) levels ($p = 0.013$) than control horses. Phylogenetic analysis showed high similarity between nucleotide sequences of EqHV in Austrian horses and EqHV circulating in other regions. Despite frequently detected evidence of EqHV infection in Austrian horses, no viral RNA was found in mosquitoes. It is therefore unlikely that mosquitoes are vectors of this flavivirus.

Keywords: arbovirus; flavivirus; hematophagous arthropod; hepacivirus A; hepatitis; insects; mosquito-borne virus; virus transmission

---

1. Introduction

EqHV is one of 14 species belonging to the genus *Hepacivirus* in the family Flaviviridae [1]. This hepatotropic virus, also referred to as canine hepacivirus, non-primate hepacivirus and hepacivirus A, represents the closest related genetic homologue of hepatitis C virus (HCV) [1,2]. It is one of the novel viral agents, which has been associated with hepatitis in horses in recent years. EqHV infection typically results in subclinical hepatitis and transient, mild increases in liver-specific plasma biochemistry parameters [3].

Prevalence studies have demonstrated a global distribution of EqHV. The virus has been detected in horse populations across six continents, in countries including the USA, Brazil, South Africa, New Zealand, Korea, Japan, China, Scotland, France, as well as Austria's neighboring countries Italy, Germany and Hungary [3–16]. However, apart from a single case of vertical transmission [17], natural routes of EqHV transmission remain elusive. Based on the frequent detection of EqHV RNA (prevalence up to 34.1%) [16], EqHV antibodies (prevalence up to 83.7%) [4] and the high EqHV prevalence in certain geographic regions and breeds [3,4,12,13], vertical transmission is unlikely to be the only route of natural infection. Phylogenetic clustering of EqHV isolates from individual horses within their respective herds also suggests a horizontal route of transmission [17]. Young horses subjected to intensive management practices appear to be particularly at risk [18,19]. HCV is known to spread by venereal transmission [20]. The spread of EqHV by the venereal route has been implicated in studies, which found the frequent occurrence of EqHV in a cohort of broodmares and breeding stallions [13] and a high frequency of EqHV RNA in horses bred for reproduction purposes [18]. The venereal transmission of EqHV remains speculative. However, comparable to HCV, experimental and iatrogenic transmission of EqHV by means of infected blood and blood products have been demonstrated [21–23].

Many known flaviviruses are horizontally transmitted between hematophagous arthropods and vertebrate hosts [24]. Examples include dengue virus, yellow fever virus (YFV), Japanese encephalitis virus (JEV), Zika virus (ZIKV), tick-borne encephalitis virus (TBEV), West Nile virus (WNV) and Usutu virus (USUV). Mosquito-borne viruses are transmitted by a vast range of mosquito species, depending primarily on the vector-competence of the mosquito species, the geographical region and susceptible vertebrate host species [24].

The primary aim of this study was to investigate whether various mosquito species, present in areas of EqHV endemicity in horses, carry EqHV nucleic acid and may transmit the virus horizontally between horses. Mosquitoes were collected across Austria and analyzed for EqHV RNA. Concurrently, the occurrence of EqHV was investigated—for the first time—in the horse population of Austria. The geographical locations of analyzed mosquito pools and study horses' properties of origin were plotted on a map to determine proximity and compare EqHV statuses. Additionally, liver-specific plasma biochemistry parameters were compared between EqHV RNA-positive horses and EqHV RNA-negative control horses. Sequencing and phylogenetic analyses of Austrian EqHV strains were performed.

2. Materials and Methods

2.1. Study Design and Population

In this cross-sectional study, serum and plasma samples were collected for surveillance purposes from 386 horses in eastern Austria between July and October 2017. Sampled horses included patients of the University of Veterinary Medicine Vienna (Vetmeduni) Equine Clinic ($n = 58$), teaching horses of the Vetmeduni ($n = 50$) and privately owned, clinically unremarkable horses enrolled voluntarily ($n = 278$). The sample population consisted of various breeds and included 156 mares, 187 geldings, 42 stallions and one horse with the sex undisclosed. The horses' ages ranged from 1 to 31 years (median age = 12.17 years). The geographic locations were recorded for the properties of origin of the horses.

Considering an estimated population of 120 000 horses in Austria, the sample size was calculated. The expected prevalence of horses positive for EqHV RNA was set to 3.6%, which is the average prevalence of two surveillance studies in horse populations of various breeds and ages, performed in Germany (n = 433 horses; RNA prevalence = 2.5%) [3] and Italy (n = 1932 horses; RNA prevalence = 4.7%) [6], respectively. Given a confidence interval of 95%, a sample size of 54 horses was required. The expected prevalence of EqHV antibody-positive horses was set to 31.4%—the antibody prevalence detected in the same German surveillance study [3]. Given a confidence interval of 95%, a sample size of 331 horses was required. Hence, the investigation of 386 horses was considered to be sufficient to estimate the prevalence of EqHV RNA and antibodies in Austrian horses. Sample size calculations were performed using Epitools Ausvet (epitools.ausvet.com.au/).

Data collection was approved by the Ethics Committee of the University of Veterinary Medicine Vienna, the Austrian Federal Ministry of Labor, Social Affairs, Health and Consumer Protection and the Austrian Federal Ministry of Education, Science and Research (study reference number BMWF-68.205/0125-WF/V/3b/2017).

2.2. Mosquito Collection

4039 mosquitoes collected in 2017 were pooled in 430 pools according to location, time of trapping and mosquito species. The mosquitoes were trapped across Austria and included both native and invasive species. In addition, 1004 mosquitoes were trapped in the 21st Viennese district in August 2017, during a targeted mosquito and virus surveillance program following the identification of an USUV RNA-positive blood donor from that district [25]. These mosquitoes were pooled in 104 pools. The mosquito collection of the 21st district was included in the study because a high number of investigated horses originated from this district, in which the Vetmeduni is located. A small number of mosquitoes from northeast Italy (n = 295; 32 pools; mainly invasive species), collected between June and October 2017, were also included in the analysis. In total, 5338 mosquitoes in 566 pools were investigated for the presence of EqHV nucleic acid. The following species were represented: *Aedes albopictus*, *Aedes japonicus*, *Aedes koreicus*, *Anopheles claviger*, *Anopheles maculipennis*, *Anopheles plumbeus*, *Culex hortensis*, *Culex pipiens*, *Culiseta annulata* and *Ochlerotatus geniculatus*.

2.3. Laboratory Analysis

2.3.1. Detection of EqHV RNA in Mosquito Pools

Mosquito preparation was conducted as described previously [26]. From each mosquito pool homogenate, 140 µL was processed by automated nucleic acid extraction employing a QIAamp 12 Viral RNA Mini QIAcube Kit (Qiagen, Germany) according to the manufacturer's instructions. Extracted RNA was transcribed into complementary deoxyribonucleic acid (cDNA) by applying the PrimeScriptRTMaster Mix Kit (TaKaRa, Kusatsu, Japan). All cDNA samples were stored at −20 °C before being used for RT-qPCR. For the SYBR Green based RT-qPCR, the SYBR®Premix Ex Taq™ II kit (TaKaRa, Kusatsu, Japan) was used, in combination with the previously described primers targeting the 5′ untranslated region (5′UTR) [5]. A standard curve for the quantification of RNA copies was assessed by serial dilution of a plasmid containing the EqHV 5′UTR based on the EqHV isolate NPHV-NZP-1 (JQ434001.1). Measurement of fluorescence was conducted with a LightCycler 480 (Roche, Mannheim, Germany). The limit of detection for this RT-qPCR assay was determined to be 7.86 viral copies/µL of homogenate. RT-qPCR analysis of mosquito pool homogenates was performed by the Department of Molecular and Medical Virology, Ruhr-University Bochum, Germany.

2.3.2. Detection of EqHV RNA in Horse Serum

Viral RNA was extracted from 200 µL of each horse´s serum using the QIAamp 96 Virus QIAcube HT Kit (Qiagen, Germany) according to the manufacturer´s instructions. For screening, EqHV-specific nucleic acids were detected by RT-qPCR using primers and probes as described previously [5]. The RT-qPCR´s limit of quantification was established as 1.57 genome equivalents (GEs) per µL in the PCR reactions, which equates 7.85 × 10^2 GEs per mL of horse serum. For absolute quantification, a serial dilution of a defined DNA plasmid standard was run in parallel with the positive samples. RT-qPCR analysis of horse serum samples was performed by the Institute of Virology, University of Veterinary Medicine Vienna, Austria.

2.3.3. Detection of Anti-EqHV Non-Structural Protein 3 (NS3)-Specific Antibodies in Horse Serum

Serum samples were frozen between −20 °C and −80 °C prior to analysis for anti-EqHV non-structural protein 3 (NS3)-specific antibodies and shipped on dry ice to the laboratory in Germany. Samples were analyzed in duplicate for the presence of anti-EqHV NS3-specific antibodies, using the LIPS assay as described previously [27]. Relative light units (RLU) were measured with a plate luminometer (LB 960 XS3; Berthold, Bad Wildbad, Germany). The threshold value, above which samples were regarded as antibody-positive, was calculated for each plate by using the mean value plus three standard deviations (SDs) of an EqHV-negative horse serum sample.

2.3.4. Plasma Biochemistry

Plasma samples were frozen between −20 °C and −80 °C prior to analysis for GLDH, gamma-glutamyl transferase (GGT), bile acids and albumin concentrations at the Central Laboratory of the University of Veterinary Medicine Vienna. These parameters were measured in plasma samples of all EqHV RT-qPCR-positive horses, as well as in plasma samples of EqHV RT-qPCR-negative control horses ($n = 45$). The control horses were randomly selected from the group of privately owned, clinically unremarkable horses enrolled voluntarily. The laboratory's reference ranges were used for data analysis.

2.4. Data Analysis

Plasma biochemistry results were assessed for normality using the Shapiro Wilk test. The independent samples *t*-test was used to compare normally distributed parameters. The Mann–Whitney U test was used to compare non-normally distributed parameters. A 95% confidence interval was set and $p \leq 0.05$ was considered statistically significant. Statistical analysis was performed using IBM SPSS Statistics 24 and the applicable figure was generated using Graph Pad Prism 5.

2.5. Sequencing and Phylogenetic Analyses

Sixteen EqHV RNA-positive samples were used for sequence analyses. Nested PCRs targeting parts of the 5′UTR, NS3 and NS5B domains were conducted as previously described [6,10] and sent for Sanger sequencing. Previously published complete genome sequences of EqHV and additional sequences of six isolates, which originated in France [12], were retrieved from the GenBank database. All sequences were aligned with MUSCLE and primer sequences were deleted. Phylogenetic analyses were conducted by using the maximum-likelihood method by MEGA7 [28] based on the general time reversible model [29]. Gamma distributed with invariant sites (G + I) was set for rates among sites with a number of six discrete gamma categories. Bootstrap replicates were set to 500. All sequences were uploaded to the NCBI database with accession numbers (MN475754 - MN475783).

3. Results

3.1. Detection of EqHV RNA in Mosquito Pools

No EqHV RNA was detected in any of the mosquito pools analyzed. The geographic locations of mosquito collection sites are indicated in Figure 1.

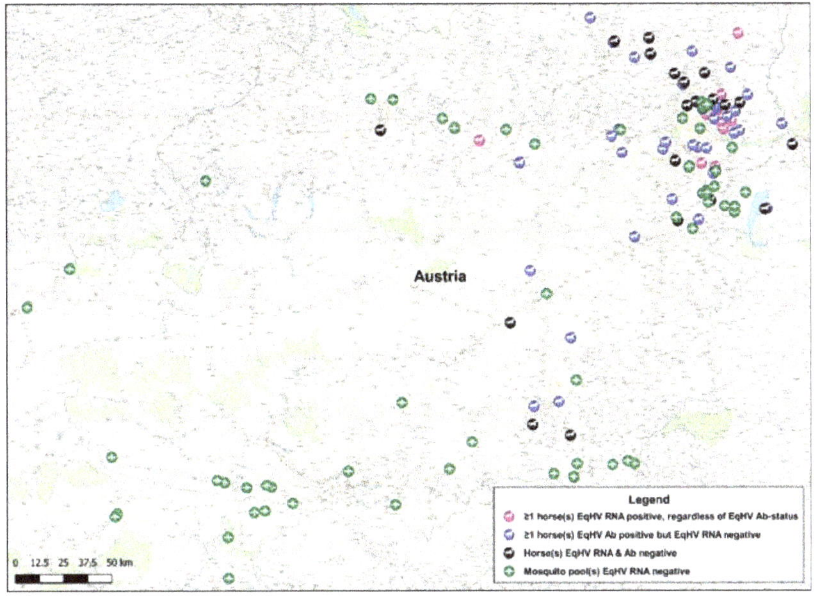

Figure 1. Geographical locations of sampling sites. A map of Austria indicating mosquito collection sites and the properties of origin of sampled horses. Green mosquito-icons each represents the collection site of a minimum of one mosquito pool. Magenta horse-icons each represents a property where at least one horse tested equine hepacivirus (EqHV) RNA-positive, regardless of the EqHV antibody-status of the horses on these properties. Blue horse-icons each represents a property where all horses tested EqHV RNA-negative, but at least one horse tested EqHV antibody-positive. Black horse-icons each represents a property where all horses tested EqHV RNA-negative, as well as EqHV antibody-negative.

3.2. Detection of EqHV RNA in Horse Serum

EqHV RNA was detectable in 4.15% of serum samples (16/386). Viral loads in 11 of the samples ranged from 1×10^3 to 9×10^6 GEs per mL serum. Five horses showed RT-qPCR signals, which were below the assay's level of quantification ($<7.9 \times 10^2$ GEs/mL serum). The 16 RT-qPCR-positive horses originated from 12 different properties, with between one and four RT-qPCR-positive horses per property (Table 1). The geographic locations of these properties are indicated in Figure 1.

Table 1. Information pertaining to the 12 properties where EqHV RNA-positive horses were identified, including the EqHV infection-state of the sampled populations.

Property		EqHV Infection-State						
		Abs-/RNA-		Abs+/RNA-		Abs+/RNA+		Abs-/RNA+
Property ID	No. of Horses Sampled	n	%	n	%	n	%	n	%
1	12	4	33.33	7	58.33	1	8.33	0	0.00
2	10	4	40.00	5	50.00	1	10.00	0	0.00
3	20	13	65.00	6	30.00	1	5.00	0	0.00
4	21	7	33.33	10	47.62	3	14.29	1	4.76
5	14	2	14.29	11	78.57	1	7.14	0	0.00
6	32	16	50.00	14	43.75	1	3.13	1	3.13
7	3	1	33.33	1	33.33	1	33.33	0	0.00
8	20	13	65.00	6	30.00	1	5.00	0	0.00
9	13	7	53.85	5	38.46	1	7.69	0	0.00
10	50	34	68.00	15	30.00	0	0.00	1	2.00
11	1	0	0.00	0	0.00	1	100	0	0.00
12	1	0	0.00	0	0.00	1	100	0	0.00
All other	189	105	55.56	84	44.44	0	0.00	0	0.00
Total	**386**	**206**	**53.37**	**164**	**42.49**	**13**	**3.37**	**3**	**0.78**

3.3. Detection of Anti-EqHV NS3-Specific Antibodies in Horse Serum

The antibody prevalence in the study population was 45.9%, with 177 of the 386 samples antibody-positive. Based on the horses' EqHV infection-state, the samples could be assigned to four categories: antibody-negative and RNA-negative (Abs–/RNA–); antibody-positive and RNA-negative (Abs+/RNA–); antibody-positive and RNA-positive (Abs+/RNA+) and antibody-negative and RNA-positive (Abs–/RNA+; Table 1). Two hundred and six samples (53.37%) tested negative for both EqHV antibodies and RNA. One hundred and sixty-four samples (42.49%) contained only antibodies, but no RNA. Thirteen samples (3.37%) contained both antibodies and RNA. Three samples (0.78%) contained only RNA, but no antibodies.

3.4. Plasma Biochemistry

EqHV RT-qPCR-positive horses (n = 16) had significantly higher plasma GLDH concentrations (p = 0.013) compared to EqHV RT-qPCR-negative control horses (n = 45; Table 2, Figure S1). All three of the EqHV RT-qPCR-positive horses which had GLDH levels above the laboratory reference range, were also EqHV antibody-positive (Abs+/RNA+). Two of these three horses were privately owned, clinically unremarkable horses enrolled voluntarily and one was admitted to the Vetmeduni Equine Clinic for follow-up examination, after a traumatic fracture of the fourth metatarsal bone. The horse had received tetanus antiserum two months before sample collection, following the acute trauma. Clinical findings did not give any indication for elevation of GLDH concentrations in this patient. All three of the horses which were acutely infected with EqHV (Abs–/RNA+), had GLDH levels well below the upper normal limit of the laboratory reference range. GGT, bile acids and albumin concentrations did not differ significantly between EqHV RT-qPCR-positive horses and EqHV RT-qPCR-negative control horses (p > 0.05; Table 2).

Table 2. Specifications of the liver-specific biochemistry parameters glutamate dehydrogenase (GLDH), gamma-glutamyl transferase (GGT), bile acids and albumin measured in plasma samples of all EqHV RT-qPCR-positive horses (n = 16), as well as EqHV RT-qPCR-negative control horses (n = 45).

Parameter (Reference Range)	GLDH (<13 U/L)		GGT (<30 U/L)		Bile acids (<20 umol/L)		Albumin (2.4–4.5 g/dL)	
EqHV RT-qPCR Status	Positive	Negative	Positive	Negative	Positive	Negative	Positive	Negative
n	16	45	16	45	16	45	16	45
Range (min–max)	1.81–67.45	1.44–42.54	4–61	1–42	3–19	2–11	2.69–3.25	1.99–4.41
Normal distribution	No	No	No	No	No	No	Yes	Yes
Median	4.84	2.78	14.5	12	5	5	3.04	2.98
Mean	12.98	4.83	16.13	13.33	5.81	5.53	3.03	3.04
Standard deviation	20.27	6.83	13.27	7.92	3.94	2.16	0.16	0.52
Parametric/ Nonparametric test	The Mann–Whitney U test		The Mann–Whitney U test		The Mann–Whitney U test		Independent samples t-test	
p-value	p = 0.013 *		p = 0.434		p = 0.659		p = 0.855	

* EqHV RT-qPCR-positive horses had significantly higher plasma GLDH concentrations (p < 0.05) compared to EqHV RT-qPCR-negative control horses.

3.5. Sequence and Phylogenetic Analyses

For a molecular characterization of the EqHV positive samples, three nested PCRs were performed to target different conserved regions of the viral genome. Most likely as a result of variation in the viral load of samples, we were able to recover 12/16 partial 5′UTR sequences, 7/16 partial NS3 sequences and 11/16 partial NS5B sequences (Table 3).

Table 3. The newly identified sequences which were submitted to NCBI and assigned with accession numbers. n.a.—not available.

Isolate	Accession Numbers		
	5′UTR	**NS3**	**NS5B**
N40-17	MN475754	MN475766	MN475773
N87-17	MN475755	MN475767	MN475774
N107-17	MN475756	MN475768	MN475775
N147-17	MN475757	MN475769	MN475776
N154-17	MN475758	MN475770	MN475777
N201-17	MN475759	n.a.	MN475778
N234-17	MN475760	MN475771	MN475779
N235-17	MN475761	MN475772	MN475780
N265-17	MN475762	n.a.	MN475781
N351-17	MN475763	n.a.	MN475782
N364-17	MN475764	n.a.	MN475783
N403-17	MN475765	n.a.	n.a.

Phylogenetic analysis based on partial NS5B sequences revealed only minor genetic distances between Austrian samples and previously published samples (Figure 2). Similarly, phylogenetic trees based on the 5′UTR (Figure S2) and on NS3 (Figure S3) visualize only minor differences between the sequences.

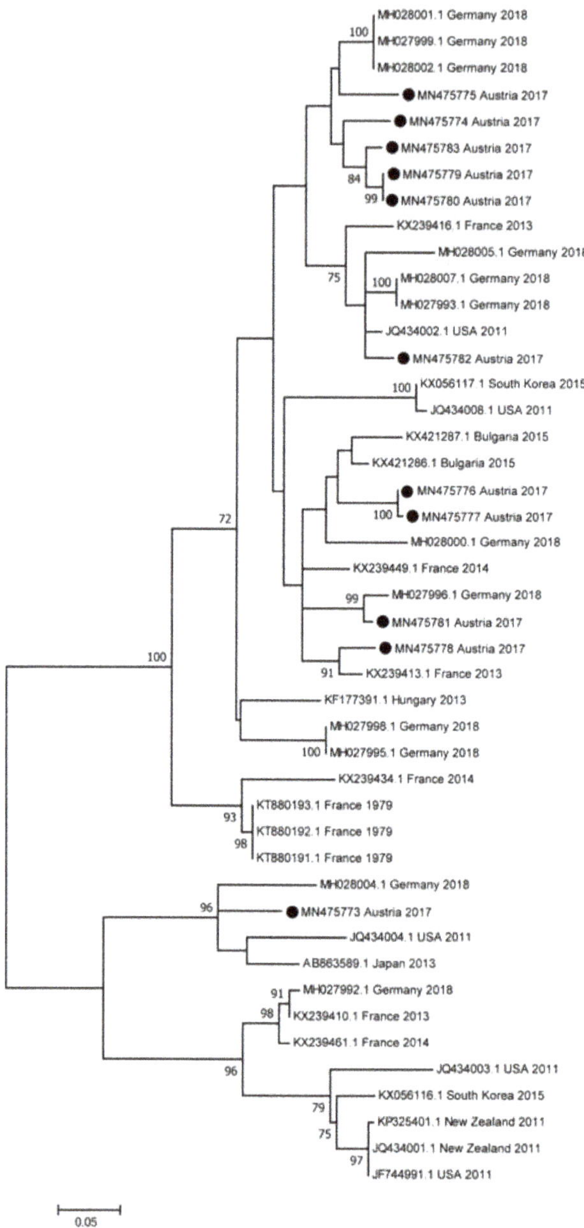

Figure 2. Maximum-likelihood phylogeny based on partial NS5B sequences of EqHV. In addition to sequences obtained from Austrian horses, the phylogenetic tree contains previously published complete genome sequences of EqHV retrieved from the GenBank database, as well as sequences of six isolates, which originated in France [12]. Different samples are identified with their accession number, country of origin and year of sampling or, if not applicable, year of publication. The analysis involved 45 nucleotide sequences. All positions containing gaps and missing data were eliminated, whereby a total of 258 positions were included in the final dataset. Bootstrap values <70% are not shown. The scale bar represents the number of substitutions per site. Black circles indicate samples obtained in this study.

4. Discussion

The EqHV carrier-state of a large population of native and invasive mosquito species, collected across a large geographic area—the entire Austria and a small part of Italy—was investigated. This study also represents the first surveillance for EqHV in the horse population of Austria. The EqHV antibody prevalence (45.9%) and RNA prevalence (4.15%) detected in this population of Austrian horses were within previously reported ranges [4,16,30]. Despite the frequently detected evidence of EqHV infection in the equine study population, no indication of viral nucleic acid was found in mosquito populations throughout Austria and northeast Italy.

Mosquito surveillance programs were initiated in Austria in 2011, with the aim of detecting invasive mosquito species [31,32] and investigating the variety of viruses, which mosquitoes may carry and transmit to vertebrate hosts. Particularly the flaviviruses WNV and USUV have been detected [26,33–37]. For the current study, nucleic acid extracts from the entire 2017 mosquito-collection were investigated for the presence of EqHV RNA. The vast majority of mosquitoes were collected between June and October 2017, coinciding with the time of blood collection from the horses (July to October 2017). All mosquito species native to Austria were represented in the collection, as well as a large number of individuals from the invasive species *Aedes japonicus*. *A. japonicus* is rapidly dispersing in Austria [31], and was shown to be a flavivirus vector [35].

Mosquito-borne flaviviruses replicate within the invertebrate vector during the extrinsic incubation period—the time between uptake of an infectious blood meal by a female mosquito and subsequent viral transmission to the host [24]. A small percentage of infected female mosquitoes typically overwinter in frost-free areas, resulting in survival of the flavivirus [38]. Vertical transmission of flaviviruses is also frequently observed in infected female mosquitoes [26]. During the course of evolution, certain mosquito species developed "vector competence" for certain viruses. It was therefore essential to investigate a spectrum of mosquito species for the presence of EqHV RNA, since it is not known which mosquito species could be competent vectors for EqHV transmission. A large number of mosquito pools (n = 566), consisting of 5338 individuals from various species, representing five mosquito genera (*Aedes* spp., *Anopheles* spp., *Culex* spp., *Culiseta* spp. and *Ochlerotatus* spp.) contained no EqHV RNA. Based on these findings, mosquitoes are unlikely to play a role as vectors of EqHV.

Comparable numbers of Austrian mosquitoes enabled us to determine the minimal infection rate (MIR) of WNV in previous investigations [36]. These results were in accordance with Czech and Hungarian studies, in which WNV-positive mosquitoes were detected in four of 650 pools [39] and in three of 645 pools [40], respectively. However, in areas with documented human or animal flavivirus infections, significantly higher numbers of mosquitoes were infested, as shown in [26] and [36] for WNV, and in [35] for USUV. We are therefore confident that the number of investigated mosquitoes was sufficient to also detect other flaviviruses.

Similar to HCV, transmission by means of infected blood and blood products has been demonstrated for EqHV [21,23]. Experimental infections of horses have been performed by intravenous administration of various volumes (range from 5 mL to 500 mL) of serum or plasma with varying viral loads (range from 3.92×10^3 RNA copies/mL to 7.78×10^6 RNA copies/mL) [21,23]. Considering this proven route of EqHV transmission, it remains to be determined whether EqHV can be mechanically transmitted by other hematophagous insects. The family Tabanidae (horse flies) is of particular importance as mechanical vectors of viruses such as equine infectious anemia (EIA) [41]. Painful bites, large lesions, persistent feeding behavior and the volume of blood left on their large mouthparts favors mechanical transmission [42]. The volume of blood retained on the mouthparts of a horse fly is approximately 5 to 25 nL [41,43]. Viral load in the blood of the host plays an important role in successful mechanical transmission of viruses. The minimum infectious dose of EqHV has not yet been determined.

Natural and experimental EqHV infections typically result in transient elevation of concentrations of liver-specific enzymes, indicative of hepatic inflammation. Increased GLDH, sorbitol dehydrogenase (SDH), GGT and aspartate aminotransferase (AST) levels have been reported [3,21,23,44]. The increased plasma GLDH concentrations in EqHV RT-qPCR-positive horses are consistent with hepatocellular damage observed on liver histopathology of EqHV-infected horses [21]. Although plasma GLDH concentrations were significantly higher in EqHV RNA-positive horses compared to EqHV RNA-negative horses, the GLDH values of the majority of EqHV RNA-positive individuals were still within the laboratory reference range. Therefore, potential cases of EqHV should not only be monitored for liver-specific enzyme concentrations above references values, but also for deviations of enzyme concentrations from the individual horse's baseline values. The timing of sampling is also of importance, with previous studies demonstrating that elevations in concentration of liver-specific enzymes typically do not occur during the acute phase of infection, but rather coincides with seroconversion [3,21]. Our findings are in accordance with this data.

In accordance with findings from other regions, the phylogenetic analyses revealed high similarity at nucleotide sequence-level of EqHV circulating in Austria, compared to other countries [4,6,10].

5. Conclusions

Despite evidence of EqHV infection being frequently detected in Austrian horses, no indication of viral nucleic acid was found in mosquito populations throughout the country. Consequently, mosquitoes are unlikely to play a role as vectors of this flavivirus. Nucleotide sequence differences between EqHV in Austria compared to EqHV of other origins are minor.

Supplementary Materials: The following are available online at http://www.mdpi.com/1999-4915/11/11/1014/s1, Figure S1: Plasma GLDH concentrations of EqHV RT-qPCR-positive horses (n = 16) and EqHV RT-qPCR-negative control horses (n = 45), Figure S2: Maximum-likelihood phylogeny based on partial 5′UTR sequences of EqHV, Figure S3: Maximum-likelihood phylogeny based on partial NS3 sequences of EqHV.

Author Contributions: Conceptualization, M.B., N.N., E.S., J.M.V.C.; methodology, M.B., P.d.H., A.A., T.R., J.K., N.N., E.S., J.-M.V.C.; validation, A.A., T.R., N.N., E.S., J.-M.V.C.; formal analysis, M.B., A.A., B.T., J.K., N.N., E.S., J.-M.V.C.; investigation, M.B., P.d.H., J.K., N.N.; resources, P.d.H., A.A., T.R., B.T., N.N., E.S., J.-M.V.C.; data curation, M.B., A.A., B.T., J.K., N.N., E.S.; writing—original draft preparation, M.B.; writing—review and editing, P.d.H., A.A., T.R., B.T., J.K., N.N., E.S., J.-M.V.C.; visualization, M.B., B.T., J.K., N.N.; supervision, E.S., J.-M.V.C.; project administration, M.B., J.M.V.C.

Funding: This research received no external funding.

Acknowledgments: We gratefully acknowledge Bernhard Seidel (Technical Office of Ecology and Landscape Assessment, Persenbeug, Austria) for trapping and determination of the mosquitoes; Alexander Tichy (University of Veterinary Medicine Vienna) for assistance with statistical analysis of data; Peter D. Burbelo (NIH, Maryland, USA) for providing the Renilla-luciferase-NS3 fusion plasmid. We would like to thank all members of the Department of Molecular and Medical Virology at the Ruhr-University Bochum and especially Rosemarie Bohr, Regina Bütermann, Monika Kopytkowski, Klaus Sure, and Ute Wiegmann-Misiek for technical support. Open Access Funding by the University of Veterinary Medicine Vienna.

Conflicts of Interest: The authors declare no conflict of interest.

References

1. Smith, D.B.; Becher, P.; Bukh, J.; Gould, E.A.; Meyers, G.; Monath, T.; Muerhoff, A.S.; Pletnev, A.; Rico-Hesse, R.; Stapleton, J.T.; et al. Proposed update to the taxonomy of the genera *Hepacivirus* and *Pegivirus* within the Flaviviridae family. *J. Gen. Virol.* **2016**, *97*, 2894–2907. [CrossRef] [PubMed]

2. Pfaender, S.; Brown, R.J.; Pietschmann, T.; Steinmann, E. Natural reservoirs for homologs of hepatitis C virus. *Emerg. Microbes Infect.* **2014**, *3*, e21. [CrossRef] [PubMed]

3. Pfaender, S.; Cavalleri, J.M.; Walter, S.; Doerrbecker, J.; Campana, B.; Brown, R.J.; Burbelo, P.D.; Postel, A.; Hahn, K.; Anggakusuma Riebesehl, N.; et al. Clinical course of infection and viral tissue tropism of hepatitis C virus-like nonprimate hepaciviruses in horses. *Hepatology* **2015**, *61*, 447–459. [CrossRef] [PubMed]

4. Badenhorst, M.; Tegtmeyer, B.; Todt, D.; Guthrie, A.; Feige, K.; Campe, A.; Steinmann, E.; Cavalleri, J.M.V. First detection and frequent occurrence of Equine Hepacivirus in horses on the African continent. *Vet. Microbiol.* **2018**, *223*, 51–58. [CrossRef]

5. Burbelo, P.D.; Dubovi, E.J.; Simmonds, P.; Medina, J.L.; Henriquez, J.A.; Mishra, N.; Wagner, J.; Tokarz, R.; Cullen, J.M.; Iadarola, M.J.; et al. Serology-enabled discovery of genetically diverse hepaciviruses in a new host. *J. Virol.* **2012**, *86*, 6171–6178. [CrossRef]

6. Elia, G.; Lanave, G.; Lorusso, E.; Parisi, A.; Cavaliere, N.; Patruno, G.; Terregino, C.; Decaro, N.; Martella, V.; Buonavoglia, C. Identification and genetic characterization of equine hepaciviruses in Italy. *Vet. Microbiol.* **2017**, *207*, 239–247. [CrossRef]

7. Gemaque, B.S.; Junior Souza de Souza, A.; do Carmo Pereira Soares, M.; Malheiros, A.P.; Silva, A.L.; Alves, M.M.; Gomes-Gouvea, M.S.; Pinho, J.R.; Ferreira de Figueiredo, H.; Ribeiro, D.B.; et al. Hepacivirus infection in domestic horses, Brazil, 2011–2013. *Emerg. Infect. Dis.* **2014**, *20*, 2180–2182. [CrossRef]

8. Kim, H.-S.; Moon, H.-W.; Sung, H.W.; Kwon, H.M. First identification and phylogenetic analysis of equine hepacivirus in Korea. *Infect. Genet. Evol.* **2017**, *49*, 268–272. [CrossRef]

9. Lu, G.; Sun, L.; Xu, T.; He, D.; Wang, Z.; Ou, S.; Jia, K.; Yuan, L.; Li, S. First Description of Hepacivirus and Pegivirus Infection in Domestic Horses in China: A Study in Guangdong Province, Heilongjiang Province and Hong Kong District. *PLoS ONE* **2016**, *11*, e0155662. [CrossRef]

10. Lyons, S.; Kapoor, A.; Sharp, C.; Schneider, B.S.; Wolfe, N.D.; Culshaw, G.; Corcoran, B.; McGorum, B.C.; Simmonds, P. Nonprimate hepaciviruses in domestic horses, United Kingdom. *Emerg. Infect. Dis.* **2012**, *18*, 1976–1982. [CrossRef]

11. Matsuu, A.; Hobo, S.; Ando, K.; Sanekata, T.; Sato, F.; Endo, Y.; Amaya, T.; Osaki, T.; Horie, M.; Masatani, T.; et al. Genetic and serological surveillance for non-primate hepacivirus in horses in Japan. *Vet. Microbiol.* **2015**, *179*, 219–227. [CrossRef] [PubMed]

12. Pronost, S.; Hue, E.; Fortier, C.; Foursin, M.; Fortier, G.; Desbrosse, F.; Rey, F.A.; Pitel, P.H.; Richard, E.; Saunier, B. Prevalence of Equine Hepacivirus Infections in France and Evidence for Two Viral Subtypes Circulating Worldwide. *Transbound. Emerg. Dis.* **2017**, *64*, 1884–1897. [CrossRef] [PubMed]

13. Reichert, C.; Campe, A.; Walter, S.; Pfaender, S.; Welsch, K.; Ruddat, I.; Sieme, H.; Feige, K.; Steinmann, E.; Cavalleri, J.M.V. Frequent occurrence of nonprimate hepacivirus infections in Thoroughbred breeding horses—A cross-sectional study for the occurrence of infections and potential risk factors. *Vet. Microbiol.* **2017**, *203*, 315–322. [CrossRef] [PubMed]

14. Reuter, G.; Maza, N.; Pankovics, P.; Boros, A. Non-primate hepacivirus infection with apparent hepatitis in a horse—Short communication. *Acta. Vet. Hung.* **2014**, *62*, 422–427. [CrossRef]

15. Schlottau, K.; Fereidouni, S.; Beer, M.; Hoffmann, B. Molecular identification and characterization of nonprimate hepaciviruses in equines. *Arch. Virol.* **2019**, *164*, 391–400. [CrossRef]

16. Tanaka, T.; Kasai, H.; Yamashita, A.; Okuyama-Dobashi, K.; Yasumoto, J.; Maekawa, S.; Enomoto, N.; Okamoto, T.; Matsuura, Y.; Morimatsu, M.; et al. Hallmarks of hepatitis C virus in equine hepacivirus. *J. Virol.* **2014**, *88*, 13352–13366. [CrossRef]

17. Gather, T.; Walter, S.; Todt, D.; Pfaender, S.; Brown, R.J.; Postel, A.; Becher, P.; Moritz, A.; Hansmann, F.; Baumgaertner, W.; et al. Vertical transmission of hepatitis C virus-like non-primate hepacivirus in horses. *J. Gen. Virol.* **2016**, *97*, 2540–2551. [CrossRef]

18. Figueiredo, A.S.; Lampe, E.; de Albuquerque, P.P.L.F.; Chalhoub, F.L.L.; de Filippis, A.M.B.; Villar, L.M.; Cruz, O.G.; Pinto, M.A.; de Oliveira, J.M. Epidemiological investigation and analysis of the NS5B gene and protein variability of non-primate hepacivirus in several horse cohorts in Rio de Janeiro state, Brazil. *Infect. Genet. Evol.* **2018**, *59*, 38–47. [CrossRef]

19. Figueiredo, A.S.; de Moraes, M.V.D.S.; Soares, C.C.; Chalhoub, F.L.L.; de Filippis, A.M.B.; Dos Santos, D.R.L.; de Almeida, F.Q.; Godoi, T.L.O.S.; de Souza, A.M.; Burdman, T.R.; et al. First description of Theiler's disease-associated virus infection and epidemiological investigation of equine pegivirus and equine hepacivirus coinfection in Brazil. *Transbound. Emerg. Dis.* **2019**, *66*, 1737–1751. [CrossRef]

20. Terrault, N.A.; Dodge, J.L.; Murphy, E.L.; Tavis, J.E.; Kiss, A.; Levin, T.R.; Gish, R.G.; Busch, M.P.; Reingold, A.L.; Alter, M.J. Sexual transmission of hepatitis C virus among monogamous heterosexual couples. The HCV partners study. *Hepatology* **2013**, *57*, 881–889. [CrossRef]

21. Pfaender, S.; Walter, S.; Grabski, E.; Todt, D.; Bruening, J.; Romero-Brey, I.; Gather, T.; Brown, R.J.P.; Hahn, K.; Puff, C.; et al. Immune protection against reinfection with nonprimate hepacivirus. *PNAS* **2017**, *114*, E2430–E2439. [CrossRef] [PubMed]

22. Postel, A.; Cavalleri, J.-M.V.; Pfaender, S.; Walter, S.; Steinmann, E.; Fischer, N.; Feige, K.; Haas, L.; Becher, P. Frequent presence of hepaci and pegiviruses in commercial equine serum pools. *Vet. Microbiol.* **2016**, *182*, 8–14. [CrossRef] [PubMed]

23. Ramsay, J.D.; Evanoff, R.; Wilkinson, T.E.; Divers, T.J.; Knowles, D.P.; Mealey, R.H. Experimental transmission of equine hepacivirus in horses as a model for hepatitis C virus. *Hepatology* **2015**, *61*, 1533–1546. [CrossRef] [PubMed]

24. Huang, Y.-J.S.; Higgs, S.; Horne, K.M.; Vanlandingham, D.L. Flavivirus-mosquito interactions. *Viruses* **2014**, *6*, 4703–4730. [CrossRef] [PubMed]

25. Bakonyi, T.; Jungbauer, C.; Aberle, S.W.; Kolodziejek, J.; Dimmel, K.; Stiasny, K.; Allerberger, F.; Nowotny, N. Usutu virus infections among blood donors, Austria, July and August 2017—Raising awareness for diagnostic challenges. *Euro. Surveill.* **2017**, *22*. [CrossRef] [PubMed]

26. Kolodziejek, J.; Seidel, B.; Jungbauer, C.; Dimmel, K.; Kolodziejek, M.; Rudolf, I.; Hubálek, Z.; Allerberger, F.; Nowotny, N. West Nile virus positive blood donation and subsequent entomological investigation, Austria, 2014. *PLoS ONE* **2015**, *10*, e0126381. [CrossRef]

27. Burbelo, P.D.; Ching, K.H.; Klimavicz, C.M.; Iadarola, M.J. Antibody profiling by Luciferase Immunoprecipitation Systems (LIPS). *J. Vis. Exp.* **2009**, *32*. [CrossRef]

28. Kumar, S.; Stecher, G.; Tamura, K. MEGA7: Molecular Evolutionary Genetics Analysis Version 7.0 for Bigger Datasets. *Mol. Biol. Evol.* **2016**, *33*, 1870–1874. [CrossRef]

29. Nei, M.; Kumar, S. *Molecular Evolution and Phylogenetics*; Oxford University Press: New York, NY, USA, 2000.

30. Lyons, S.; Kapoor, A.; Schneider, B.S.; Wolfe, N.D.; Culshaw, G.; Corcoran, B.; Durham, A.E.; Burden, F.; McGorum, B.C.; Simmonds, P. Viraemic frequencies and seroprevalence of non-primate hepacivirus and equine pegiviruses in horses and other mammalian species. *J. Gen. Virol.* **2014**, *95*, 1701–1711. [CrossRef]

31. Seidel, B.; Nowotny, N.; Bakonyi, T.; Allerberger, F.; Schaffner, F. Spread of *Aedes japonicus japonicus* (Theobald, 1901) in Austria, 2011-2015, and first records of the subspecies for Hungary, 2012, and the principality of Liechtenstein, 2015. *Parasite Vectors* **2016**, *9*, 356. [CrossRef]

32. Seidel, B.; Montarsi, F.; Huemer, H.P.; Indra, A.; Capelli, G.; Allerberger, F.; Nowotny, N. First record of the Asian bush mosquito, *Aedes japonicus japonicus*, in Italy: Invasion from an established Austrian population. *Parasite Vectors* **2016**, *9*, 284. [CrossRef] [PubMed]

33. Bakonyi, T.; Ferenczi, E.; Erdélyi, K.; Kutasi, O.; Csörgő, T.; Seidel, B.; Weissenböck, H.; Brugger, K.; Bán, E.; Nowotny, N. Explosive spread of a neuroinvasive lineage 2 West Nile virus in Central Europe, 2008/2009. *Vet. Microbiol.* **2013**, *165*, 61–70. [CrossRef] [PubMed]

34. Camp, J.V.; Bakonyi, T.; Soltész, Z.; Zechmeister, T.; Nowotny, N. *Uranotaenia unguiculata* Edwards, 1913 are attracted to sound, feed on amphibians, and are infected with multiple viruses. *Parasite Vectors* **2018**, *11*, 456. [CrossRef] [PubMed]

35. Camp, J.V.; Kolodziejek, J.; Nowotny, N. Targeted surveillance reveals native and invasive mosquito species infected with Usutu virus. *Parasite Vectors* **2019**, *12*, 46. [CrossRef]

36. Kolodziejek, J.; Jungbauer, C.; Aberle, S.W.; Allerberger, F.; Bagó, Z.; Camp, J.V.; Dimmel, K.; de Heus, P.; Kolodziejek, M.; Schiefer, P.; et al. Integrated analysis of human-animal-vector surveillance: West Nile virus infections in Austria, 2015–2016. *Emerg. Microbes Infect.* **2018**, *7*, 25. [CrossRef]

37. Pachler, K.; Lebl, K.; Berer, D.; Rudolf, I.; Hubalek, Z.; Nowotny, N. Putative new West Nile virus lineage in *Uranotaenia unguiculata* mosquitoes, Austria, 2013. *Emerg. Infect. Dis.* **2014**, *20*, 2119–2122. [CrossRef]

38. Rudolf, I.; Betášová, L.; Blažejová, H.; Venclíková, K.; Straková, P.; Šebesta, O.; Mendel, J.; Bakonyi, T.; Schaffner, F.; Nowotny, N.; et al. West Nile virus in overwintering mosquitoes, central Europe. *Parasite Vectors* **2017**, *10*, 452. [CrossRef]

39. Bakonyi, T.; Ivanics, E.; Erdélyi, K.; Ursu, K.; Ferenczi, E.; Weissenböck, H.; Nowotny, N. Lineage 1 and 2 strains of encephalitic West Nile virus, central Europe. *Emerg. Infect. Dis.* **2006**, *12*, 618–623. [CrossRef]

40. Rudolf, I.; Bakonyi, T.; Šebesta, O.; Mendel, J.; Peško, J.; Betášová, L.; Blažejová, H.; Venclíková, K.; Straková, P.; Nowotny, N.; et al. West Nile virus lineage 2 isolated from *Culex modestus* mosquitoes in the Czech Republic, 2013: Expansion of the European WNV endemic area to the North? *Euro. Surveill.* **2014**, *19*, 20867. [CrossRef]

41. Issel, C.J.; Foil, L.D. Equine infectious anaemia and mechanical transmission. Man and the wee beasties. *Rev. Sci. Tech. OIE* **2015**, *34*, 513–523. [CrossRef]

42. Carn, V.M. The role of dipterous insects in the mechanical transmission of animal viruses. *Brit. Vet. J.* **1996**, *152*, 377–393. [CrossRef]

43. Scoles, G.A.; Miller, J.A.; Foil, L.D. Comparison of the efficiency of biological transmission of *Anaplasma marginale* (Rickettsiales. Anaplasmataceae) by *Dermacentor andersoni* Stiles (Acari: Ixodidae) with mechanical transmission by the horse fly, *Tabanus fuscicostatus* Hine (Diptera: Muscidae). *J. Med. Entomol.* **2008**, *45*, 109–114. [CrossRef] [PubMed]
44. Gather, T.; Walter, S.; Pfaender, S.; Todt, D.; Feige, K.; Steinmann, E.; Cavalleri, J.M.V. Acute and chronic infections with nonprimate hepacivirus in young horses. *Vet. Res.* **2016**, *47*, 97. [CrossRef] [PubMed]

Communication

Further Evidence for in Utero Transmission of Equine Hepacivirus to Foals

Stephane Pronost [1,2,*], Christine Fortier [1,2], Christel Marcillaud-Pitel [3], Jackie Tapprest [4], Marc Foursin [5], Bertrand Saunier [6], Pierre-Hugues Pitel [1], Romain Paillot [1,2] and Erika S. Hue [1,2]

[1] LABÉO Frank Duncombe, 14280 Saint-Contest, France; christine.fortier@laboratoire-labeo.fr (C.F.); Pierre-Hugues.PITEL@laboratoire-labeo.fr (P.-H.P.); Romain.PAILLOT@laboratoire-labeo.fr (R.P.); erika.hue@laboratoire-labeo.fr (E.S.H.)
[2] BIOTARGEN EA7450, UNICAEN, NORMANDIE UNIV, 14000 Caen, France
[3] RESPE, 14280 Saint-Contest, France; c.marcillaud-pitel@respe.net
[4] Laboratory for Animal Health in Normandy, French Agency for Food, Environmental and Occupational Health & Safety (ANSES), 14430 Goustranville, France; jackie.tapprest@anses.fr
[5] Clinique équine de la Boisrie, 61500 Chailloué, France; marc.foursin@orange.fr
[6] Structural Virology Unit—CNRS UMR 3569, Institut Pasteur, 75015 Paris, France; bertrand.saunier@pasteur.fr
* Correspondence: stephane.pronost@laboratoire-labeo.fr; Tel.: +33-2-31-47-19-54

Received: 26 September 2019; Accepted: 3 December 2019; Published: 5 December 2019

Abstract: (1) Background: Equine hepacivirus (EqHV), also referred to as non-primate hepacivirus (NPHV), infects horses—and dogs in some instances—and is closely related to hepatitis C virus (HCV) that has infected up to 3% of the world's human population, causing an epidemic of liver cirrhosis and cancer. EqHV also chronically infects the liver of horses, but does not appear to cause serious liver damages. Previous studies have been looking to identify route(s) of EqHV transmission to and between horses. (2) Methods: In this retrospective study, we sought to evaluate the prevalence of vertical transmission taking place in utero with measuring by quantitative RT-PCR the amounts of EqHV genome in samples from 394 dead foals or fetuses, paired with the allantochorion whenever available. (3) Results: Detection of EqHV in three foals most likely resulted from a vertical transmission from the mares to the fetuses, consistent with the in utero transmission hypothesis. In support of this observation, the presence of EqHV genome was found for the first time in two of the allantochorions. (4) Conclusions: As seemingly benign viruses could turn deadly (e.g., Zika flavivirus) and EqHV happens to have infected a significant proportion of the world's horse herds, EqHV infectious cycle should be further clarified.

Keywords: non-primate hepacivirus; equine hepacivirus; in utero transmission; horse; fetuses

1. Introduction

Since Burbelo et al. [1] first reported the infection of horses by an equine hepacivirus (EqHV) in 2012, the presence of this non-primate hepacivirus (NPHV) has been described in the equine populations of the five continents [1–9]. Depending on the geographic locations, the prevalence of EqHV measured by quantitative reverse transcriptase PCR (RT-PCR) varies from less than 1 to over 10 percent [2]. Among all hepaciviruses recently discovered in different animal species, EqHV displays the highest genomic homology to hepatitis C virus, which chronically infects humans in the liver. The study of hepaciviral infections in equids could shed some light on the physiopathology of HCV [10]; conversely, knowledge on HCV could also help investigating EqHV pathogenicity and elucidating its transmission route(s) (Figure 1).

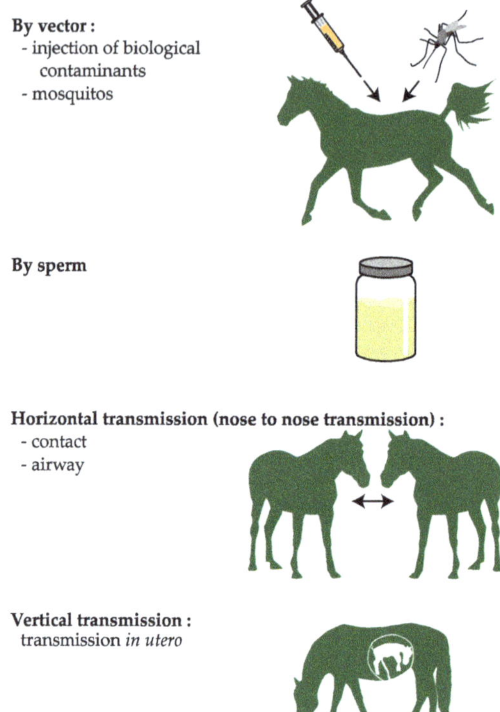

Figure 1. Hypotheses about the different modes of transmission by equine hepacivirus (EqHV) (according to [11]).

A recent study performed in an area where EqHV is endemic in horses failed to detect the virus in a sample of over 5000 mosquitoes, making the latter an unlikely vector of this virus [12]. Other studies have reported the detections of EqHV-specific antibodies and/or of viral genome in serum, tissue samples and, lastly, in cerebrospinal fluid [13]. Detection and replication of EqHV genome in different organs of adult horses, such as liver, spleen, cerebellum, and lungs were also reported following the experimental infections [9,14,15]. Despite recent progress, EqHV tissue tropism remains largely uncharacterized in non-experimental conditions and little is known about the presence of the EqHV in equine fetuses. By comparison, most HCV infections in young children result from vertical transmission [16]. The contamination could take place during pregnancy or upon delivery [17,18], but its exact timing and mechanisms are not fully understood. Vertical transmission of EqHV in horses has started to be investigated only recently. In one study, the presence of EqHV viral RNA was measured in serum samples from 20 mare-foal pairs [19]. Evidence of transmission to the foal was reported for only one mare, and no potential route of transmission of the virus, such as intrauterine or postpartum transfer was identified [19].

The purpose of the present study is, by measuring the amount of viral genome in samples of a large population of foals deceased during the perinatal period (aborted fetuses, stillborn foals, death occurring during the first week of life), to evaluate the incidence of EqHV vertical transmission by the intrauterine route. Genomic sequence analyses showed that infections of paired mare and dead foal were caused by identical EqHV strains, demonstrating the existence of an in utero vertical transmission.

2. Materials and Methods

The study and all animal work involved received ethical approval from the LABÉO Frank Duncombe ethical advisor (LFD-CE-07/2012, 2012). Samples were collected by equine veterinary practitioners according to a high standard of veterinary care.

2.1. Necrospy and Histological Analysis

Specimens from foals deceased during the perinatal period were collected from French stud farms according to a standard protocol [20] and shipped at +4 °C to the Anses' Necropsy Center (Laboratory for Animal Health in Normandy, France). As this is a retrospective study, information about the time lapse between delivery and autopsy of the fetuses may not be accurate; nevertheless, practitioner's good practices include a reasonable routing time. Most samples were intended for routine bacteriological cultures, equine herpesvirus 1 (EHV-1), and equine viral arteritis (EVA) PCR analysis and histopathological analysis [21]. For owners who wished necropsy be performed on dead foals' placentas, the latter were shipped using same guidelines: the samples were transported at +4 °C by a carrier and delivered within half an hour to LABÉO sample center; the samples were processed by the analytical services upon receipt.

In total, 394 tissue samples from aborted fetuses/neonatal foal deaths collected between January 2013 and December 2016 were used for this retrospective study. More information was available for 201 samples which allows to characterize the population: 159 fetuses with a median age of 8.75 months (7.25–9.50) and 42 foals with a median age of 24 h (4–72). In 2016, no sample was tested positive out of 130. During that year, 35 serum samples from mare who had aborted were also obtained and were all found negative. Serum was also obtained from foals of Cases #1 and #3 (cf. lower).

2.2. Molecular Detection and Characterisation of the EqHV Strains

2.2.1. Nucleic Acid Extraction and Quantitative RT-PCR

RNA was extracted from 25 mg of homogenized organ tissue (liver + lung mixture, or allantochorion) with RNeasy Mini Kit (Qiagen, Courtaboeuf, France) or from 140 µL of serum with QIAmp viral RNA Mini kit (Qiagen, Courtaboeuf, France) according to the manufacturer's instructions, and eluted in a final volume of 50 µl elution buffer (Qiagen, Courtaboeuf, France). After performing the originally planned tests, the extracted RNAs were stored at −80 °C in a temperature-monitored freezer until further use.

Quantitative RT-PCR was performed with One Step Prime Script RT-PCR kit (Takara, Ozyme, France) according to the manufacturer's instructions and adapted from Burbelo et al. [1] on a StepOne™ Real-Time PCR system (Life Technologies, Saint-Aubin, France) [2]. Quantitative RT-PCR were performed with primers Qanti-5UF1, Qanti-5UR1, and probe 5′-FAM-CCACGAAGGAAGGCGGGGGC-BHQ1-3′ [1] and with a second pair of primers (Sau5UF 5′-TCGAGGGAGCTGRAATTCGT-3′, Sau5UR 5′-GCCCTCGCAAGCATCCTATC-3′), as previously described [2]. Thermal cycling proceeded at 42 °C for 5 min, 95 °C for 10 s, followed by 45 cycles: 95 °C for 5 s and 60 °C for 34 s. Fluorescence was measured at the end of each annealing/elongation step (60 °C). Data were analyzed using the StepOne™ software, version 2.2.2 (Life Technologies, Saint-Aubin, France). The limit of detection was 3.5×10^2 genome copies/mL for serum and 7.7×10^3 genome copies/g for tissue.

2.2.2. Sequencing and Phylogenic Analysis

Sequencing of 5′-untranslated region (5′UTR), non-structural protein 3 (NS3), and non-structural protein 5B (NS5B) amplimers was performed by Biofidal (Vaulx-en-Velin, France) with a Phusion Hot start II (Fisher Scientific, Illkirch, France) [2]. All sequences were deposited in GenBank (accession numbers MN229470-MN229484, KX239312-KX239466, KT175006-KT175040, as described in Supplementary Table S1). The concatenated sequences were obtained by joining: (i) 5′UTR, NS3, and NS5B regions (598 bp) for strains used in Figure 4A,B; and, (ii) 5′UTR and NS5B regions (472 bp) for strains used in Figure 4C.

Phylogenetic trees of nucleotide sequence alignments were created using the neighbor-joining method based on the Jukes-Cantor model of MEGA5 [22], as previously described [2]. Finally, all NS5B sequences used to build the tree were converted into a Nexus format using EMBOSS Seqret [23] and a median joining network was built using PopART (Population Analysis with Reticulate Trees) software, version 1.7 [24,25].

3. Results

3.1. Prevalence of EqHV in Perinatal Foal Deaths

Among 394 tissue samples collected from aborted fetuses, stillborn foals, or neonatal foal deaths samples, 3 (0.76%) were found positive for EqHV by quantitative RT-PCR (Table 1); 1/84 in 2013, 1/131 in 2014, 1/49 in 2015, and none out of 130 in 2016. The positive results were confirmed by nucleotide sequencing. The prevalence of EqHV viral RNA detected in this study contrasts with that of 6% previously observed in serum samples of French horses [2], suggesting that most transmissions occur via other paths, as reported in other studies [19].

Table 1. Features of the three cases, for which EqHV genome was detected in foal and/or allantochorion samples, in France between 2013 and 2016. (n.a. = not applicable).

Cases (Year)	Subjects	Sampling	Life Status (/Birth)	Viral Loads	EqHV Strains
Case #1 (2013 & 2015)	Foal #1 (4 days)	Liver+Lung	Neonatal death (4 days)	2.4×10^7 copies/g	FR-Eq73_Liver-Lung/FR/2013
	Foal #2 (2 months)	Serum	Alive (2 months)	1.1×10^7 copies/mL	FR-Eq84_Serum/FR/2015
Case #2 (2014)	Foal (2 days)	Liver+Lung	Neonatal death (2 days)	2.3×10^7 copies/g	FR-Eq69_Liver-Lung/FR/2014
	Mare (5 years)	Allantochorion	n.a.	1.5×10^4 copies/g	FR-Eq69_Allanto/FR/2014
		Serum	Alive (10 months)	7.7×10^7 copies/ml	FR-Eq74_Serum/FR/2015
Case #3 (2015)	Foal (2 months)	Serum	Alive (50 days)	Negative	n.a.
	Mare (7 years)	Allantochorion	n.a.	3.9×10^4 copies/g	FR-Eq70_Allanto/FR/2015
		Serum	Alive	1.8×10^3 copies/mL	FR-Eq85_Serum/FR/2015

In Case #1, EqHV genome was first detected in liver + lung samples from foal #1 (FR-Eq73 Liver-Lung) with a viral load of 2.3×10^7 copies/g. Necropsy and histological analyses concluded to the lack of evidence for an overt viral infection; i.e., no significant lesion was observed in these two organs. This mare foaled again in 2015 (foal #2). No sample was available from the mare, but the presence of EqHV genome (FR-Eq84 Serum) was detected in serum from this foal.

In Case #2, EqHV was detected in both liver + lung necropsy sample and in the allantochorion (FR-Eq69 Liver-lung and FR-Eq69 Allanto in 2014) with a viral load of 2.4×10^7 and 1.5×10^4 copies/g, respectively. Necropsy and histological analyses concluded to the lack of lesion reminding a viral infection; no overt lesion was observed in these two organs. The mare (FR-Eq74) was not tested at the time of foaling but 10 months later and the serum sample was found positive (7.7×10^7 copies/mL).

In Case #3, 3.9×10^4 copies/g of EqHV genome (FR-Eq70 in 2015) was detected in the allantochorion from a mare that foaled normally. Necropsy and histological analyses of the allantochorion concluded to a lack of histopathological signs of a viral infection; no obvious lesion was observed. Two months later, serum from the foal was tested negative while the maternal serum sample was still weakly positive with a viral load of 1.8×10^3 copies/mL (FR-Eq85 in 2015).

Overall, the viral load measured in serum samples (1.8 to 7.7×10^7 copies/mL) or liver + lung biopsies (2.3 to 2.4×10^7 copies/g) were about 10^3 higher than in allantochorion (1.5 to 3.9×10^4 copies/g).

3.2. Vertical Transmission of EqHV from Mare to Foal

The 5′UTR, NS3, and NS5B regions were sequenced in all EqHV positive samples, with the exception of the two allantochorions, in which amplification of NS3 region failed.

Sequences from amplified NS5B region were obtained for the 7 samples reported in Table 1 and were incorporated in a phylogenetic tree including 46 other strains: (i). 20 that were selected to obtain a representative selection of EqHV strains from France in space and time according to Pronost et al. (2016, [2]); or, (ii). 26 samples from other countries (Supplementary Table S1) representing EqHV genetic diversity in the world so far. Albeit indirectly in Case #1, the NS5B phylogenetic tree strongly suggests a link between EqHV strains in samples from three mares and their respective foals or allantochorion (Figure 2). All EqHV strains from Cases #1 to #3 belong to subtype 1 [2] and these data were confirmed with the median joining network analysis (Supplementary Figure S1).

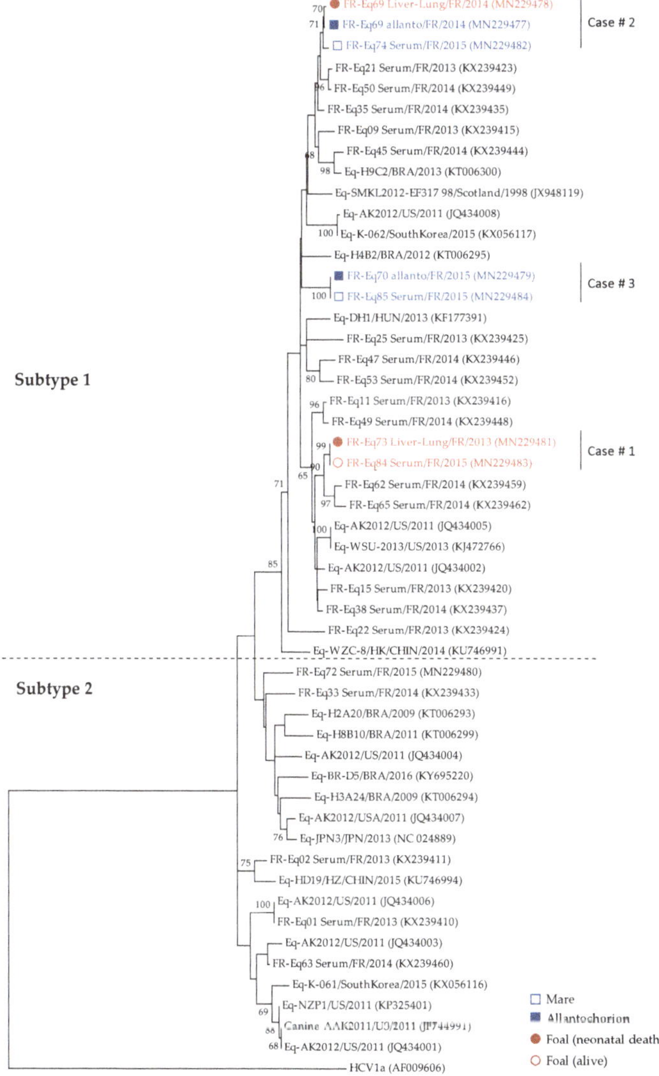

Figure 2. Phylogenetic analysis of equine hepacivirus NS5B sequences identified in horses. Neighbor-joining tree of partial nucleotide sequences from NS5B (259 bp) and corresponding region of a hepatitis C virus (HCV) genome (genotypes 1a). The tree was constructed with Jukes–Cantor model. A bootstrap was performed with a replicate rate of 500 (values ≥ 65% shown on branches).

For Case #1, no sample was directly available from the mare, but samples from her foals #1 (liver + lung) and #2 (serum) were analyzed; the two EqHV strains identified (FR-Eq73 and FR-Eq84) clustered with two other strains (FR-Eq62 and FR-Eq65). Alignment of the 260-bp NS5B sequences (Figure 3A) and the concatenated sequences (5′UTR + NS3 + NS5B; Figure 4A) show a total homology between the two strains. Sequence homologies with the two strains (FR-Eq62 and FR-Eq65) clustering in the phylogenetic tree (Figure 2) vary from 94.2% (15 nt difference) to 95.4% (12 nt) in the NS5B region and from 97% (18 nt) to 97.7% (14 nt) in the concatenated region. As a comparison, NS5B homology between French strains (with FR-Eq69 Liver-lung as reference strain) and strains isolated in other countries varied from 79% to 100%.

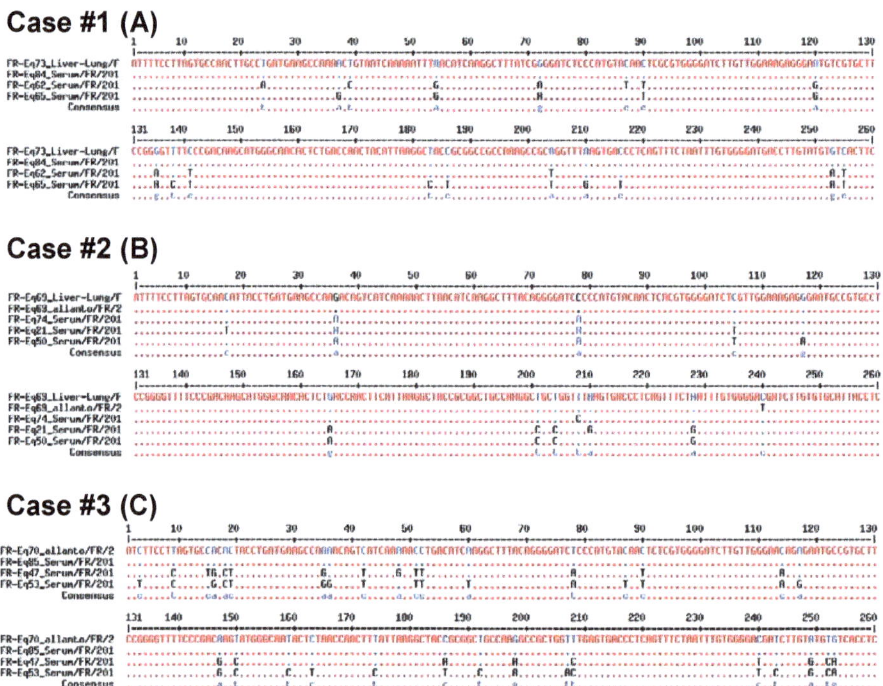

Figure 3. Alignment of NS5B sequences for samples Case #1 (**A**), Case #2 (**B**) and Case #3 (**C**) with the closest sequences identified by phylogenetic analysis (cf. Figure 2).

For Case #2, three different samples were analyzed: liver and lung biopsies from the foal (FR-Eq69 Liver-lung), the allantochorion (FR-Eq69 Allanto) and the maternal serum (FR-Eq74 Serum) drawn 10 months after foaling. Phylogenetic analysis evidenced only one base difference (99.6% homology) between the foal (liver + lung) and the allantochorion in the NS5B region (Figure 3B, nucleotide in position 240). No concatenated sequence was available for these two strains because NS3 sequencing failed for the allantochorion sample. These data suggest that these strains are closely related, since the number of discrepancies between analyzed regions is consistent with that of the mutations usually observed between isolates. Sequence homology with the strain obtained from the mare, ten months after abortion, varies from 98.8% (3 nt) in the NS5B region to 99% (6 nt) in the concatenated region when compared with the strain detected in the foal sample (FR-Eq69/liver-lung/FR/2014) (Figure 4B). Compared to this strain, the closest strains in the NS5B tree (FR-Eq21 and FR-Eq50; Figure 2) display a homology of respectively 96.5% (9 nt) and 96.9% (8 nt) in the NS5B region and 98.2% (11 nt) and 98.2% (11 nt) in the concatenated sequence (Figure 4B).

For Case #3, three samples were obtained: serum and allantochorion from the mare and serum from her foal. The EqHV sequences in the allantochorion (FR-Eq70) and the serum (FR-Eq85) samples were analyzed (the foal serum was negative). No differences were observed in the NS5B sequences (Figure 3C) nor in the concatenated 5′UTR+NS5B sequences (Figure 4C). Compared to the sequences of the two closest strains (FR-Eq47 and FR-Eq53), sequence homologies varied respectively from 91.5% (22 nt) to 88.1% (31 nt) in the NS5B region and from 92.6% (35 nt) to 91.1% (42 nt) in the concatenated region.

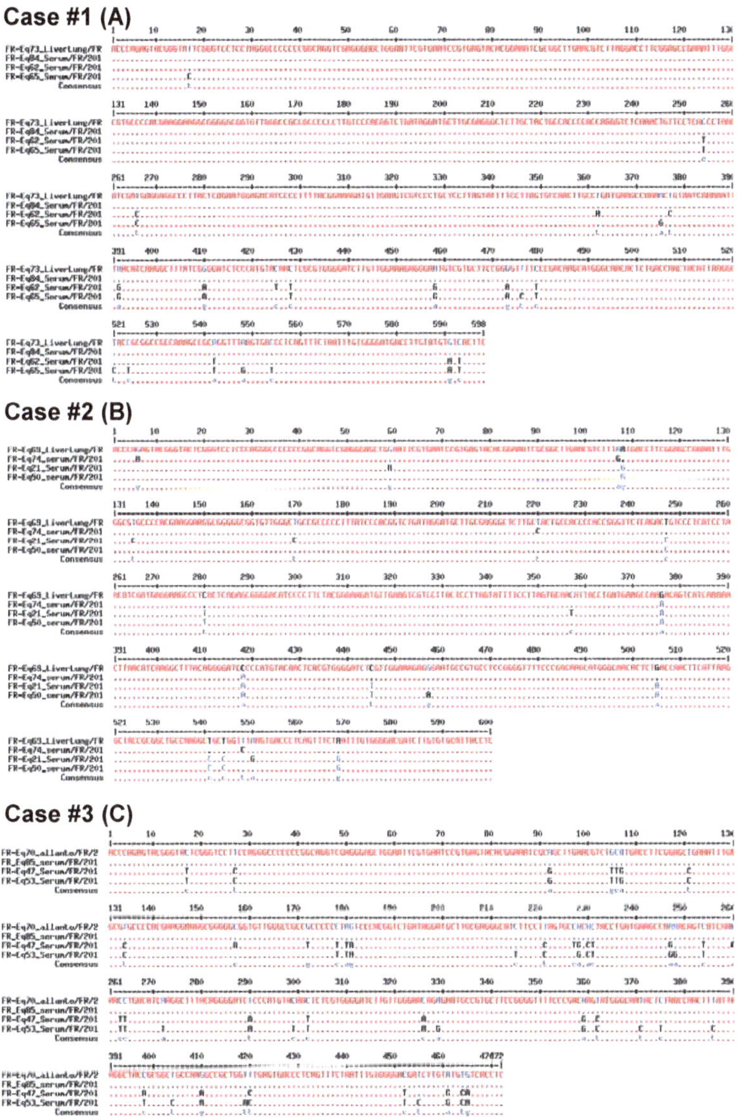

Figure 4. Alignment of concatenated sequences for samples Case #1 (**A**), Case #2 (**B**) and Case #3 (**C**) with those of the closest sequences identified by phylogenetic analysis (sequence numbers in Supplementary Table S1). Concatenated sequences were obtained from 5′UTR, NS5B, and NS3 sequences for A and B and from 5′UTR and NS5B for C.

4. Discussion

From a phylogenetic point of view, EqHV (or NPHV) is very closely related to HCV. Therefore, transmission paths reported for HCV in humans—or in chimpanzees prior to the international ban on experiments involving great apes—have also been partially investigated in horses (Figure 1). The parenteral transmission of NPHV was experimentally demonstrated by Ramsay et al. and by Scheel et al. [14,15]. The works of Postel et al. and, more recently, of Lu et al. described the presence of equine hepacivirus in biological products from horses [26–28]. These observations highlight the risk of contamination when these products are injected as is in horses, as observed until the end of the 1980s for HCV with human blood transfusion, leading to an epidemic that has infected up to 3% of the world's human population [29]. EqHV is also believed to be transmitted directly by blood, like reported for two more-distantly-related equine pegiviruses, TDAV (Theiler's disease-associated virus), and EPgV (equine pegivirus) [30]. Other hypotheses, previously described for other viruses, cannot be totally excluded, such as infections transmitted by mosquitos or, less unlikely, medical treatment with contaminated blood products or instruments [1,9,19,26,31]. Yet, plasma and antitoxin inoculations are unlikely to account for the high seroprevalence of NPHV and EPgV in horses [32], suggesting that other modes of transmission may exist.

In addition to the parenteral route, a mother-to-child (vertical) transmission has been observed in 5% of human hepatitis C cases [16,18]. In spite a high prevalence of EqHV infections worldwide [1–9], such occurrence appears infrequent in horses. One case of vertical transmission from a mare to her colt was reported in 2015 [19]. This study was carried out on 20 gestating mares, 4 of whom were infected with EqHV at the time of delivery. The presence of EqHV genome was also detected in umbilical cord blood and in the serum of one of the foals. To the authors' knowledge, this was the first report suggesting the possibility of in utero infection by EqHV. The rate of transversal infections was surprisingly high in the stud farm where this study took place; hence, we sought to further evaluate the prevalence of EqHV vertical transmissions in samples collected between 2013 and 2016 from several stud farms in France.

Among almost 400 cases of foal perinatal death, for which samples had been harvested in our laboratory, only three new cases of possible vertical transmission were identified, confirming an anticipated low incidence of new EqHV infections by this route. In Case #2, the presence of EqHV genome in three different biological compartments: organs of the foal (liver + lungs), the allantochorion, and the serum from the mare, could support this interpretation. The same 5'UTR sequences were obtained from organs of the foal and allantochorion, while only one-base difference was identified between NS5B sequences (FR-Eq69). The viral load in the allantochorion was very low and no NS3 sequence was obtained, but in our hands NS5B sequence is the most discriminating between strains [2]. Unfortunately, no blood sample was drawn from the mare at the time of delivery, but serum had been obtained 10 months later (FR-Eq74).

In HCV-infected individuals, quasispecies are defined as a group of similar-yet-not-identical viral genetic variants evolving at a rate between 0.8 and 2 milli-substitutions per nucleotide per site per year and overall presenting less than 5% nucleotide difference between genomes [20,27,28]. However, Gather et al. found only one-nucleotide change between EqHV genomes in maternal serum, umbilical cord blood and serum from her foal [19]. This could result from a relatively short delay between EqHV transmission and foal delivery. Alternatively, EqHV quasispecies in horses could drift at a slower pace than HCV quasispecies in humans. Therefore, in Case #2, it is far from clear whether the delay in the mare's sampling would entirely explain a 5-nucleotide difference with her foal's samples.

However, the most compelling argument for in utero transmission comes from the high viral loads detected in tissue samples from the neonate foal. First, it is highly improbable that a transmission upon delivery would produce so many genome copies in only two days [14,15]. Second, a simple contamination of the foal samples by maternal blood is also very unlikely, both during gestation and upon delivery. Thus, in the latter case, the virion concentration of maternal origin should be much more diluted than observed. In the former case, allantochorion anatomy is such that antibodies barely cross the chorionic barrier, if at all [33]; let alone viral particles of probably 50–70 nm in diameter. This comes in contrast to women, whose syncytiochorial placenta is bathed in maternal blood with potentially easy transfer of related HCV (and virus specific antibodies) from the maternal circulation to the syncytiotrophoblast. On the contrary, the equine epitheliochorial placenta has six layers of maternal and fetal tissues between the two blood circulations; hence, EqHV would be unlikely crossing this barrier, unless a receptor required for viral entry or acting as virion carrier was expressed on the allantochorion [34]. Were it nevertheless the case, given the worldwide prevalence of EqHV infections in mares, the number of neonate foals contaminated should be much higher than what is observed. Therefore, unless a hypothetical receptor isoform is involved, variants of the virus could promote EqHV in utero transmission. Finally, if the virus replicated within the allantochorion itself, as our results suggest, viral loads in the fetus would not any longer result from a contamination by maternal virions. Instead, the allantochorion would become a likely source of transmission to the fetus. The most likely explanation for Case #2 is, therefore, an in utero transmission of EqHV to the fetus.

In utero transmission to the fetus is also supported by the results obtained in Case #1. As no maternal sample was available, a direct link could not be established between EqHV identified in the three animals. Yet, the sequences of two EqHV-positive samples, one from a dead foal (liver + lung), born in 2013, and one from an apparently-healthy foal (serum), born later in 2015, were genetically clustered (FR-Eq73 and FR-Eq84). A 100% sequence homology was even found between the two isolates, which strongly suggested a similar source of transmission over a two-year period, here from the mare to her two offspring. In Case #3, the presence of a low EqHV viral load in allantochorion was detected, which displayed a 100% sequence homology with that identified in the mare's serum (5'UTR and NS5B region). A serum sample from her foal could be analyzed only two months after foaling, with a negative result. As the virus load in allantochorion was low, a hypothesis is that the foal cleared its infection within two months. The mare also presenting with a weak viral load upon delivering was perhaps clearing her own infection; if so, her colostrum probably contributed to the foal's quick recovery. In the absence of serological data, this cannot be confirmed, but a link between virus load and the risk of fetal infection has recently been discussed for HCV [35].

Our data are in agreement with Gather et al. [19], who suggested for the first time the possibility of in utero transmission of EqHV. Yet, in their study, all four placentas recovered from the EqHV-positive mares were negative by quantitative RT-PCR. The lack of EqHV RNA detection in placenta could result from a very low viremia, a low test sensitivity or the region of sampling given the anatomical heterogeneity of horse allantochorion [33]. In previous studies, we have developed a new EqHV quantitative RT-PCR method with specific primers designed on the basis of the first available equine NPHV genome [2]. Similarly, most studies, including that of Gather et al. [19], use primers from Burbelo et al. [1], initially designed to detect NPHV from different species. At parturition, placentitis was observed in one case, for which no EqHV genome was detected in the foal, suggesting to the authors that vertical transmission of NPHV occurred without an infection and inflammation of the placental tissue itself. In our study, the absence of histological inflammation in two placentas of EqHV-positive foals is in agreement with this interpretation, with the difference that it also establishes for the first time the presence of the virus genome in placenta. These findings suggest that, in the fetus like in adult horses, the presence of the virus is not associated to overt macroscopic or histological lesions, independently of the viral load. Nevertheless it takes sometimes several years before the consequences of what is at first considered a benign infection are identified. This outlines the importance of identifying as many transmission routes as possible.

Routes of transmission identified for other members of the *Flaviviridæ* family infecting horses and humans account for most observed infections. Thus, West Nile virus (WNV) is transmitted by mosquitoes; even if in rare occurrences, additional routes of transmission have been reported (for a review, see [36]). For example, the first case of in utero WNV transmission was reported in a woman in 2002 [37]. Abortion cases because of Japanese Encephalitis virus and severe Dengue infections were also reported after in utero transmission [38,39]. Mother-to-child transmission of WNV via breast milk has also been described (for a review, see [36]). Lastly, an epidemic of fetal microcephaly developing during Zika virus infections of pregnant women has raised serious concerns in several parts of the world [40]. Occupational exposure were also reported by different studies, which led to important safety implications for persons who work in these area.

Recently, our team completed a study on nasopharyngeal swabs and detected the presence of EqHV genomes in 4 of the 93 samples analyzed [41]. It did not enable establishing whether the transmission involved the respiratory tract, yet pointed at a possible role of the oropharyngeal sphere, as recently suggested by Altan et al. with the detection of EqHV genomes in a pool of four swabs [29]. Other works aim to elucidate the chain of transmission, such as a vector (e.g., mosquitoes) or, even if not yet reported, sperm during insemination or natural breeding. Additional studies are necessary to confirm or refute an involvement of these pathways during the transmission of the virus.

Supplementary Materials: The following are available online at http://www.mdpi.com/1999-4915/11/12/1124/s1. Table S1: Characteristics of the different strains used for the phylogenetic tree of NS5B sequences. Figure S1: Median joining network based on the same nucleotide sequences as for the NS5B phylogenetic tree (Figure 2).

Author Contributions: Here are described the contribution of the authors for this research article conceptualization, S.P. and E.S.H.; investigation, C.F. and E.S.H.; resources, C.F., M.F., J.T., and C.M.-P.; writing—original draft preparation, S.P., R.P., and E.S.H.; writing—review and editing, B.S., J.T., and R.P.; supervision, R.P. and S.P.; project administration, S.P., C.M.-P., and C.F.; funding acquisition, S.P. and P.-H.P.

Funding: This research was funded by the RESPE, LABÉO and CENTAURE European project co-funded by Normandy County Council, European Union in the framework of the ERDF-ESF operational program 2014–2020.

Acknowledgments: The authors thank the Virology and Molecular Biology Unit of the animal health department of LABÉO for its help in the conservation of the samples.

Conflicts of Interest: The authors declare no conflict of interest.

References

1. Burbelo, P.D.; Dubovi, E.J.; Simmonds, P.; Medina, J.L.; Henriquez, J.A.; Mishra, N.; Wagner, J.; Tokarz, R.; Cullen, J.M.; Iadarola, M.J.; et al. Serology-Enabled Discovery of Genetically Diverse Hepaciviruses in a New Host. *J. Virol.* **2012**, *86*, 6171–6178. [CrossRef] [PubMed]
2. Pronost, S.; Hue, E.; Fortier, C.; Foursin, M.; Fortier, G.; Desbrosse, F.; Rey, F.A.; Pitel, P.-H.; Richard, E.; Saunier, B. Prevalence of Equine Hepacivirus Infections in France and Evidence for Two Viral Subtypes Circulating Worldwide. *Transbound. Emerg. Dis.* **2016**, *64*, 1884–1897. [CrossRef]
3. Tanaka, T.; Kasai, H.; Yamashita, A.; Okuyama-Dobashi, K.; Yasumoto, J.; Maekawa, S.; Enomoto, N.; Okamoto, T.; Matsuura, Y.; Morimatsu, M.; et al. Hallmarks of Hepatitis C Virus in Equine Hepacivirus. *J. Virol.* **2014**, *88*, 13352–13366. [CrossRef] [PubMed]
4. Gemaque, B.S.; Junior Souza de Souza, A.; do Carmo Pereira Soares, M.; Malheiros, A.P.; Silva, A.L.; Alves, M.M.; Gomes-Gouvêa, M.S.; Pinho, J.R.R.; Ferreira de Figueiredo, H.; Ribeiro, D.B.; et al. Hepacivirus infection in domestic horses, Brazil, 2011–2013. *Emerg. Infect. Dis.* **2014**, *20*, 2180–2182. [CrossRef] [PubMed]
5. Lyons, S.; Kapoor, A.; Sharp, C.; Schneider, B.S.; Wolfe, N.D.; Culshaw, G.; Corcoran, B.; McGorum, B.C.; Simmonds, P. Nonprimate Hepaciviruses in Domestic Horses, United Kingdom. *Emerg. Infect. Dis.* **2012**, *18*, 1976–1982. [CrossRef]
6. Badenhorst, M.; Tegtmeyer, B.; Todt, D.; Guthrie, A.; Feige, K.; Campe, A.; Steinmann, E.; Cavalleri, J.M.V. First detection and frequent occurrence of Equine Hepacivirus in horses on the African continent. *Vet. Microbiol.* **2018**, *223*, 51–58. [CrossRef]
7. Figueiredo, A.S.; Lampe, E.; do Espírito-Santo, M.P.; do Amaral Mello, F.C.; de Almeida, F.Q.; de Lemos, E.R.S.; Godoi, T.L.O.S.; Dimache, L.A.G.; dos Santos, D.R.L.; Villar, L.M. Identification of two phylogenetic lineages of equine hepacivirus and high prevalence in Brazil. *Vet. J.* **2015**, *206*, 414–416. [CrossRef]
8. Matsuu, A.; Hobo, S.; Ando, K.; Sanekata, T.; Sato, F.; Endo, Y.; Amaya, T.; Osaki, T.; Horie, M.; Masatani, T.; et al. Genetic and serological surveillance for non-primate hepacivirus in horses in Japan. *Vet. Microbiol.* **2015**, *179*, 219–227. [CrossRef]

9.	Pfaender, S.; Cavalleri, J.M.V.; Walter, S.; Doerrbecker, J.; Campana, B.; Brown, R.J.P.; Burbelo, P.D.; Postel, A.; Hahn, K.; Anggakusuma; et al. Clinical course of infection and viral tissue tropism of hepatitis C virus-like nonprimate hepaciviruses in horses. *Hepatology* **2015**, *61*, 447–459. [CrossRef]

10.	Scheel, T.K.H.; Simmonds, P.; Kapoor, A. Surveying the global virome: Identification and characterization of HCV-related animal hepaciviruses. *Antivir. Res.* **2015**, *115*, 83–93. [CrossRef]

11.	Pronost, S.; Fortier, C.; Hue, E.; Desbrosse, F.; Foursin, M.; Fortier, G.; Saunier, B.; Pitel, P.-H. Hépacivirus, pégivirus, TDAV: Une nouvelle triade de virus hépatiques chez le cheval? *Pratique Vétérinaire Équine* **2018**, *197*, 24–31.

12.	Badenhorst, M.; de Heus, P.; Auer, A.; Rümenapf, T.; Tegtmeyer, B.; Kolodziejek, J.; Nowotny, N.; Steinmann, E.; Cavalleri, J.-M.V. No Evidence of Mosquito Involvement in the Transmission of Equine Hepacivirus (Flaviviridae) in an Epidemiological Survey of Austrian Horses. *Viruses* **2019**, *11*, 1014. [CrossRef] [PubMed]

13.	Li, L.; Giannitti, F.; Low, J.; Keyes, C.; Ullmann, L.S.; Deng, X.; Aleman, M.; Pesavento, P.A.; Pusterla, N.; Delwart, E. Exploring the virome of diseased horses. *J. Gen. Virol.* **2015**, *96*, 2721–2733. [CrossRef] [PubMed]

14.	Ramsay, J.D.; Evanoff, R.; Wilkinson, T.E.; Divers, T.J.; Knowles, D.P.; Mealey, R.H. Experimental transmission of equine hepacivirus in horses as a model for hepatitis C virus. *Hepatology* **2015**, *61*, 1533–1546. [CrossRef] [PubMed]

15.	Scheel, T.K.H.; Kapoor, A.; Nishiuchi, E.; Brock, K.V.; Yu, Y.; Andrus, L.; Gu, M.; Renshaw, R.W.; Dubovi, E.J.; McDonough, S.P.; et al. Characterization of nonprimate hepacivirus and construction of a functional molecular clone. *Proc. Natl. Acad. Sci. USA* **2015**, *112*, 2192–2197. [CrossRef] [PubMed]

16.	Indolfi, G.; Easterbrook, P.; Dusheiko, G.; El-Sayed, M.H.; Jonas, M.M.; Thorne, C.; Bulterys, M.; Siberry, G.; Walsh, N.; Chang, M.-H.; et al. Hepatitis C virus infection in children and adolescents. *Lancet Gastroenterol. Hepatol.* **2019**, *4*, 477–487. [CrossRef]

17.	Benova, L.; Mohamoud, Y.A.; Calvert, C.; Abu-Raddad, L.J. Vertical Transmission of Hepatitis C Virus: Systematic Review and Meta-analysis. *Clin. Infect. Dis.* **2014**, *59*, 765–773. [CrossRef]

18.	Tovo, P.-A.; Calitri, C.; Scolfaro, C.; Gabiano, C.; Garazzino, S. Vertically acquired hepatitis C virus infection: Correlates of transmission and disease progression. *World J. Gastroenterol.* **2016**, *22*, 1382–1392. [CrossRef]

19.	Gather, T.; Walter, S.; Todt, D.; Pfaender, S.; Brown, R.J.P.; Postel, A.; Becher, P.; Moritz, A.; Hansmann, F.; Baumgaertner, W.; et al. Vertical transmission of hepatitis C virus-like non-primate hepacivirus in horses. *J. Gen. Virol.* **2016**, *97*, 2540–2551. [CrossRef]

20.	Rooney, J.R. *Autopsy of the Horse. Technique and Interpretation*; The Williams & Wilkins Company: Baltimore, MD, USA, 1970.

21.	Laugier, C.; Foucher, N.; Sevin, C.; Leon, A.; Tapprest, J. A 24-Year Retrospective Study of Equine Abortion in Normandy (France). *J. Equine Vet. Sci.* **2011**, *31*, 116–123. [CrossRef]

22.	Tamura, K.; Peterson, D.; Peterson, N.; Stecher, G.; Nei, M.; Kumar, S. MEGA5: Molecular Evolutionary Genetics Analysis Using Maximum Likelihood, Evolutionary Distance, and Maximum Parsimony Methods. *Mol. Biol. Evol.* **2011**, *28*, 2731–2739. [CrossRef] [PubMed]

23.	Madeira, F.; Park, Y.M.; Lee, J.; Buso, N.; Gur, T.; Madhusoodanan, N.; Basutkar, P.; Tivey, A.R.N.; Potter, S.C.; Finn, R.D.; et al. The EMBL-EBI search and sequence analysis tools APIs in 2019. *Nucleic Acids Res.* **2019**, *47*, W636–W641. [CrossRef] [PubMed]

24.	Leigh, J.W.; Bryant, D. Popart: Full-feature software for haplotype network construction. *Methods Ecol. Evol.* **2015**, *6*, 1110–1116. [CrossRef]

25.	Bandelt, H.J.; Forster, P.; Röhl, A. Median-joining networks for inferring intraspecific phylogenies. *Mol. Biol. Evol.* **1999**, *16*, 37–48. [CrossRef]

26.	Postel, A.; Cavalleri, J.-M.V.; Pfaender, S.; Walter, S.; Steinmann, E.; Fischer, N.; Feige, K.; Haas, L.; Becher, P. Frequent presence of hepaci and pegiviruses in commercial equine serum pools. *Vet. Microbiol.* **2016**, *182*, 8–14. [CrossRef]

27.	Lu, G.; Huang, J.; Yang, Q.; Xu, H.; Wu, P.; Fu, C.; Li, S. Identification and genetic characterization of hepacivirus and pegivirus in commercial equine serum products in China. *PLoS ONE* **2017**, *12*, e0189208. [CrossRef]

28.	Pfaender, S.; Walter, S.; Grabski, E.; Todt, D.; Bruening, J.; Romero-Brey, I.; Gather, T.; Brown, R.J.; Hahn, K.; Puff, C. Immune protection against reinfection with nonprimate hepacivirus. *Proc. Natl. Acad. Sci. USA* **2017**, *114*, E2430–E2439. [CrossRef]

29. Altan, E.; Li, Y.; Sabino-Santos, G., Jr.; Sawaswong, V.; Barnum, S.; Pusterla, N.; Deng, X.; Delwart, E. Viruses in Horses with Neurologic and Respiratory Diseases. *Viruses* **2019**, *11*, 942. [CrossRef]

30. Paim, W.P.; Weber, M.N.; Cibulski, S.P.; da Silva, M.S.; Puhl, D.E.; Budaszewski, R.F.; Varela, A.P.M.; Mayer, F.Q.; Canal, C.W. Characterization of the viral genomes present in commercial batches of horse serum obtained by high-throughput sequencing. *Biologicals* **2019**, *61*, 1–7. [CrossRef]

31. Pybus, O.G.; Thézé, J. Hepacivirus cross-species transmission and the origins of the hepatitis C virus. *Curr. Opin. Virol.* **2016**, *16*, 1–7. [CrossRef]

32. Divers, T.J.; Barton, M.H. Chapter 13: Disorders of the Liver. In *Equine Internal Medicine*, 4th ed.; Reed, S., Baylyd, W., Sellon, D., Saunders, Sellon, D., Eds.; Elsevier: Saint-louis, MO, USA, 2017; pp. 843–887, ISBN 978-0-323-44552-8.

33. LeBlanc, M.M. Immunologic considerations. In *Equine Clinical Neonatalogy*; Lea & Febiger: Philadelphia, PA, USA, 1990; pp. 275–294. ISBN 978-0-8121-1184-2.

34. Pfaender, S.; Walter, S.; Todt, D.; Behrendt, P.; Doerrbecker, J.; Wölk, B.; Engelmann, M.; Gravemann, U.; Seltsam, A.; Steinmann, J.; et al. Assessment of cross-species transmission of hepatitis C virus-related non-primate hepacivirus in a population of humans at high risk of exposure. *J. Gen. Virol.* **2015**, *96*, 2636–2642. [CrossRef] [PubMed]

35. Steininger, C.; Kundi, M.; Jatzko, G.; Kiss, H.; Lischka, A.; Holzmann, H. Increased risk of mother-to-infant transmission of hepatitis C virus by intrapartum infantile exposure to maternal blood. *J. Infect. Dis.* **2003**, *187*, 345–351. [CrossRef] [PubMed]

36. Gould, L.H.; Fikrig, E. West Nile virus: A growing concern? *J. Clin. Investig.* **2004**, *113*, 1102–1107. [CrossRef] [PubMed]

37. Centers for Disease Control and Prevention (CDC). Intrauterine West Nile virus infection—New York, 2002. *MMWR Morb. Mortal. Wkly. Rep.* **2002**, *51*, 1135–1136.

38. Chaturvedi, U.C.; Mathur, A.; Chandra, A.; Das, S.K.; Tandon, H.O.; Singh, U.K. Transplacental infection with Japanese encephalitis virus. *J. Infect. Dis.* **1980**, *141*, 712–715. [CrossRef] [PubMed]

39. Chye, J.K.; Lim, C.T.; Ng, K.B.; Lim, J.M.; George, R.; Lam, S.K. Vertical transmission of dengue. *Clin. Infect. Dis.* **1997**, *25*, 1374–1377. [CrossRef]

40. Khaiboullina, S.F.; Ribeiro, F.M.; Uppal, T.; Martynova, E.V.; Rizvanov, A.A.; Verma, S.C. Zika Virus Transmission Through Blood Tissue Barriers. *Front. Microbiol.* **2019**, *10*. [CrossRef]

41. Pronost, S.; Hue, E.; Fortier, C.; Foursin, M.; Fortier, G.; Desbrosse, F.; Rey, F.; Pitel, P.-H.; Saunier, B. Identification of equine hepacivirus infections in France: Facts and Physiopathological insights. *J. Equine Vet. Sci.* **2016**, *39*, S22. [CrossRef]

Article

Characterization of Equine Parvovirus in Thoroughbred Breeding Horses from Germany

Toni Luise Meister [1], Birthe Tegtmeyer [2], Yannick Brüggemann [1], Harald Sieme [3], Karsten Feige [3], Daniel Todt [1], Alexander Stang [1], Jessika-M.V. Cavalleri [4,*,†] and Eike Steinmann [1,*,†]

[1] Department of Molecular and Medical Virology, Faculty of Medicine, Ruhr-University Bochum, 44801 Bochum, Germany; Toni.meister@rub.de (T.L.M.); Yannick.brueggemann@rub.de (Y.B.); Daniel.todt@rub.de (D.T.); Alexander.stang@rub.de (A.S.)

[2] Institute for Experimental Virology, TWINCORE Centre for Experimental and Clinical Infection Research; a Joint Venture between the Medical School Hannover (MHH) and the Helmholtz Centre for Infection Research (HZI), 30625 Hannover, Germany; Birthe.tegtmeyer@twincore.de

[3] Institute of Virology; University of Veterinary Medicine Hannover, 30559 Hannover, Germany; Harald.sieme@tiho-hannover.de (H.S.); Karsten.feige@tiho-hannover.de (K.F.)

[4] Department for Companion Animals and Horses, University of Veterinary Medicine, 1210 Vienna, Austria

* Correspondence: Jessika.cavalleri@vetmeduni.ac.at (J.-M.V.C.); Eike.steinmann@rub.de (E.S.); Tel.: +49-234-32-23189 (E.S.)

† These authors contributed equally to the work.

Received: 4 October 2019; Accepted: 17 October 2019; Published: 18 October 2019

Abstract: An equine parvovirus-hepatitis (EqPV-H) has been recently identified in association with equine serum hepatitis, also known as Theiler's disease. The disease was first described by Arnold Theiler in 1918 and is often observed with parenteral use of blood products in equines. However, natural ways of viral circulation and potential risk factors for transmission still remain unknown. In this study, we investigated the occurrence of EqPV-H infections in Thoroughbred horses in northern and western Germany and aimed to identify potential risk factors associated with viral infections. A total of 392 Thoroughbreds broodmares and stallions were evaluated cross-sectionally for the presence of anti-EqPV-H antibodies and EqPV-H DNA using a luciferase immunoprecipitation assay (LIPS) and a quantitative PCR, respectively. In addition, data regarding age, stud farm, breeding history, and international transportation history of each horse were collected and analysed. An occurrence of 7% EqPV-H DNA positive and 35% seropositive horses was observed in this study cohort. The systematic analysis of risk factors revealed that age, especially in the group of 11–15-year-old horses, and breeding history were potential risk factors that can influence the rate of EqPV-H infections. Subsequent phylogenetic analysis showed a high similarity on nucleotide level within the sequenced Thoroughbred samples. In conclusion, this study demonstrates circulating EqPV-H infections in Thoroughbred horses from central Europe and revealed age and breeding history as risk factors for EqPV-H infections.

Keywords: equine parvovirus-hepatitis; Germany; risk factors; transmission

---

1. Introduction

Equine serum hepatitis (i.e., Theiler's disease (TD)) is a serious and potentially life-threatening disease and one of the most common causes of acute hepatitis and liver failure in horses [1]. Specific treatment options are still lacking. TD was first reported in 1918 by Sir Arnold Theiler, after he observed signs of liver disease in animals vaccinated against African horse sickness with a combination of live virus and equine antiserum [2]. Similar to historical outbreaks of human posttransfusion hepatitis, multiple outbreaks of Theiler´s disease have been observed following parenterally administered equine serum products [3–6]. An incidence between 1.4%–18% of fulminant hepatitis among horses receiving an equine biological product has been reported [2,7]. Given the association between prior treatment with equine serum/plasma and the appearance of Theiler's disease an etiologic role for a contaminating toxin or infectious agent has been suggested [8]. However, the exact pathogenic agent remained unknown for nearly a century.

Parvoviridae comprise a large family of non-enveloped DNA-viruses, which is currently subdivided into eight genera collectively known as parvoviruses. Members of this family have been described to infect a wide array of hosts, including humans, domestic, and wild animals [9–11]. Parvovirus infections have been associated with various severe and fatal diseases affecting the respiratory, gastrointestinal, and haematological systems and further potentially causing abortions [6,12–14]. Most recently, a novel equine Parvovirus (equine parvovirus-hepatitis virus (EqPV-H)) was isolated from serum and liver tissue of a horse that died of TD following administration of tetanus antitoxin (TAT) [15]. Administration of TAT contaminated with EqPV-H further resulted in seroconversion and acute hepatitis in experimentally infected horses, indicating that EqPV-H might be the causative agent of TD [15,16]. A recent study further reported a high prevalence of EqPV-H among commercial equine serum pools, indicating the necessity of careful risk assessment for medical and research applications [17]. However, despite its association with equine diseases, EqPV-H has not gained much attention of equine veterinarians and its worldwide prevalence and epidemiology remain poorly investigated.

Here, we examined the prevalence of EqPV-H among Thoroughbred breeding horses in northern and western Germany to identify potential risk factors for EqPV-H infections. A total of 392 serum samples from Thoroughbred broodmares and stallions were analysed for the presence of anti-EqPV-H antibodies, DNA, and viral sequences, respectively. Furthermore, an analysis of risk factors potentially affecting the prevalence of EqPV-H infections was performed to investigate natural routes of virus transmission.

2. Materials and Methods

2.1. Serum Sample Collection

A total of 392 serum samples from Thoroughbreds stabled on stud farms in northern and western Germany (Lower Saxony, North Rhine-Westphalia, Hamburg, Schleswig-Holstein) representing more than 25% of all registered breeding horses at The German Thoroughbred Studbook Authority (Cologne) were collected and processed between 2012 and 2015 [18]. All samples were then stored at −80 °C until further analysis regarding the presence of EqPV-H.

2.2. Detection of EqPV-H DNA

Viral DNA was extracted with a viral DNA Kit from Qiagen (Cat. No. 1048147, Hilden, Germany) according to the manufacturer's recommendations. DNA samples were stored at −20 °C until further analysis. A probe-based quantitative polymerase chain reaction (qPCR) was used with primers and probe designed and provided by Dr. Amit Kapoor as described previously [15]. A serial dilution of a plasmid containing the EqPV-H VP1 sequence was generated as standard row for the quantification of EqPV-H within the samples tested. qPCR measurements were performed using the LightCycler 480 real-time PCR system (Roche, Mannheim, Germany).

2.3. Detection of Anti-EqPV-H Antibodies

Samples were analysed for the presence of anti-EqPV-H-VP1 antibodies using the previously described luciferase immunoprecipitation system (LIPS) [19–21]. The EqPV-H-LIPS antigen VP1 was produced as described by Divers et al. [15]. Following the LIPS assay, relative light units (RLU) were determined using a plate luminometer (LB 960 XS3; Berthold, Bad Wildbad, Germany). To calculate sensitivity the mean RLU plus three standard deviations (SD) of an EqPV-H negative horse serum was defined as a cut-off limit. A potential cross-reactivity between the LIPS and other related parvoviruses could not be excluded.

2.4. Data Collection and Study Design

Three different groups regarding the state of EqPV-H infection were distinguished: Seropositive and EqPV-H DNA positive (DNA$^+$/AB$^+$), seronegative and EqPV-H DNA negative (DNA$^-$/AB$^-$), and seropositive and EqPV-H DNA negative (DNA$^-$/AB$^+$). Information regarding gender, age, state of reproduction, breeding, and international transportation history of the study population were received earlier from the Association for breeding and racing of Thoroughbreds (Cologne, Germany). Furthermore, the established criteria were subdivided into different groups. Age groups of similar size were created: 3–6-year-old horses ($n = 78$), 7–10-year-old horses ($n = 113$), 11–15-year-old horses ($n = 129$), and 16–29-year-old horses ($n = 72$). Similarly, horses were assigned to groups regarding the breeding history (0–4 breeding years) and the stock size of the stud farms (1–9 horses, 10–39 horses and >40 horses). Additional information about the transportation history was sorted according to the target country (France, Great Britain, Ireland, Germany, etc.). The data were determined by the time of sample collection between 2012–2015.

2.5. Sequencing and Phylogeny

For sequence analysis, a PCR (I) was designed within the NS1 of EqPV-H [17]. PCR was performed using the expand high fidelity PCR system (Roche Diagnostics, Basel, Switzerland) as described before [17]. PCR products were visualised on a 2% agarose gel, excised, and purified using a Monarch® DNA gel extraction kit (New England Biolabs, Ipswich, Massachusetts, United States). Purified products were then sent for Sanger sequencing using the applicable PCR primers. Highlighter plot analysis [22] displaying nucleotide exchanges in EqPV NS1 in the screened cohort compared to a previously published strain from Europe (MK792434). Labelling of nucleotides according to IUPAC code. Length of bar scale of neighbour-joining tree indicating number of nucleotide exchanges. The input multiple sequence alignment was created with Mega X.

3. Results

3.1. Frequent Occurrence of EqPV-H among German Thoroughbreds

We first investigated the frequency of EqPV-H infections among Thoroughbreds from the north and west of Germany. A total of 392 serum samples from Thoroughbreds originating from Lower Saxony and North Rhine-Westphalia were collected during an annual fertility monitoring and tested for the presence of anti-EqPV-H-VP1 antibodies and EqPV-H DNA, respectively (Figure 1A). A total of 28 samples were tested positive for EqPV-H DNA (7.14%) via qPCR. We further evaluated the samples for the presence of anti-VP1-antibodies using a previously described LIPS [17]. An elevated serum titre of anti EqPV H VP1 antibodies above the detection limit was present in 136 (34.69%) horses. Based on these findings the individual horses were assigned to the following groups: DNA$^-$/AB$^-$; DNA$^-$/AB$^+$ and DNA$^+$/AB$^+$ (Figure 1B). Interestingly, no sample was positive for EqPV-H DNA only, indicating no acute EqPV-H infections within the examined cohort.

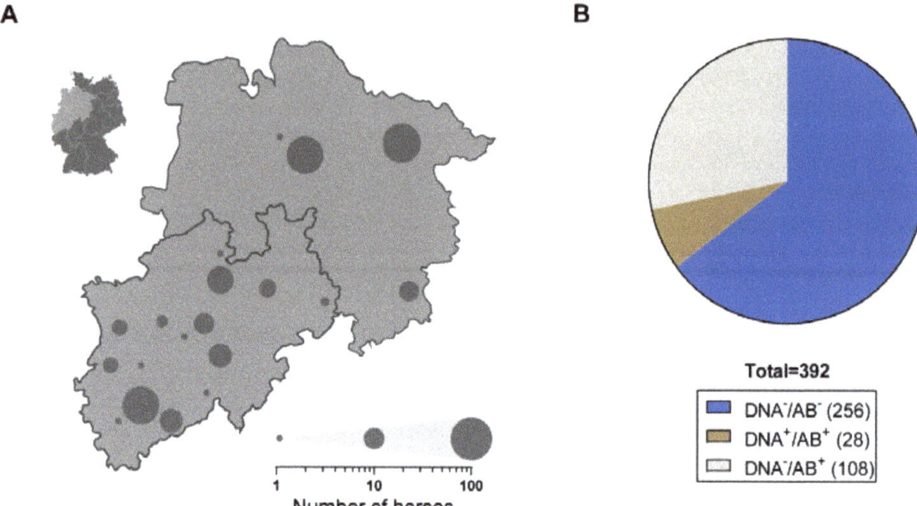

Figure 1. (**A**) Map of Germany highlighting the location of sampling (insert upper left). Serum samples were collected from 392 Thoroughbreds in North Rhine-Westphalia and Lower Saxony. Grey circles highlight the sample location and circle size is scaled to relative the number of examined horses (see legend lower right). (**B**) Serum samples were analysed for the presence of anti-equine parvovirus-hepatitis (anti-EqPV-H) VP1 AB and EqPV-H DNA performing luciferase immunoprecipitation system (LIPS) and qPCR, respectively. Individual horses were assigned to three different groups: Seronegative and EqPV-H DNA negative (DNA$^-$/AB$^-$; 65.3%), seropositive and EqPV-H DNA negative (DNA$^-$/AB$^+$; 27.55%), and seropositive and EqPV-H DNA positive (DNA$^+$/AB$^+$; 7.14%).

3.2. Viral Characteristics of EqPV-H-Positive Horses

Next, we characterized the distribution of viral loads in the study cohort showing three high-titre samples with copy numbers above 5×10^4 DNA copies/mL (Figure 2A). For the anti-EqPV-H-VP1 antibodies, titres up to a 600-fold increase over the limit of detection were observed (Figure 2B). Interestingly, DNA$^+$/AB$^+$ samples showed significantly higher titres of anti-VP1-antibodies as compared to the DNA$^-$/AB$^+$ horses (Figure 2C). The 28 EqPV-H positive samples were further confirmed in a gel electrophoresis of the qPCR products (Figure 2D). The highest viral loads were detected for samples 13, 14, 15, 20, and 23, respectively (Figure 2E). The highest anti-VP1-antibody levels were observed for samples 10, 15, 16, 18, and 23 (Figure 2F).

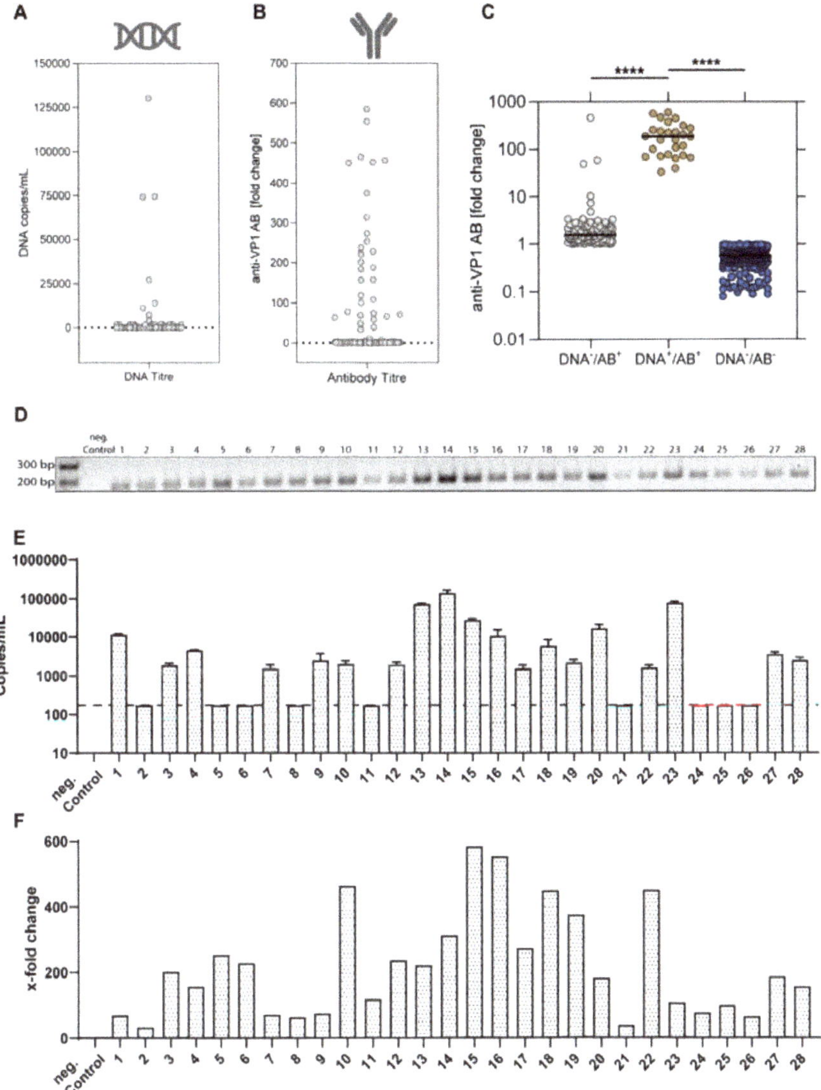

Figure 2. (**A**) Viral loads of EqPV-H (DNA copies/mL) were determined via qPCR ($n = 392$). Horses with viral load of >175 copies/mL were considered EqPV-H DNA positive. (**B**) EqPV-H VP1 antibodies were detected using a LIPS assay and the relative increase of RLU compared to an EqPV-H-negative control sample was calculated ($n = 392$). (**C**) Comparison of relative EqPV-H VP1 antibody levels between different groups: (Median values; DNA⁻/AB⁻, $n = 256$; DNA⁻/AB⁺, $n = 108$; DNA⁺/AB⁺, $n = 28$). Statistical significance was determined using a one-way ANOVA with Dunnett's post-hoc test (**** $p < 0.0001$). (**D**) Agarose gel electrophoresis of the qPCR product from DNA positive serum samples ($n = 28$) and an EqPV-H-negative control sample. (**E**) Viral loads of EqPV-H as determined by qPCR are displayed in DNA copies/mL. The dotted line indicates the limit of detection (mean +/− SD, $n = 3$). (**F**) Evaluation of serum titres for anti-EqPV-H antibodies in EqPV-H DNA positive serum samples determined by LIPS. RLU was normalised to the cut-off limit and is displayed as x-fold change.

3.3. Age and Breeding History are Potential Risk Factors of EqPV-H Infection in Thoroughbreds

We next performed an analysis using data on age, breeding history, stock size of the stud farm, and transportation to a foreign country to determine factors, which might be involved in promoting EqPV-H and would identify natural routes of transmission. All results of the descriptive analysis are shown in Figure 3, with detailed information provided in Table 1. By analysing the impact of the age on EqPV-H infection, it became apparent that with increasing age, especially between 11–15 years, the proportion of EqPV-H DNA positive and seropositive horses [DNA$^+$/AB$^+$] was elevated (Figure 3A). In the group of 3–6-year-old horses approximately 14% were tested DNA positive and seropositive as compared to 43% for the group of 11–15 and 25% for the group of 16–29-year-old horses (Figure 3A). Likewise, a slight increase in the fraction of seropositive horses was observed upon increasing breeding history (Figure 3B). Regarding the classification of the stud farm based on their stock size, no difference between the different groups could be noted (Figure 3C). More horses were recruited from farms with more than 40 horses (Table 1). Most of the sampled and analysed horses stayed in Germany and other countries including Great Britain (n = 27), Ireland (n = 33), and France (n = 26). Importantly, travelling history to another country was not associated with an increased risk for EqPV-H infection (Figure 3D and Table 1).

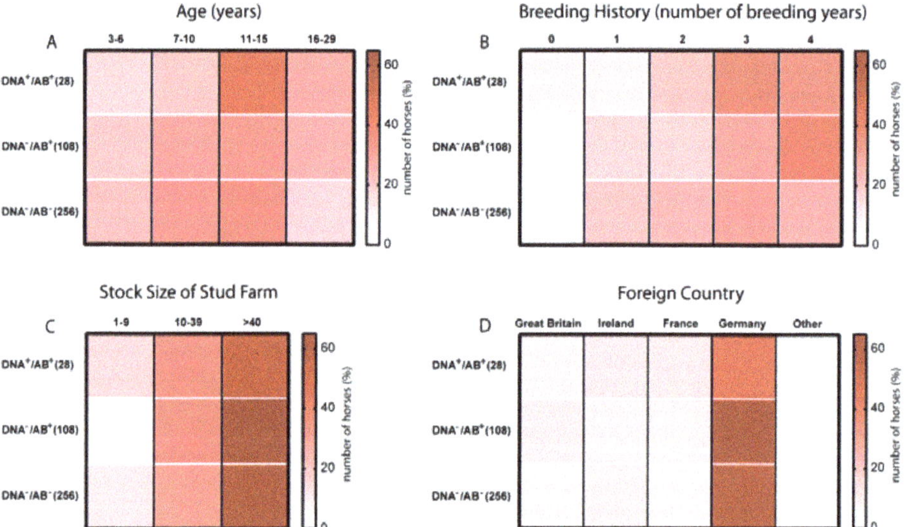

Figure 3. Heatmaps displaying the occurrence of anti-EqPV-H VP1 AB and EqPV-H DNA with regard to potential risk factors. The three groups based on the EqPV-H infection status were further classified based on (**A**) age, (**B**) breeding history, (**C**) stock size of the stud farm, and (**D**) foreign country. The colouring indicates the number of horses (in %) classified in the specific category. Each row adds up to 100%.

Table 1. General information regarding gender, age, breeding history, stock size, travel background, and EqHV-coinfection of the equine serum samples collected from Thoroughbreds in North Rhine-Westphalia and Lower Saxony.

Variables (*n*)	State of EqPV-H Infection					
	DNA$^-$/AB$^+$ (108)		DNA$^+$/AB$^+$ (27) *		DNA$^-$/AB$^-$ (256)	
	n	%	*n*	%	*n*	%
GENDER **						
Mare (380)	106	98.15	26	92.86	248	96.88
Stallion (10)	2	1.85			8	3.12
AGE						
3–6 years (78)	20	18.52	4	14.29	54	21.09
7–10 years (113)	30	27.78	5	17.86	78	30.47
11–15 years (129)	32	29.6	12	42.86	85	33.20
16–29 years (72)	26	24.07	7	25	39	15.23
NUMBER OF BREEDING YEARS						
Not covered (9)	2	1.85	2	7.14	5	1.95
1 breeding year (68)	17	15.74	2	7.14	49	19.14
2 breeding year (80)	17	15.74	5	17.86	58	22.66
3 breeding year (113)	29	26.85	10	35.71	74	28.91
4 breeding year (112)	41	37.96	9	32.14	62	24.22
STOCK SIZE						
1–9 horses (34)	4	3.70	4	14.29	26	10.16
10–39 horses (119)	35	32.41	9	32.14	75	29.30
>40 horses (239)	69	63.89	15	53.57	155	60.55
FOREIGN COUNTRY **						
Great Britain (27)	10	9.26	2	7.14	15	5.86
Ireland (33)	8	7.41	3	10.71	22	8.59
France (26)	8	7.41	3	10.71	15	5.856
Germany (227)	67	62.04	12	42.86	148	57.81
Other (4)	1	0.93	0	0	3	1.172
EqHV-COINFECTION **						
RNA$^-$/AB$^+$ (196)	60	55.57	12	44.45	124	48.44
RNA$^+$/AB$^+$ (41)	8	7.41	4	14.82	29	11.33
RNA$^-$/AB$^-$ (152)	40	37.04	10	37.04	102	39.84

* Data regarding potential risk factors could only be obtained for 27 of the 28 DNA$^+$/AB$^+$ horses. ** Data regarding gender, foreign country, and EqHV-coinfection could not be obtained for all the 392 horses.

In a recent study, we observed a frequent occurrence of equine hepacivirus (EqHV) infections among the same cohort of Thoroughbreds with approximately 62% seropositive horses for EqHV antibodies [18]. We surveyed whether an EqPV-H co-infection potentially favours an EqHV infection. However, a similar prevalence of approximately 60% seropositivity for EqHV was observed among all groups of horses, indicating that an EqHV infection does not predispose individual horses for an EqPV-H infection (Table 1 and Figure S1). Overall, we observed an age dependent state of EqPV-H infection and our results further suggest a slightly higher prevalence among horses with an extended breeding history. In contrast, no correlation between the stock size of the stud farm and transportation history to a foreign country could be observed indicating that neither acts as a predisposing factor for an EqPV-H infection.

3.4. Sequence and Phylogenetic Analysis of EqPV-H Detected in Thoroughbreds

We next performed a molecular characterization of the EqPV-H DNA positive samples. A PCR (I) was designed to amplify a region within the NS1 gene of EqPV-H and the obtained amplicons were subjected to conventional Sanger sequencing. The obtained sequences were submitted to the GenBank database with the accession numbers indicated in Table 2. Due to very low viral loads sequencing reactions could not be successfully performed for all samples (20/28 sequences for PCRI). The obtained sequences were highly similar to a previously described EqPV-H sequence (Figure 4). As indicated by the highlighter plot and the length of the tree branches, a high grade of genomic conservation is found within the analysed cohort. Sequence identity with the previously described European strain is ranging from 95% to 98% (Figure 4). In accordance with previous findings these results indicate a high degree of conservation and genomic stability between the world-wide circulating strains and low genetic variability of the EqPV-H strain.

Table 2. The newly identified specimens were submitted to the National Centre for Biotechnology Information (NCBI) and were assigned to the following accession numbers. The sample ID refers to Figure 2D–F.

Sample ID	EqPV Sequence Name	NCBI Accession Number
1	EqPV-H/VB1_518-1104nt	MN184860
3	EqPV-H/VB3_518-1104nt	MN184861
4	EqPV-H/VB4_518-1104nt	MN184862
10	EqPV-H/VB10_518-1104nt	MN184863
11	EqPV-H/VB11_518-1104nt	MN184864
17	EqPV-H/VB17_518-1104nt	MN184865
5	EqPV-H/VB5_518-1104nt	MN184866
19	EqPV-H/VB19_518-1104nt	MN184867
20	EqPV-H/VB20_518-1104nt	MN184868
23	EqPV-H/VB23_518-1104nt	MN184869
12	EqPV-H/VB12_518-1104nt	MN184870
14	EqPV-H/VB14_518-1104nt	MN184871
24	EqPV-H/VB24_518-1104nt	MN184872
25	EqPV-H/VB25_518-1104nt	MN184873
27	EqPV-H/VB27_518-1104nt	MN184874
15	EqPV-H/VB15_518-1104nt	MN184875

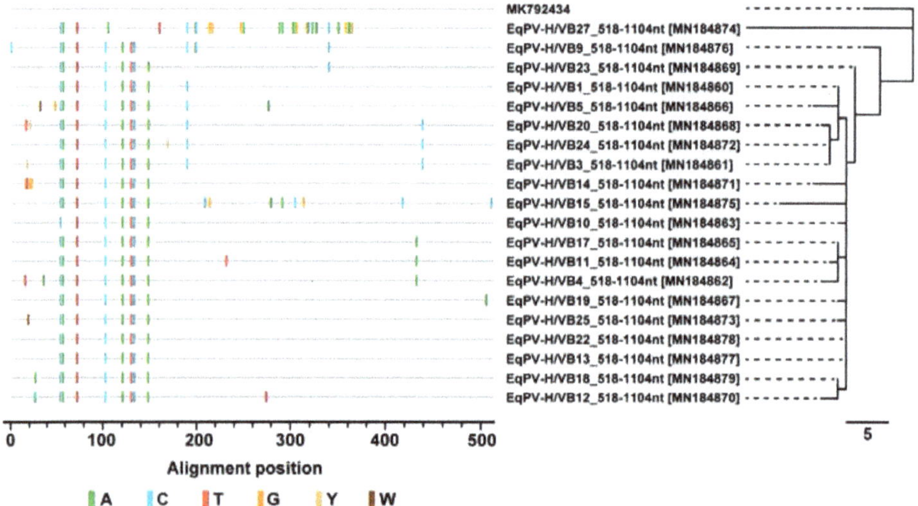

Figure 4. Highlighter plot displaying nucleotide exchanges in EqPV NS1 in the screened cohort compared to a previously published strain from Europe (MK792434). Labelling of nucleotides according to the International Union of Pure and Applied Chemistry (IUPAC) code. Length of bar scale of neighbour-joining tree indicating number of nucleotide exchanges. The input multiple sequence alignment was created with Mega X.

4. Discussion

In this cross-sectional study, we examined the prevalence of EqPV-H among almost 400 Thoroughbreds in northern and western Germany representing the first analysed European country. Based on the information from the German Association for breeding and racing of Thoroughbreds (Cologne), a total of 1450 brood mares and about 80 stallions were registered as breeding horses in 2014. Thus, we investigated about a quarter of the actively breeding Thoroughbreds in Germany and the results can be considered representative for the region. We observed a frequent occurrence of EqPV-H infections with 34.69% seropositive and 7.14% viraemic horses. The phylogenetic analysis demonstrated a high level of conservation in NS1 sequences in comparison with the world-wide circulating strains implying low levels of genetic variability. A detection of EqPV-H DNA with concurrent absence of anti-EqPV-H-VP1 antibodies was not observed for any sample, indicating no acute infection at the time of sampling. Of note, a cross-reaction in the LIPS assay with other potential parvoviruses in the horse samples cannot fully be excluded. Previously, in the first report describing the identification of EqPV-H a PCR prevalence in serum of 13% (13/100) was reported in the USA [15], and in China the average prevalence was 11.9% (17/143) in racehorses [11]. Further prevalence studies with larger cohorts from various continents are required for a more detailed epidemiology of EqPV-H.

We also performed a descriptive comparison between potential risk factors affecting EqPV-H prevalence. Iatrogenic transmission of EqPV-H has been demonstrated by Divers et al. through the administration of equine origin blood products [15], which should be confirmed ideally with a clonal viral EqPV-H inoculum in future studies. Of note, the virus can also be transmitted to horses that did not have such treatments [8]. The relative number of viraemic and seropositive horses in our analysis was increased with advanced age, especially between 11 and 15 years of age, but also in the group of 16–29-year-old horses compared to DNA$^-$/AB$^-$ and DNA$^-$/AB$^+$ horses. Possible reasons can be persistent humoral immunity or high risk of re-infection with longer living times. Due to the study design, horses younger than three years were not included. Therefore, it is impossible to draw a conclusion whether young foals in which the immune system is still maturing are at an increased risk of infection. Stage of reproduction was additionally accompanied by slightly higher fraction of EqPV-H infections. However, given that an extended breeding history mostly involves an advanced age of the examined horse this observation might be due to an indirect correlation. Neither stock size of the stud farm, transportation history to a foreign country, or an EqHV infection could be identified as a predisposing factor.

Of note, our previous investigation of the same study cohort revealed that younger Thoroughbreds transported abroad were at a higher risk to get infected with EqHV than horses at the same age staying in Germany. It is described that transportation of horses [23–25] and a changing environment as well as disruption of established social groups [26] can lead to a stress-induced immunosuppression, which could favour viral transmission. This risk factor seems therefore not to be highly involved in the susceptibility of horses to EqPV-H infections. Furthermore, equine hepacivirus co-infection was not a determinant for increased EqPV-H infections. Recently, a high rate of co-infections of these two equine viruses was also described in the prevalence report by Lu et al. from China [11]. As an additional way of viral spreading, we also investigated the potential of vertical transmission of EqPV-H in a single case. Postnatal serum samples were taken from a foal after delivery from an EqPV-H DNA positive and anti-VP1-positive mare (Figure S2). Transfer of EqPV-H-specific antibodies to the foal via colostrum could be observed, but no viremia was detected. The specific antibody transfer was expected, since the colostrum represents the important route of antibody transfer in neonatal foals, whereas the placental barrier hampers intrauterine transfer of macromolecules such as immunoglobulins from the mare to the foetus [27].

In conclusion, these data revealed that Thoroughbreds from central Europe have a frequent occurrence of anti-VP1 antibodies and EqPV-H DNA suggesting circulating EqPV-H infections in the horse population, endemic herds, or persistent shedding. In addition, age and multiple breeding were identified as potential risk factors for EqPV-H transmission.

Supplementary Materials: The following are available online at http://www.mdpi.com/1999-4915/11/10/965/s1, Figure S1: Correlation between EqPV-H and EqHV status of thoroughbreds. The EqPV-H status is displayed on the abscissa with the EqHV status shown in four shades of grey. Data for EqHV RNA and seroprevalence were obtained during a previous study [18]; Figure S2: Postnatal serum samples from a foal and an EqPV-H-positive mare were evaluated for the presence of EqPV-H VP1 antibodies and DNA at the indicated times using a luciferase immunoprecipitation (LIPS) assay and the relative increase of RLU compared to an EqPV-H-negative control sample was calculated.

Author Contributions: Conceptualization, J.-M.V.C. and E.S.; Methodology, T.L.M., B.T., H.S., D.T., and A.S.; Formal analysis, Y.B., H.S., K.F., D.T., A.S., J.-M.V.C., and E.S.; Investigation, T.L.M. and B.T.; Resources, H.S., K.F., J.-M.V.C., and E.S.; Data curation, T.L.M. and A.S.; Writing—original draft preparation, T.L.M., Y.B., and E.S.; Writing—review and editing, T.L.M., B.T., Y.B., H.S., K.F., D.T., A.S., J.-M.V.C., and E.S.; Visualization, T.L.M. and Y.B.; Supervision, J.-M.V.C. and E.S.; Project administration, J.-M.V.C. and E.S.

Funding: This research received no external funding.

Acknowledgments: We are grateful to Amit Kapoor for providing us with the qPCR protocol and also thank Peter D. Burbelo (NIH, Maryland, USA) for providing the Renilla-luciferase-VP1 fusion plasmid. We would like to thank all members of the Department of Molecular and Medical Virology at the Ruhr-University Bochum and especially Rosemarie Bohr, Regina Bütermann, Monika Kopytkowski, Klaus Sure, and Ute Wiegmann-Misiek for technical support. We additionally acknowledge support by the DFG Open Access Publication Funds of the Ruhr-Universität Bochum.

Conflicts of Interest: The authors declare no conflict of interest.

References

1. Sturgeon, B. Theiler's disease. *Vet. Rec.* **2017**, *180*, 14–15. [CrossRef] [PubMed]

2. Theiler, A. Acute Liver-Atrophy and Parenchymatous Hepatitis in Horses. In *5th and 6th Repts. of the Director of Veterinary Research*; Dept. of Agriculture, Union of South Africa: Pretoria, South Africa, 1918; pp. 7–164.

3. Panciera, R.J. Serum hepatitis in the horse. *J. Am. Vet. Med. Assoc.* **1969**, *155*, 408–410. [PubMed]

4. Guglick, M.A.; MacAllister, C.G.; Ely, R.W.; Edwards, W.C. Hepatic disease associated with administration of tetanus antitoxin in eight horses. *J. Am. Vet. Med. Assoc.* **1995**, *206*, 1737–1740. [PubMed]

5. Aleman, M.; Nieto, J.E.; Carr, E.A.; Carlson, G.P. Serum Hepatitis Associated with Commercial Plasma Transfusion in Horses. *J. Vet. Intern. Med.* **2005**, *19*, 120–122. [CrossRef] [PubMed]

6. Chandriani, S.; Skewes-Cox, P.; Zhong, W.; Ganem, D.E.; Divers, T.J.; van Blaricum, A.J.; Tennant, B.C.; Kistler, A.L. Identification of a previously undescribed divergent virus from the Flaviviridae family in an outbreak of equine serum hepatitis. *Proc. Natl. Acad. Sci. USA* **2013**, *110*, E1407–E1415. [CrossRef]

7. Tomlinson, J.E.; Tennant, B.C.; Struzyna, A.; Mrad, D.; Browne, N.; Whelchel, D.; Johnson, P.J.; Jamieson, C.; Löhr, C.V.; Bildfell, R.; et al. Viral testing of 10 cases of Theiler's disease and 37 in-contact horses in the absence of equine biologic product administration, A prospective study (2014–2018). *J. Vet. Intern. Med.* **2019**, *33*, 258–265. [CrossRef]

8. Tomlinson, J.E.; van de Walle, G.R.; Divers, T.J. What Do We Know About Hepatitis Viruses in Horses? The Veterinary clinics of North America. *Equine Pract.* **2019**, *35*, 351–362. [CrossRef]

9. Cotmore, S.F.; Agbandje-McKenna, M.; Chiorini, J.A.; Mukha, D.V.; Pintel, D.J.; Qiu, J.; Soderlund-Venermo, M.; Tattersall, P.; Tijssen, P.; Gatherer, D.; et al. The family Parvoviridae. *Arch. Virol.* **2014**, *159*, 1239–1247. [CrossRef]

10. Palinski, R.M.; Mitra, N.; Hause, B.M. Discovery of a novel Parvovirinae virus, porcine parvovirus 7, by metagenomic sequencing of porcine rectal swabs. *Virus Genes* **2016**, *52*, 564–567. [CrossRef]

11. Lu, G.; Sun, L.; Ou, J.; Xu, H.; Wu, L.; Li, S. Identification and genetic characterization of a novel parvovirus associated with serum hepatitis in horses in China. *Emerg. Microbes Infect.* **2018**, *7*, 170. [CrossRef]

12. Wong, F.C.; Spearman, J.G.; Smolenski, M.A.; Loewen, P.C. Equine Parvovirus: Initial Isolation and Partial Characterization. *Can. J. Comp. Med.* **1985**, *49*, 50–54. [PubMed]

13. Li, L.; Giannitti, F.; Low, J.; Keyes, C.; Ullmann, L.S.; Deng, X.; Aleman, M.; Pesavento, P.A.; Pusterla, N.; Delwart, E. Exploring the virome of diseased horses. *J. Gen. Virol.* **2015**, *96*, 2721–2733. [CrossRef] [PubMed]

14. Tomlinson, J.E.; Kapoor, A.; Kumar, A.; Tennant, B.C.; Laverack, M.A.; Beard, L.; Delph, K.; Davis, E.; Schott Ii, H.; Lascola, K.; et al. Viral testing of 18 consecutive cases of equine serum hepatitis, A prospective study (2014–2018). *J. Vet. Intern. Med.* **2019**, *33*, 251–257. [CrossRef] [PubMed]

15. Divers, T.J.; Tennant, B.C.; Kumar, A.; McDonough, S.; Cullen, J.; Bhuva, N.; Jain, K.; Chauhan, L.S.; Scheel, T.K.H.; Lipkin, W.I.; et al. New Parvovirus Associated with Serum Hepatitis in Horses after Inoculation of Common Biological Product. *Emerg. Infect. Dis.* **2018**, *24*, 303–310. [CrossRef]
16. Divers, T.J.; Tomlinson, J.E. Theiler's disease. *Equine Vet. Educ.* **2019**, *19*, 120. [CrossRef]
17. Meister, T.L.; Tegtmeyer, B.; Postel, A.; Cavalleri, J.-M.V.; Todt, D.; Stang, A.; Steinmann, E. Equine Parvovirus-Hepatitis Frequently Detectable in Commercial Equine Serum Pools. *Viruses* **2019**, *11*, 461. [CrossRef]
18. Reichert, C.; Campe, A.; Walter, S.; Pfaender, S.; Welsch, K.; Ruddat, I.; Sieme, H.; Feige, K.; Steinmann, E.; Cavalleri, J.M.V. Frequent occurrence of nonprimate hepacivirus infections in Thoroughbred breeding horses—A cross-sectional study for the occurrence of infections and potential risk factors. *Vet. Microbiol.* **2017**, *203*, 315–322. [CrossRef]
19. Burbelo, P.D.; Ching, K.H.; Klimavicz, C.M.; Iadarola, M.J. Antibody profiling by Luciferase Immunoprecipitation Systems (LIPS). *J. Vis. Exp. JoVE* **2009**, e1549. [CrossRef]
20. Burbelo, P.D.; Dubovi, E.J.; Simmonds, P.; Medina, J.L.; Henriquez, J.A.; Mishra, N.; Wagner, J.; Tokarz, R.; Cullen, J.M.; Iadarola, M.J.; et al. Serology-enabled discovery of genetically diverse hepaciviruses in a new host. *J. Virol.* **2012**, *86*, 6171–6178. [CrossRef]
21. Pfaender, S.; Cavalleri, J.M.V.; Walter, S.; Doerrbecker, J.; Campana, B.; Brown, R.J.P.; Burbelo, P.D.; Postel, A.; Hahn, K.; Anggakusuma Riebesehl, N.; et al. Clinical course of infection and viral tissue tropism of hepatitis C virus-like nonprimate hepaciviruses in horses. *Hepatol. (Baltim. Md.)* **2015**, *61*, 447–459. [CrossRef]
22. Keele, B.F.; Giorgi, E.E.; Salazar-Gonzalez, J.F.; Decker, J.M.; Pham, K.T.; Salazar, M.G.; Sun, C.; Grayson, T.; Wang, S.; Li, H.; et al. Identification and characterization of transmitted and early founder virus envelopes in primary HIV-1 infection. *Proc. Natl. Acad. Sci. USA* **2008**, *105*, 7552–7557. [CrossRef] [PubMed]
23. Grandin, T. Assessment of stress during handling and transport. *J. Anim. Sci.* **1997**, *75*, 249–257. [CrossRef] [PubMed]
24. Smith, B.L.; Jones, J.H.; Hornof, W.J.; Miles, J.A.; Longworth, K.E.; Willits, N.H. Effects of road transport on indices of stress in horses. *Equine Vet. J.* **1996**, *28*, 446–454. [CrossRef] [PubMed]
25. Schmidt, A.; Biau, S.; Möstl, E.; Becker-Birck, M.; Morillon, B.; Aurich, J.; Faure, J.-M.; Aurich, C. Changes in cortisol release and heart rate variability in sport horses during long-distance road transport. *Domest. Anim. Endocrinol.* **2010**, *38*, 179–189. [CrossRef]
26. Alexander, S.L.; Irvine, C.H. The effect of social stress on adrenal axis activity in horses: The importance of monitoring corticosteroid-binding globulin capacity. *J. Endocrinol.* **1998**, *157*, 425–432. [CrossRef]
27. Chucri, T.M.; Monteiro, J.M.; Lima, A.R.; Salvadori, M.L.B.; Kfoury, J.R.; Miglino, M.A. A review of immune transfer by the placenta. *J. Reprod. Immunol.* **2010**, *87*, 14–20. [CrossRef]

Communication

Identification of a Novel Equine Papillomavirus in Semen from a Thoroughbred Stallion with a Penile Lesion

Ci-Xiu Li [1], Wei-Shan Chang [1], Katerina Mitsakos [2], James Rodger [3], Edward C. Holmes [1,*] and Bernard J. Hudson [2]

[1] Marie Bashir Institute for Infectious Diseases and Biosecurity, Charles Perkins Centre, School of Life & Environmental Sciences and Sydney Medical School, The University of Sydney, Sydney, NSW 2006, Australia
[2] Royal North Shore Hospital, Clinical Microbiology & Infectious Diseases, Reserve Road, St Leonards, NSW 2065, Australia
[3] Vets & Veterinary Surgeons, Jerry Plains Veterinary Hospital, 10 Pagan Street, Jerry Plains, NSW 2330, Australia
* Correspondence: edward.holmes@sydney.edu.au; Tel.: +61-2-9351-5591

Received: 9 July 2019; Accepted: 1 August 2019; Published: 4 August 2019

Abstract: Papillomaviruses (PVs) have been identified in a wide range of animal species and are associated with a variety of disease syndromes including classical papillomatosis, aural plaques, and genital papillomas. In horses, 13 PVs have been described to date, falling into six genera. Using total RNA sequencing (meta-transcriptomics) we identified a novel equine papillomavirus in semen taken from a thoroughbred stallion suffering a genital lesion, which was confirmed by nested RT-PCR. We designate this novel virus *Equus caballus papillomavirus 9* (EcPV9). The complete 7656 bp genome of EcPV9 exhibited similar characteristics to those of other horse papillomaviruses. Phylogenetic analysis based on concatenated E1-E2-L2-L1 amino acid sequences revealed that EcPV9 clustered with EcPV2, EcPV4, and EcPV5, although was distinct enough to represent a new viral species within the genus *Dyoiotapapillomavirus* (69.35%, 59.25%, and 58.00% nucleotide similarity to EcPV2, EcPV4, and EcPV5, respectively). In sum, we demonstrate the presence of a novel equine papillomavirus for which more detailed studies of disease association are merited.

Keywords: equine papillomaviruses; horse; genital wart; phylogeny; evolution

---

1. Introduction

Equine papillomaviruses are small, non-enveloped viruses that comprise a circular and double-stranded DNA genome of up to approximately 8 kb in length. These viruses are associated with a variety of equine diseases including classical viral papillomatosis, genital papillomatosis, and aural and genital plaques. To date, 13 species of papillomavirus have been documented to infect the Equidae (horses and donkeys): *Bos taurus papillomavirus 1* (BPV1) [1], *2* (BPV2) [1], and *13* (BPV13) [2], *Equus caballus papillomaviruses 1–8* (EcPV1–8) [3–8], and *Equus asinus papillomaviruses 1–2* (EaPV1–2) [9]. These viruses are further classified into six genera based on the level of nucleotide sequence diversity in the L1 gene. Notably, these viruses have been isolated from a variety of lesions (aural plaques, genital masses, verrucous) and almost all seem to be associated with distinct pathologies (Table S1). Herein, we describe the detection of a novel papillomavirus in a 12-year-old thoroughbred Australian stallion using bulk RNA sequencing (meta-transcriptomics).

During the 2018 southern hemisphere serving season, the stallion experienced difficulty covering mares, primarily manifest as apparent pain on ejaculation. A wart-like lesion, 1 cm in circumference, was observed at the tip of the penis consistent with a genital papillomavirus lesion (Figure S1). Further endoscopy and ultrasound excluded neoplasia with no evidence of further internal lesions.

2. Materials and Methods

Sample Collection, RNA Sequencing and Virus Discovery

Two sets of urine and semen samples were collected from the stallion for microbiological investigation, one set placed into a standard specimen container and the other stored in RNA later. Because the nature of any causative pathogen was unknown, we employed a meta-transcriptomic approach as this is able to detect any microbial species (i.e., bacteria, eukaryotes, viruses) as long as sufficient expressed RNA is present. Accordingly, total RNA was extracted using the RNeasy plus universal kit (QIAGEN, Chadstone Centre, Victoria, Australia), with RNA sequencing libraries then constructed with the SMARTer Stranded total RNA-seq kit (TaKaRa, Clayton, Victoria, Australia). RNA sequencing of 100 bp pair-end libraries on the Illumina NovaSeq platform yielded 84.54 Gb of data (Table S2). All sequencing reads have been uploaded onto the NCBI Sequence Read Archive (SRA) database under BioProject PRJNA552109.

RNA sequencing reads were quality trimmed and horse reads were subsequently removed by mapping to the horse genome. To identify potential viral transcripts, non-horse reads from each library were compared against the non-redundant nucleotide (nt) and non-redundant protein (nr) databases using Blastn and Diamond blastx, respectively, with e-value thresholds of 1×10^{-10} and 1×10^{-4} [10], and were then annotated by taxonomy. Reads from the virus-positive library were *de novo* assembled using Megahit v1.1 [11,12]. Virus-associated contigs were extracted and assembled using Geneious 11.1.5 [13], followed by subsequent blast analysis against the NCBI nt database using BLASTn as further confirmation.

3. Results and Discussion

3.1. Identification of a Novel Equine Papillomavirus

A 7605 bp genome sequence of a papilloma-like virus was identified in one semen library. Prediction of open reading frames (ORFs) was performed using the ORF Finder tool at NCBI (https://www.ncbi.nlm.nih.gov/orffinder/). A conserved domain search (https://pave.niaid.nih.gov/#analyze/l1_taxonomy_tool) revealed that the L1 protein of the new virus exhibited the highest nucleotide and amino acid identities with EcPV2, at 69.35% and 70.44%, respectively, indicative of a novel papillomavirus. A novel *Equus caballus papillomavirus 8* (EcPV8) associated with viral plaques, viral papillomas, and squamous cell carcinoma has been recently described [14]. We therefore refer to the novel equine papillomavirus described here as *Equus caballus papillomavirus 9* (EcPV9, GenBank accession number MN117918), in accordance with current guidelines for the classification of papillomaviruses [15]. To obtain the full virus genome and to verify the sequence obtained from the deep sequencing and assembly processes, overlapping primers were designed and nested RT-PCR was performed. This resulted in the determination of a circular genome of 7656 bp in length. Remapping of the sequence reads from this library revealed a maximum coverage of 3419X (Figure 1), corresponding to an abundance of 152.97 RPM (reads mapped per million input reads).

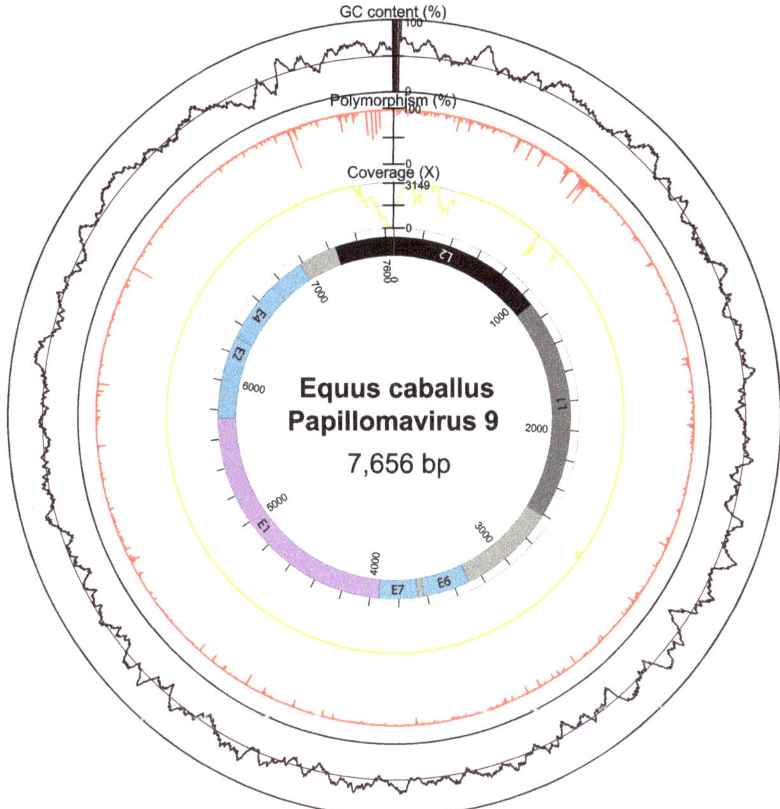

Figure 1. Genome organization of *Equus caballus papillomavirus 9* (EcPV9). The external circles of the metadata ring indicate the percentage GC content (brown), percentage nucleotide polymorphism (red), and read coverage (yellow). The inner gray circle represents the genome, with colored regions showing the predicted open reading frames (ORFs).

3.2. Genomic Properties and Evolutionary Relationships of EcPV9

The genome of EcPV9 has a GC content of 52.9% and the classic papillomavirus ORFs were identified, encoding five early (E1, E2, E4, E6, and E7) and two late (L1 and L2) proteins, consistent with other PVs (Figure 2). The predicted nucleotide and amino acid features are summarized in Figure 2 and Table S3. The noncoding region (NCR) framed by the L1 stop and the E6 start codons comprised 680 nt and exhibited one polyadenylation site (AATAAA) at nt 34. One E1 (TAGATCATTGTTAACAAC) (nt 580), two SP1 (GGCGGG) (nt 5021 and 5084), and three NF1 (CGGAA) (nt 2247, 3144 and 5728) binding sites were also predicted, although the AP1 (TGANTCA) binding site is absent (Figure 2; Table S3). In addition, 16 typical E2 binding sites were identified comprising 8 true consensus sequences (ACC-N6-GGT) and 8 putative consensus sequences (2 ACC-N4-GGT, 4 ACC-N5-GGT, 2 ACC-N7-GGT) (Figure 2; Table S3). Such A/T rich N regions are commonplace in E2.

Two zinc-binding domain(s) (CXXC-X29-CXXC) were found in E6 (nt 708 and 937; amino acids 10 and 85) and one in E7 (nt 1201; aa 50), separated by 29 amino acids (Figure S2). No PDZ binding domain (XS/TXV/L) was located at the C-terminus of the predicted EcPV9 E6 protein sequence (Figure S3), which has been reported as a characteristic feature of high risk (i.e., pathogenic) HPV types in comparison to low risk HPVs [16]. Notably, it was previously reported that a PDZ binding domain (XS/TXV/L) was located at the C-terminus of the predicted EcPV-2 E6 protein sequence [8], which was not observed here

(Figure S3). No putative pRB binding site (retinoblastoma tumor suppressor-binding domain) (LXCXE) was identified in the putative EcPV9 E7 protein, consistent with all equine and dyoiotapapillomaviruses determined to date [17,18], and the putative E4 protein showed a typical high proline content (12.8%, 18P/141 aa).

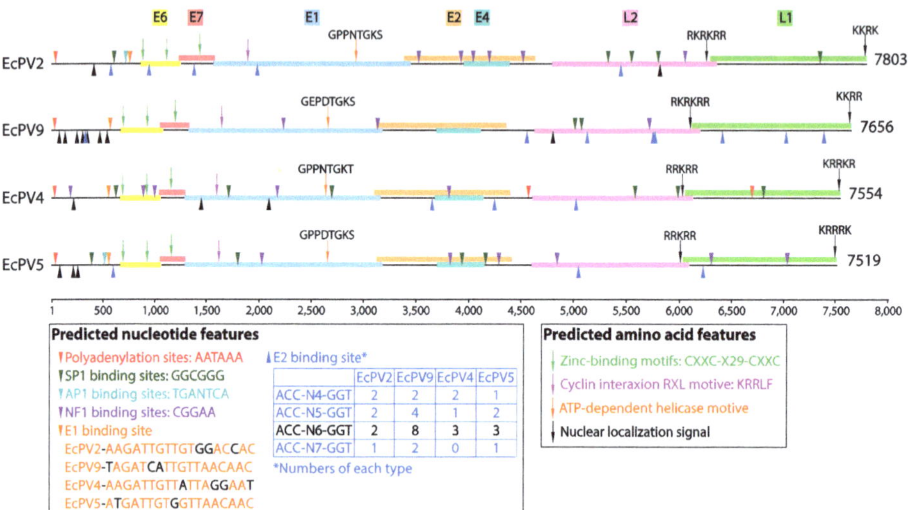

Figure 2. The genome of EcPV9 and three equine papillomaviruses from the genus *Dyoiotapapillomavirus*. Open reading frames (ORFs) are shown at the top and colored according to their putative function. Predicted nucleotide features are shown above and below each genome, using triangles and colored according to different features. Predicted amino acid features are shown above each genome, using down arrows and colored according to the feature in question. Sequences of E1 binding sites are shown in the left box at the bottom with variable nucleotides shown in black type. True consensus E2 binding sites (ACC-N6-GGT) are colored black, while other putative E2 binding sites (ACC-N4-GGT, ACC-N5-GGT, ACC-N7-GGT) are shown in blue, and the numbers of E2 binding sites within each papillomavirus are shown in the left bottom box. Sequences of the ATP-dependent helicase motive and nuclear localization signal are shown above each arrow.

We next performed pairwise alignments based on both nucleotide (nt) and amino acid (aa) sequences for the seven viral ORFs. In the case of L1, which is currently used for classification, EcPV9 shared 69.35%, 59.25%, and 58.00% nucleotide similarity with EcPV2, EcPV4, and EcPV5, respectively (Table 1). For E1, EcPV9 shared 62.20%, 51.12%, and 53.81% nucleotide similarity with EcPV2, EcPV4, and EcPV5, respectively. Among the other ORFs, the nucleotide identities with EcPV2, EcPV4, and EcPV5 were between 40.23% and 59.49%, with equivalent amino acid identities between 27.18% and 70.44% (Table 1).

To determine the evolutionary relationships of EcPV9, we inferred a phylogenetic tree based on the concatenated alignment of four coding sequences (E1, E2, L2, and L1). Amino acid sequences (concatenated E1-E2-L2-L1) of 13 equine PVs, as well as the type species of each of the 52 PV genera, were aligned using the E-INS-I algorithm in the MAFFT v7 package [19]. A phylogenetic tree was then estimated using the maximum likelihood method in PhyML 3.0 [20], incorporating the LG+Γ model of amino acid substitution, a SPR branch-swapping algorithm, and 1000 bootstrap replications. This analysis revealed that EcPV9 is clearly related to *Dyoiota* PVs—EcPV2, EcPV4, and EcPV5 (Figure 3). Hence, this evolutionary analysis demonstrates that EcPV9 is a novel species within the genus *Dyoiotapapillomavirus*, yet most closely related to EcPV2, classified as *Dyoiotapapillomavirus 1*.

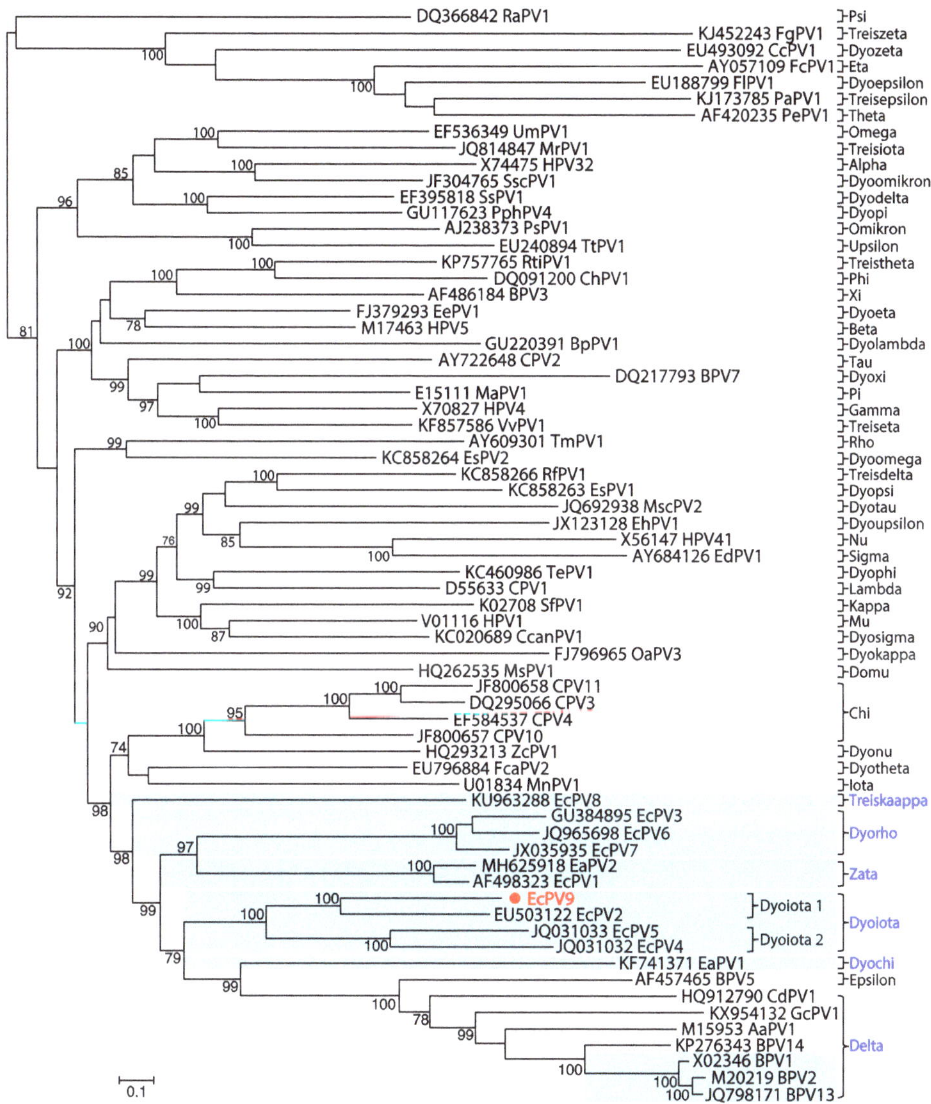

Figure 3. Phylogenetic relationships of EcPV9 to other papillomaviruses based on an analysis of concatenated E1, E2, L2, and L1 amino acids sequences. EcPV9 is shown in red. The names of reference sequences, that contain both the GenBank accession number and the virus species name, are shown in black. Those papillomaviruses associated with horses are shown in a light blue background. The names of previously defined genera are shown to the right of the phylogenies. The tree is mid-point rooted for clarity and nodes supported by >70% of bootstrap replicates are indicated.

Table 1. Comparison of sequence similarity based on nucleotide and amino acid sequences of 7 ORFs with EcPV9 to EcPV2, EcPV4, and EcPV5.

EcPV9		EcPV2 (%)	EcPV4 (%)	EcPV5 (%)
E6	nt	52.85	42.93	44.17
	aa	41.86	35.16	34.88
E7	nt	40.23	43.02	50.74
	aa	33.04	27.18	33.98
E1	nt	62.20	51.12	53.81
	aa	58.63	44.97	46.56
E2	nt	51.05	49.54	48.77
	aa	42.79	35.31	36.38
E4	nt	49.32	40.83	42.19
	aa	40.14	33.33	32.24
L2	nt	59.49	46.60	49.63
	aa	59.17	42.55	45.07
L1	nt	69.35	59.25	58.00
	aa	70.44	57.74	59.88

3.3. Disease Association

As no biopsy samples could be taken from this case, it is not possible to confidently determine its significance in the observed pathologies. Nevertheless, the novel EcPV described here was extracted from semen samples, collected when a wart-like lesion was visible on the tip of the penis (Figure S1), and hence compatible with a disease syndrome caused by a papillomavirus. In addition, it was notable that EcPV9 exhibited greatest sequence similarity with EcPV2, a major aetiologic agent of equine squamous cell carcinoma (SCC) disease [8], again compatible with the idea that EcPV9 might also be associated with papillomavirus-related malignancies in horses. Finally, our meta-transcriptomic analysis identified no other likely microbial pathogen in any of the samples analyzed from this stallion.

In conclusion, we report the identification of a novel equine papillomavirus (genus *Dyoiotapapillomavirus*) in a thoroughbred Australian stallion suffering a genital papilloma ("wart"), highlighting the broad diversity of these viruses in horses. Further investigation of the clinical impact of this virus on horse health is clearly merited.

Supplementary Materials: The following are available online at http://www.mdpi.com/1999-4915/11/8/713/s1, Figure S1: Photograph of the penile lesion, Figure S2: Amino acid sequence alignment of the E6(A) and E7(B) proteins from EcPV2, EcPV9, EcPV4, and EcPV5, Figure S3: Amino acid sequence alignment of the C-terminal ends of E6 proteins derived from HPV types associated with genital infections and 4 PV types from the genus *Dyoiotapapillomavirus*, Table S1: Equine associated papillomaviruses and their respective associated disease/syndrome, Table S2: Information on the RNA sequencing libraries generated here, Table S3: Genomic nucleotide and amino acid features of the genus *Dyoiotapapillomavirus*.

Author Contributions: Conceptualization, E.C.H. and B.J.H.; methodology, C.-X.L., K.M., J.R., E.C.H., and B.J.H.; laboratory analysis, C.-X.L., K.M., and J.R.; software, C.-X.L. and W.-S.C.; formal analysis, C.-X.L.; data curation, C.-X.L. and W.-S.C.; writing—original draft preparation, C.-X.L.; writing—review and editing, E.C.H., K.M., and B.J.H.; visualization, C.-X.L.; project administration, K.M., E.C.H., and B.J.H.; funding acquisition, E.C.H.

Funding: E.C.H. is funded by an ARC Australian Laureate Fellowship (FL170100022).

Conflicts of Interest: The authors declare no conflict of interest.

References

1. Nasir, L.; Brandt, S. Papillomavirus associated diseases of the horse. *Vet. Microbiol.* **2013**, *167*, 159–167. [CrossRef] [PubMed]

2. Lunardi, M.; de Alcântara, B.K.; Otonel, R.A.; Rodrigues, W.B.; Alfieri, A.F.; Alfieri, A.A. Bovine papillomavirus type 13 DNA in equine sarcoids. *J. Clin. Microbiol.* **2013**, *51*, 2167–2171. [CrossRef] [PubMed]

3. Torres, S.M.; Koch, S.N. Papillomavirus-associated diseases. *Vet. Clin. N. Am. Equine. Pract.* **2013**, *29*, 643–655. [CrossRef] [PubMed]

4. Sykora, S.; Brandt, S. Papillomavirus infection and squamous cell carcinoma in horses. *Vet. J.* **2017**, *223*, 48–54. [CrossRef] [PubMed]

5. Lange, C.E.; Tobler, K.; Ackermann, M.; Favrot, C. Identification of two novel equine papillomavirus sequences suggests three genera in one cluster. *Vet. Microbiol.* **2011**, *149*, 85–90. [CrossRef] [PubMed]

6. Lange, C.E.; Vetsch, E.; Ackermann, M.; Favrot, C.; Tobler, K. Four novel papillomavirus sequences support a broad diversity among equine papillomaviruses. *J. Gen. Virol.* **2013**, *94*, 1365–1372. [CrossRef] [PubMed]

7. Linder, K.E.; Bizikova, P.; Luff, J.; Zhou, D.; Yuan, H.; Breuhaus, B.; Nelson, E.; Mackay, R. Generalized papillomatosis in three horses associated with a novel equine papillomavirus (EcPV8). *Vet. Dermatol.* **2018**, *29*, 72-e30. [CrossRef] [PubMed]

8. Scase, T.; Brandt, S.; Kainzbauer, C.; Sykora, S.; Bijmholt, S.; Hughes, K.; Sharpe, S.; Foote, A. *Equus caballus papillomavirus-2* (EcPV-2): An infectious cause for equine genital cancer? *Equine. Vet. J.* **2010**, *42*, 738–745. [CrossRef] [PubMed]

9. Lecis, R.; Tore, G.; Scagliarini, A.; Antuofermo, E.; Dedola, C.; Cacciotto, C.; Dore, G.M.; Coradduzza, E.; Gallina, L.; Battilani, M.; et al. *Equus asinus papillomavirus* (EaPV1) provides new insights into equine papillomavirus diversity. *Vet. Microbiol.* **2014**, *170*, 213–223. [CrossRef] [PubMed]

10. Buchfink, B.; Xie, C.; Huson, D.H. Fast and sensitive protein alignment using diamond. *Nat. Meth.* **2015**, *12*, 59–60. [CrossRef] [PubMed]

11. Li, D.; Luo, R.; Liu, C.M.; Leung, C.M.; Ting, H.F.; Sadakane, K.; Yamashita, H.; Lam, T.W. MEGAHIT v1.0: A fast and scalable metagenome assembler driven by advanced methodologies and community practices. *Methods* **2016**, *102*, 3–11. [CrossRef] [PubMed]

12. Li, D.; Liu, C.M.; Luo, R.; Sadakane, K.; Lam, T.W. MEGAHIT: An ultra-fast single-node solution for large and complex metagenomics assembly via succinct de Bruijn graph. *Bioinformatics* **2015**, *31*, 1674–1676. [CrossRef] [PubMed]

13. Kearse, M.; Moir, R.; Wilson, A.; Stones-Havas, S.; Cheung, M.; Sturrock, S.; Buxton, S.; Cooper, A.; Markowitz, S.; Duran, C.; et al. Geneious Basic: an integrated and extendable desktop software platform for the organization and analysis of sequence data. *Bioinformatics* **2012**, *28*, 1647–1649. [CrossRef] [PubMed]

14. Peters-Kennedy, J.; Lange, C.E.; Rine, S.L.; Hacket, R.P. *Equus caballus papillomavirus 8* (EcPV8) associated with multiple viral plaques, viral papillomas, and squamous cell carcinoma in a horse. *Equine Vet. J.* **2019**, *51*, 470–474. [CrossRef] [PubMed]

15. Hans-Ulrich, B.; Burk, R.D.; Chen, Z.; Van Doorslaer, K.; zur Hausen, H.; De Villiers, E.-M. Classification of papillomaviruses (PVs) based on 189 PV types and proposal and taxonomic amendments. *Virology* **2010**, *401*, 70–79.

16. Gardiol, D.; Kühne, C.; Glaunsinger, B.; Lee, S.S.; Javier, R.; Banks, L. Oncogenic human papillomavirus E6 proteins target the discs large tumour suppressor for proteasome-mediated degradation. *Oncogene* **1999**, *18*, 5487–5496. [CrossRef] [PubMed]

17. Wise-Draper, T.M.; Wells, S.I. Papillomavirus E6 and E7 proteins and their cellular targets. *Front. Biosci.* **2008**, *13*, 1003–1017. [CrossRef] [PubMed]

18. Narechania, A.; Terai, M.; Chen, C.Z.; DeSalle, R.; Burk, R.D. Lack of the canonical pRB-binding domain in the E7 ORF of artiodactyl papillomaviruses is associated with the development of fibropapillomas. *J. Gen. Virol.* **2004**, *85*, 1243–1250. [CrossRef] [PubMed]

19. Katoh, K.; Standley, D.M. Mafft multiple sequence alignment software version 7: Improvements in performance and usability. *Mol. Biol. Evol.* **2013**, *30*, 772–780. [CrossRef] [PubMed]

20. Guindon, S.; Gascuel, O. A simple, fast, and accurate algorithm to estimate large phylogenies by maximum likelihood. *Syst. Biol.* **2003**, *52*, 696–704. [CrossRef] [PubMed]

Communication

The First Detection of Equine Coronavirus in Adult Horses and Foals in Ireland

Manabu Nemoto [1,2], Warren Schofield [3] and Ann Cullinane [1,*]

1 Virology Unit, The Irish Equine Centre, Johnstown, Naas, Co. Kildare W91 RH93, Ireland; nemoto_manabu@equinst.go.jp
2 Equine Research Institute, Japan Racing Association, Shimotsuke, Tochigi 329-0412, Japan
3 Troytown Grey Abbey Equine Hospital, Green Road, Co. Kildare R51 YV04, Ireland; warrenschofield@gmail.com
* Correspondence: acullinane@irishequinecentre.ie; Tel.: +353-45-866266; Fax: +353-45-866273

Received: 27 July 2019; Accepted: 12 October 2019; Published: 14 October 2019

Abstract: The objective of this study was to investigate the presence of equine coronavirus (ECoV) in clinical samples submitted to a diagnostic laboratory in Ireland. A total of 424 clinical samples were examined from equids with enteric disease in 24 Irish counties between 2011 and 2015. A real-time reverse transcription polymerase chain reaction was used to detect ECoV RNA. Nucleocapsid, spike and the region from the p4.7 to p12.7 genes of positive samples were sequenced, and sequence and phylogenetic analyses were conducted. Five samples (1.2%) collected in 2011 and 2013 tested positive for ECoV. Positive samples were collected from adult horses, Thoroughbred foals and a donkey foal. Sequence and/or phylogenetic analysis showed that nucleocapsid, spike and p12.7 genes were highly conserved and were closely related to ECoVs identified in other countries. In contrast, the region from p4.7 and the non-coding region following the p4.7 gene had deletions or insertions. The differences in the p4.7 region between the Irish ECoVs and other ECoVs indicated that the Irish viruses were distinguishable from those circulating in other countries. This is the first report of ECoV detected in both foals and adult horses in Ireland.

Keywords: equine coronavirus; Ireland; enteric disease

---

1. Introduction

Equine coronavirus (ECoV) is a positive-stranded RNA virus and belongs to the species *Betacoronavirus 1* in the genus *Betacoronavirus* [1,2]. The clinical signs associated with ECoV infection during outbreaks in the USA [3] and Japan [4–6] were fever, anorexia, lethargy and diarrhoea. The same clinical signs were also recorded in an experimental challenge study using Japanese draft horses [7]. The main transmission route is considered to be faecal–oral [7] and ECoV is usually detected in faecal samples. However, the molecular detection of ECoV in faeces from horses with diarrhoea, does not prove causation. Coronaviruses can cause both enteric and respiratory disease in many avian and mammalian species but ECoV is less likely to be found in respiratory secretions than in faeces [8,9].

Both molecular and seroepidemiology studies suggest that ECoV may be more prevalent in the USA than in other countries [10]. ECoV was detected in samples collected from equids in 48 states of the USA [11]. In central Kentucky, approximately 30% of both healthy and diarrheic Thoroughbred foals were infected with ECoV [12]. All of the qPCR positive foals with diarrhoea were co-infected with other pathogens such as rotavirus or *Clostridium perfringens*, suggesting that there was potential for ECoV to be over-diagnosed as a causative agent in complex diseases. In contrast in Japan, although an outbreak of diarrhoea occurred among ECoV-infected draft horses at one racecourse [4–6], there have been no similar outbreaks subsequently, and all rectal swabs collected from diarrheic Thoroughbred foals were negative. Furthermore, only 2.5% of the rectal swabs collected from healthy foals in the largest Thoroughbred horse breeding region in Japan were positive for ECoV [13]. In France, 2.8% of 395 faecal samples and 0.5% of 200 respiratory samples collected in 58 counties tested positive for ECoV [9]. Similar to the reports from Japan and France, a low prevalence of ECoV was also observed in the UK [14], Saudi Arabia and Oman [15]. The objective of this study was to investigate the presence of ECoV in clinical samples submitted to a diagnostic laboratory in Ireland. The samples were tested by real-time reverse transcription polymerase chain reaction (rRT-PCR) as it has been shown to be the most sensitive diagnostic method for ECoV [16] and is routinely employed as an alternative to virus isolation in diagnostic laboratories worldwide, both for timely diagnosis and in epidemiological studies [9,10]. Virus isolation and biological characterisation were beyond the capacity of this study, which was similar in scope to that of the studies in horse populations in the USA, Europe and Asia [8,9,13,14].

2. Materials and Methods

Faecal samples and rectal swabs collected from equids with enteric disease were included in this study. Samples were collected in 24 Irish counties between 2011 and 2015. All samples (65 in 2011, 69 in 2012, 97 in 2013, 97 in 2014 and 96 in 2015) were stored at −80 °C prior to testing by rRT-PCR. Faecal samples were diluted 1:10 in phosphate buffered saline (PBS), and rectal swabs were immersed in PBS. The suspensions were clarified by centrifugation at 2000 × g for 10 min, and viral RNA was extracted from the supernatant using the LSI MagVet Universal Isolation Kit (Thermo Fisher Scientific, Cambridge, MA, USA).

The rRT-PCR assay was performed as previously described using a primer set targeting the nucleocapsid (N) gene (ECoV-380f, ECoV-522r and ECoV-436p) [3] (Table 1) and AgPath-ID One-Step RT-PCR Kit (Thermo Fisher Scientific, MA, USA) according to the manufacturer's instructions. To prove that the extraction was successful and that there was no inhibition during rRT-PCR amplification, an internal positive control primer/probe (PrimerDesign, Southampton, UK) was added to the master mix. Thermal cycling conditions were; 48 °C for 10 min and 95 °C for 10 min, followed by 40 cycles at 94 °C for 15 s and 60 °C for 45 s.

Table 1. Primers and probe for rRT-PCR and sequencing primers.

Primer Name	Sequence 5′-3′	Use	Target	Reference
ECoV-380-F ECoV-522-R ECoV-436-probe	TGGGAACAGGCCCGC CCTAGTCGGAATAGCCTCATCAC TGGGTCGCTAACAAG	PCR	Nucleocapsid	[3]
ECoV-N-F ECoV-N-R	TCAGGCATGGACACCGCATTGTT CCAGGTGCCGACATAAGGTTCAT	Sequencing	Nucleocapsid	[4]
ECoV-S1-F ECoVS1-R	CAATGCCTTTATGGCTTGGT AAACTCGGAAGGGATCTGAA	Sequencing	Spike Gene	[9]
ECoV-p4.7-F ECoV-p4.7-R	TAATCGGCCTTGCTGGTGTAGC GCTTCATCAGCAGTCCAGGTA	Sequencing	p4.7 to p12.7 genes	Oue (personal communication)

The SuperScript III One-Step RT-PCR System with Platinum Taq High Fidelity (Thermo Fisher Scientific, MA, USA) was used for sequencing analysis of two of the five ECoV samples identified. There was inadequate viral nucleic acid in the other three samples for sequencing. The primer sets used to amplify the nucleocapsid (N) gene [4], the partial spike (S) gene [9], and the region from the p4.7 to p12.7 genes of non-structural proteins (Oue, personal communication) are shown in Table 1. The RT-PCR products were sequenced commercially by GATC Biotech (Cologne, Germany). Sequence analysis was performed using the BLAST and CLUSTALW programs, and Vector NTI Advance 11.5 software (Thermo Fisher Scientific, MA, USA). Phylogenetic analysis of nucleotide sequences was conducted with MEGA software Version 5.2 [17]. A phylogenetic tree was constructed based on nucleotide sequences of the K2+G (N gene) and TN93 (S gene) using the maximum likelihood method. MEGA software was used to select the optimal substitution models. Statistical analysis of the tree was performed with the bootstrap test (1000 replicates) for multiple alignments. The complete genome sequences of NC99

(EF446615) [2], Tokachi09 (LC061272), Obihiro12-1 (LC061273) and Obihiro12-2 (LC061274) [1], the N (AB671298) and S (AB671299) genes of Obihiro2004, the N gene of Hidaka-No.61/2012 (LC054263) and Hidaka-No.119/2012 (LC054264) [13], the S gene of ECoV_FRA_2011/1 (KC178705), ECoV_FRA_2011/2 (KC178704), ECoV_FRA_2012/1 (KC178703), ECoV_FRA_2012/2 (KC178702) and ECoV_FRA_2012/3 (KC178701) [9] were used in sequence and/or phylogenetic analysis.

The accession numbers registered in GenBank/EMBL/DDBJ are as follows: the complete sequences of the N gene; 11V11708/IRL (LC149485) and 13V08313/IRL (LC149486), the partial sequences of the S gene; 11V11708/IRL (LC149487) and13V08313/IRL (LC149488) and the complete sequences from the p4.7 to p12.7 genes; 11V11708/IRL (LC149489) and13V08313/IRL (LC149490).

3. Results

Five samples (11V11708/IRL/2011 and 11V06761/IRL/2011 collected in 2011, and 13V08313/IRL/2013, 13V08314/IRL/2013 and 13V07530/IRL/2013 collected in 2013) tested positive for ECoV. All samples collected in 2012, 2014 and 2015 tested negative. Sample 11V11708/IRL/2011 was collected in November 2011 from a donkey foal in County Cork. Samples 13V08313/IRL/2013 and 13V08314/IRL/2013 were collected on the same day in April 2013 from adult horses on a farm in County Clare. At the time of sample collection, both horses were described as having mild diarrhoea for 24 to 48 h. This resolved within a week. Samples 11V06761/IRL/2011 and 13V07530/IRL/2013 were collected in County Kildare from diarrheic foals on two separate premises in March 2011 and April 2013, respectively. One six-week-old foal was the only clinical case on a public Thoroughbred stud farm with approximately 30 mares when it presented with diarrhoea. Recovery took over three weeks during which it received fluid therapy, probiotics, antiulcer medication and antibiotics. The second foal was a 14-day-old filly, which had been hospitalised with diarrhoea two days prior to sample collection. The foal responded well to supportive treatment and at the time of sample collection, the diarrhoea had resolved. The five ECoV positive samples tested negative for equine rotavirus.

The nucleotide sequences of the complete N gene, the partial S gene and the region from the p4.7 to p12.7 genes of two positive samples (11V11708/IRL/2011 and 13V08313/IRL/2013) were determined. The nucleotide identities of the N and S genes of the two Irish ECoVs were 99.8% (1338/1341 nucleotides) and 99.5% (650/653 nucleotides), respectively. The nucleotide identities of the N gene of the two Irish ECoVs and the ECoVs from other continents are summarised in Table 2.

Phylogenetic analysis was performed for the nucleotide sequences of the complete N and partial S genes (Figure 1). The analysis for the N gene showed that Irish ECoVs were independently clustered although they were closely related to Japanese viruses identified after 2009. In the phylogenetic tree of the S gene, Irish ECoVs were closely related to all other ECoVs analysed.

The length of the region from the p4.7 to p12.7 genes in the two viruses was 544 base pairs. Compared with NC99, Irish ECoVs, had a total of 37 nucleotide deletions within p4.7 and the non-coding region following the p4.7 gene. Compared with Obihiro 12-1 and 12-2, Irish ECoVs had a three-nucleotide insertion. When compared with Tokachi09, the Irish ECoVs had a 148-nucleotide insertion (see Figure S1). The p12.7 gene of the two Irish ECoVs did not have deletions or insertions, and the nucleotide identities were 98.8–99.7% between these viruses and the other ECoVs (NC99, Tokachi09, Obihiro12-1 and Obihiro12-2).

Table 2. Nucleotide sequence identities of Irish ECoVs and ECoVs from other continents.

	Nucleotide Identities (%) of Nucleocapsid Gene to:				
Name	11V11708/IRL	13V08313/IRL	NC99	Obihiro2004	Tokachi09
Accession No.	LC149485	LC149486	EF446615	AB671298	LC061272
11V11708/IRL	-	99.8	98.1	98.6	99.2
13V08313/IRL	99.8	-	98.3	98.7	99.4
Name	Obihiro12-1	Obihiro12-2	Hidaka-No.61/2012	Hidaka-No.119/2012	
Accession No.	LC061273	LC061274	LC054263	LC054264	
11V11708/IRL	99.3	99.4	99.5	99.2	
13V08313/IRL	99.6	99.6	99.7	99.4	

	Nucleotide Identities (%) of Spike Gene to:					
Name	11V11708/IRL	13V08313/IRL	NC99	Obihiro2004	Tokachi09	Obihiro12-1
Accession No.	LC149487	LC149488	EF446615	AB671299	LC061272	LC061273
11V11708/IRL	-	99.5	99.7	98.9	98.6	99.4
13V08313/IRL	99.5	-	99.5	98.8	98.5	99.2
Name	Obihiro12-2	FRA_2011/1	FRA_2011/2	FRA_2012/1	FRA_2012/2	FRA_2012/3
Accession No.	LC061274	KC178705	KC178704	KC178703	KC178702	KC178701
11V11708/IRL	99.5	99.5	98.6	99.4	99.5	99.5
13V08313/IRL	99.4	99.3	98.4	99.3	99.3	99.3

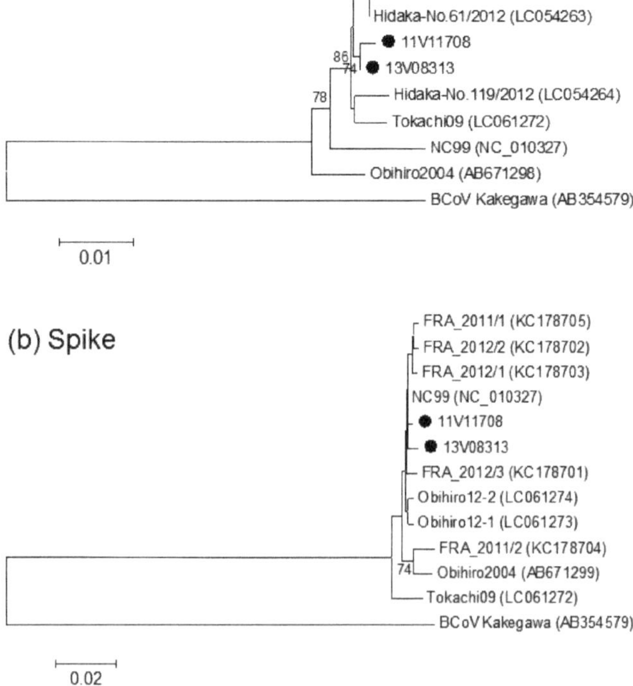

Figure 1. Phylogenetic analysis of the nucleotide sequences of the complete nucleocapsid (**a**) and the partial spike (**b**) genes of equine coronavirus. Closed circles indicate two of the equine coronaviruses detected in Ireland (11V011708 and 13V03813). The percentage bootstrap support is indicated by the value at each node; values <70% are omitted. Scale bars indicate nucleotide substitutions per site. BCoV (Bovine coronavirus) Kakegawa strain is used as an out-group.

4. Discussion

This study provides the first report of ECoV circulating in Ireland, the third European country with a significant horse industry where the virus has been detected in horses with enteric disease. However, detection of ECoV in faeces samples from horses with enteric disease does not prove causation. In this study, 424 samples collected between 2011 and 2015 from equids with enteric disease were tested, and only five samples (1.2%) were positive for ECoV. The inclusion of an internal positive control in the rRT-PCR eliminated the possibility of false negative results due to the presence of PCR inhibitors but the high content of nucleases associated with faeces samples may have caused some RNA degradation. However, this low prevalence of ECoV is similar to that identified in France [9] and among Thoroughbred foals in Japan [13].

Although ECoV has been identified on three continents, little is known about the genetic and pathogenic diversity in field viruses. In this study, sequence and phylogenetic analysis (Figure 1) demonstrated a high level of homology between viruses detected in a donkey and a horse in two provinces in Ireland in different years. This suggests that Irish ECoVs may have low genetic diversity. Compared with the ECoVs of other countries, the N, S and p12.7 genes of the two Irish viruses were highly conserved. In contrast, the region from p4.7 and the non-coding region following the p4.7 gene had deletions or insertions (Figure S1). Because of polymorphism in this region, this region could be useful for epidemiological investigation [5]. The differences in the p4.7 region between the Irish ECoVs and other ECoVs indicated that the viruses in Ireland may be distinguishable from those circulating in other countries. The positive samples were collected in November (1), March (1) and April (3) in this study. Higher case numbers are identified in the USA during the colder months (October to April) [11], and our results were consistent with the circulation period in USA. It has been reported that outbreaks mainly occurred among adult riding, racing and show horses in USA [11]. The choice of cases to include in the current study may not have been optimal for detection of ECoV as the majority of samples were from foals. However, two positive samples were collected from adult horses in a combined riding school/show jumping yard in the West of Ireland. At the time of sample collection in April 2013, the monthly mean temperatures were below long-term average and in parts of the West, were the coldest in 24 years [18]. Cold weather may have been a predisposing factor to the ECoV infection on the farm.

Two positive samples were collected from Thoroughbred foals. A faeces sample collected from one foal with severe watery diarrhoea and inappetance was positive for ECoV but a sample collected three days later tested negative. A potential difficulty in detecting ECoV from naturally infected horses has been noted previously as serial samples from seven sick horses in the USA suggested that ECoV only persisted for three to nine days in faeces [3]. In both cases, the diarrhoea may have been caused by other unidentified coinfecting pathogens as has been suggested by investigators in the USA [12].

This is the first report of ECoV detection in faeces samples from both foals and adult horses in Ireland. The viruses identified in Ireland are genetically closely related to the Japanese viruses and the results of this study give no indication of significant genetic or phenotypic diversity. In recent years, there has been an increase in awareness and testing for ECoV in the USA and elsewhere [10]. Horse breeding and racing activities in Ireland are the most prominent and important of any country on a per capita basis. There are over 50 Thoroughbred horses per 10,000 of population in Ireland, compared to between three and five for Great Britain, France and the USA [19]. Thus, an investigation of ECoV in Ireland is pertinent not only to increase awareness nationally of the epidemiology of the virus and promote discussion on its clinical importance, but also to inform the industry globally of the health status of Irish horses. Ireland exports horses all over the world. By illustration, in 2016 the country was the second biggest seller of bloodstock at public auctions second only to the USA [19].

Many questions remain with regard to the clinical significance of ECoV. The outbreak at a draft-horse racetrack in Japan in 2009 affected 132 of approximately 600 horses and resulted in non-starters and the implementation of movement restrictions [4]. However, draft horses appear to have a higher infection rate than other breeds and an outbreak of similar severity has not been reported in Thoroughbred racehorses [10,20]. The much higher incidence of ECoV positive Thoroughbred foals identified in Kentucky compared to similar populations internationally suggests an increased susceptibility to ECoV infection in that population. In the past, specific environmental factors were associated with extensive reproductive loss in the Kentucky area and to a lesser extent in other states [21], but predisposing regional factors such as differences in management, environment or husbandry have not been identified for ECoV. It has been suggested that ECoV is a coinfecting agent in foals with diarrhoea and clinical infections have predominantly been reported in adult horses with a mono-infection with EcoV [10]. There was no indication from the results of this study that coronavirus is a major cause of diarrhoea in Irish horses but the introduction of rRT-PCR as a routine diagnostic test will assist in elucidating the significance of this virus to the Irish breeding, racing and sports industries. The primary focus in future will be on testing adult horses that present with anorexia, lethargy, fever and changes in faecal character as a significant association has been demonstrated between this clinical status and molecular detection of ECoV in faeces [11].

Supplementary Materials: The following are available online at http://www.mdpi.com/1999-4915/11/10/946/s1, Figure S1: Alignment of the nucleotide sequences of the region from the p4.7 to p12.7 genes of the ECoV NC99, Tokachi09, Obihiro12-1, Obihiro12-2, 11V11708/IRL/2011 and 13V08313/IRL/2013. Minus signs (−) show missing nucleotides and asterisks (*) show conserved nucleotides.

Author Contributions: Conceptualization, M.N. and A.C.; methodology, M.N. and A.C.; formal analysis, M.N.; investigation, M.N., A.C. and W.S.; resources, A.C.; data curation, M.N.; writing—original draft preparation, M.N.; writing—review and editing, M.N. and A.C.; supervision, A.C.

Funding: Manabu Nemoto was supported by the Japan Racing Association while on sabbatical at the Irish Equine Centre. The laboratory work was funded by the Irish Equine Centre.

Acknowledgments: The authors are grateful to the horse owners and their veterinary surgeons for their support and discussions relating to the positive cases and also to Dr Marie Garvey for assistance with manuscript preparation.

Conflicts of Interest: The authors declare no conflict of interest.

References

1. Nemoto, M.; Oue, Y.; Murakami, S.; Kanno, T.; Bannai, H.; Tsujimura, K.; Yamanaka, T.; Kondo, T. Complete genome analysis of equine coronavirus isolated in Japan. *Arch. Virol.* **2015**, *160*, 2903–2906. [CrossRef] [PubMed]

2. Zhang, J.; Guy, J.S.; Snijder, E.J.; Denniston, D.A.; Timoney, P.J.; Balasuriya, U.B. Genomic characterization of equine coronavirus. *Virology* **2007**, *369*, 92–104. [CrossRef] [PubMed]

3. Pusterla, N.; Mapes, S.; Wademan, C.; White, A.; Ball, R.; Sapp, K.; Burns, P.; Ormond, C.; Butterworth, K.; Bartol, J.; et al. Emerging outbreaks associated with equine coronavirus in adult horses. *Vet. Microbiol.* **2013**, *162*, 228–231. [CrossRef] [PubMed]

4. Oue, Y.; Ishihara, R.; Edamatsu, H.; Morita, Y.; Yoshida, M.; Yoshima, M.; Hatama, S.; Murakami, K.; Kanno, T. Isolation of an equine coronavirus from adult horses with pyrogenic and enteric disease and its antigenic and genomic characterization in comparison with the NC99 strain. *Vet. Microbiol.* **2011**, *150*, 41–48. [CrossRef] [PubMed]

5. Oue, Y.; Morita, Y.; Kondo, T.; Nemoto, M. Epidemic of equine coronavirus at Obihiro Racecourse, Hokkaido, Japan in 2012. *J. Vet. Med. Sci.* **2013**, *75*, 1261–1265. [CrossRef] [PubMed]

6. Narita, M.; Nobumoto, K.; Takeda, H.; Moriyama, T.; Morita, Y.; Nakaoka, Y. Prevalence of disease with inference of equine coronavirus infection among horses stabled in a draft-horse racecourse. *J. Jpn. Vet. Med. Assoc.* **2011**, *64*, 535–539. [CrossRef]

7. Nemoto, M.; Oue, Y.; Morita, Y.; Kanno, T.; Kinoshita, Y.; Niwa, H.; Ueno, T.; Katayama, Y.; Bannai, H.; Tsujimura, K.; et al. Experimental inoculation of equine coronavirus into Japanese draft horses. *Arch. Virol.* **2014**, *159*, 3329–3334. [CrossRef] [PubMed]

8. Pusterla, N.; Holzenkaempfer, N.; Mapes, S.; Kass, P. Prevalence of equine coronavirus in nasal secretions from horses with fever and upper respiratory tract infection. *Vet. Rec.* **2015**, *177*, 289. [CrossRef] [PubMed]

9. Miszczak, F.; Tesson, V.; Kin, N.; Dina, J.; Balasuriya, U.B.; Pronost, S.; Vabret, A. First detection of equine coronavirus (ECoV) in Europe. *Vet. Microbiol.* **2014**, *171*, 206–209. [CrossRef] [PubMed]

10. Pusterla, N.; Vin, R.; Leutenegger, C.M.; Mittel, L.D.; Divers, T.J. Enteric coronavirus infection in adult horses. *Vet. J.* **2018**, *231*, 13–18. [CrossRef] [PubMed]

11. Pusterla, N.; Vin, R.; Leutenegger, C.; Mittel, L.; Divers, T. Equine coronavirus: An emerging enteric virus of adult horses. *Equine Vet. Educ.* **2016**, *28*, 216–223. [CrossRef]

12. Slovis, N.; Elam, J.; Estrada, M.; Leutenegger, C. Infectious agents associated with diarrhoea in neonatal foals in central Kentucky: A comprehensive molecular study. *Equine Vet. J.* **2014**, *46*, 311–316. [CrossRef] [PubMed]

13. Nemoto, M.; Oue, Y.; Higuchi, T.; Kinoshita, Y.; Bannai, H.; Tsujimura, K.; Yamanaka, T.; Kondo, T. Low prevalence of equine coronavirus in foals in the largest Thoroughbred horse breeding region of Japan, 2012–2014. *Acta Vet. Scand.* **2015**, *57*, 53. [CrossRef] [PubMed]

14. Bryan, J.; Marr, C.M.; Mackenzie, C.J.; Mair, T.S.; Fletcher, A.; Cash, R.; Phillips, M.; Pusterla, N.; Mapes, S.; Foote, A.K. Detection of equine coronavirus in horses in the United Kingdom. *Vet. Rec.* **2019**, *184*, 123. [CrossRef] [PubMed]

15. Hemida, M.G.; Chu, D.K.W.; Perera, R.; Ko, R.L.W.; So, R.T.Y.; Ng, B.C.Y.; Chan, S.M.S.; Chu, S.; Alnaeem, A.A.; Alhammadi, M.A.; et al. Coronavirus infections in horses in Saudi Arabia and Oman. *Transbound. Emerg. Dis.* **2017**, *64*, 2093–2103. [CrossRef] [PubMed]

16. Nemoto, M.; Morita, Y.; Niwa, H.; Bannai, H.; Tsujimura, K.; Yamanaka, T.; Kondo, T. Rapid detection of equine coronavirus by reverse transcription loop-mediated isothermal amplification. *J. Virol. Methods* **2015**, *215*, 13–16. [CrossRef] [PubMed]

17. Tamura, K.; Peterson, D.; Peterson, N.; Stecher, G.; Nei, M.; Kumar, S. MEGA5: Molecular evolutionary genetics analysis using maximum likelihood, evolutionary distance, and maximum parsimony methods. *Mol. Biol. Evol.* **2011**, *28*, 2731–2739. [CrossRef] [PubMed]

18. Met Eireann: The Irish Meteorological Service. Available online: http://www.met.ie/climate/irish-climate-monthly-summary.asp (accessed on 12 September 2019).

19. Economic Impact of Irish Breeding and Racing. 2017. Available online: https://www.hri.ie/uploadedFiles/HRI-Corporate/HRI_Corporate/Press_Office/Economic_Impact/HRI%20Report.pdf (accessed on 12 September 2019).

20. Kooijman, L.J.; James, K.; Mapes, S.M.; Theelen, M.J.; Pusterla, N. Seroprevalence and risk factors for infection with equine coronavirus in healthy horses in the USA. *Vet. J.* **2017**, *220*, 91–94. [CrossRef] [PubMed]

21. Sebastian, M.M.; Bernard, W.V.; Riddle, T.W.; Latimer, C.R.; Fitzgerald, T.D.; Harrison, L.R. Review Paper: Mare Reproductive Loss Syndrome. *Vet. Pathol.* **2008**, *45*, 710–722. [CrossRef] [PubMed]

Article

Development and Validation of a S1 Protein-Based ELISA for the Specific Detection of Antibodies against Equine Coronavirus

Shan Zhao [1], Constance Smits [2], Nancy Schuurman [1], Samantha Barnum [3], Nicola Pusterla [3], Frank van Kuppeveld [1], Berend-Jan Bosch [1], Kees van Maanen [2,*,†] and Herman Egberink [1,*,†]

[1] Virology Division, Department of Infectious Diseases & Immunology, Faculty of Veterinary Medicine, Utrecht University, Yalelaan 1, 3584CL Utrecht, The Netherlands; s.zhao@uu.nl (S.Z.); N.M.P.Schuurman@uu.nl (N.S.); f.j.m.vankuppeveld@uu.nl (F.v.K.); b.j.bosch@uu.nl (B.-J.B.)

[2] GD Animal Health, Department of Small Ruminants, Horses and Companion Animals, Arnsbergstraat 7, 7418EZ Deventer, The Netherlands; c.smits@gddiergezondheid.nl

[3] Department of Medicine and Epidemiology, School of Veterinary Medicine, University of California, Davis, One Shields Ave., Davis, CA 95616, USA; smmapes@ucdavis.edu (S.B.); npusterla@ucdavis.edu (N.P.)

* Correspondence: c.v.maanen@gddiergezondheid.nl (K.v.M.); H.F.Egberink@uu.nl (H.E.)

† These authors contributed equally to the work.

Received: 29 October 2019; Accepted: 29 November 2019; Published: 30 November 2019

Abstract: Equine coronavirus (ECoV) is considered to be involved in enteric diseases in foals. Recently, several outbreaks of ECoV infection have also been reported in adult horses from the USA, France and Japan. Epidemiological studies of ECoV infection are still limited, and the seroprevalence of ECoV infection in Europe is unknown. In this study, an indirect enzyme-linked immunosorbent assay (ELISA) method utilizing ECoV spike S1 protein was developed in two formats, and further validated by analyzing 27 paired serum samples (acute and convalescent sera) from horses involved in an ECoV outbreak and 1084 sera of horses with unknown ECoV exposure. Both formats showed high diagnostic accuracy compared to virus neutralization (VN) assay. Receiver-operating characteristic (ROC) analyses were performed to determine the best cut-off values for both ELISA formats, assuming a test specificity of 99%. Employing the developed ELISA method, we detected seroconversion in 70.4% of horses from an ECoV outbreak. Among the 1084 horse sera, seropositivity varied from 25.9% (young horses) to 82.8% (adult horses) in Dutch horse populations. Further, sera of Icelandic horses were included in this study and a significant number of sera (62%) were found to be positive. Overall, the results demonstrated that the ECoV S1-based ELISA has reliable diagnostic performance compared to the VN assay and is a useful assay to support seroconversion in horses involved with ECoV outbreaks and to estimate ECoV seroprevalence in populations of horses.

Keywords: equine coronavirus; spike S1 protein; ELISA; virus neutralization; seroprevalence

---

1. Introduction

Coronaviruses (CoVs) are enveloped, positive single-stranded RNA viruses that belong to the subfamily *Orthocoronavirinae* in the family *Coronaviridae* of the order Nidovirales. They are classified into four genera (*alpha* , *beta-*, *gamma-* and *deltacoronavirus*) and infect both mammalian and avian hosts [1,2]. Equine coronavirus (ECoV) belongs to *Betacoronavirus 1* species, within the *Embecovirus* subgenus of the *Betacoronavirus* genus, as does human coronavirus OC43, HKU1 and bovine coronavirus [3]. ECoV was isolated for the first time from a two-week-old diarrheic foal in North Carolina (USA) in 1999, suggesting the role of ECoV in causing enteric disease [4]. Since 2010, several cases of ECoV infections have also been reported in adult horses from the United States, Europe and Japan [5–9]. Equine coronavirus has been detected in fecal samples from horses with clinical signs that included anorexia, lethargy, fever and, less frequently, diarrhea, colic and neurologic deficits [10,11]. The morbidity rate varies from 10% to 83% during outbreaks. Mortality is low and has been related to endotoxemia, septicemia or hyperammonemia-associated encephalopathy [12,13]. The outbreaks in adult horses demand further studies on the pathogenesis and epidemiology of ECoV infections. For this, diagnostic assays with high sensitivity and specificity are crucial.

ECoV is known to be associated with enteric infections but can also be detected in a small percentage of horses with respiratory signs. Virus shedding can be observed in fecal samples or nasal swabs from sick horses as well as healthy horses, but with a strong association between clinical signs assumed to be related to ECoV infection and virus detection in fecal samples suggesting a possible etiological role of ECoV [10,14]. Recently, real-time quantitative PCR (qPCR) methods have been established and were shown to be able to detect ECoV in feces efficiently. However, ECoV viral nucleic acid is generally only detectable by qPCR within a limited timeframe of 3–9 days post infection, as reported from both field and experimental studies [6,7,12,15]. On the other hand, serological assays can be used to support the diagnosis of a clinical ECoV infection by showing seroconversion or a significant increase in antibody titer in paired serum samples. Serological assays are also needed to gain more insight into the transmission rate of infection within animal populations [16]. Antibodies induced by betacoronaviruses persist in blood for a longer period after infection [17,18]. The virus neutralization (VN) assay has long been used as a gold standard to confirm serological responses to coronavirus infections [19–21]. Although the VN assay is highly specific for the detection of antibodies, it is also time-consuming and laborious to perform. Alternative high-throughput serologic assays that correlate well with neutralizing antibodies are therefore needed. Severe infections of ECoV have been shown to be associated with high viral load, but mild or asymptomatic infections may occur with low levels of virus replication being negative in PCR and with variable immune responses [12]. Consequently, specific, sensitive and high-throughput serodiagnostic methods are necessary to avoid the underestimation of prevalence in surveillance studies.

The spike protein (S) of coronaviruses is the key mediator in virus cell entry and therefore the major target for neutralizing antibodies. The S ectodomain consists of two functionally interdependent subunits, S1 and S2. The N-terminal S1 subunit is responsible for receptor binding, while the C-terminal S2 subunit mediates membrane fusion [22,23]. The S1 subunit is the most variable immunogenic antigen among coronaviruses, and therefore it is an ideal candidate for the detection of CoV species-specific antibodies [24,25]. The objective of the study was to develop and validate an ELISA method for the detection of specific antibodies to ECoV and provide a tool for the diagnosis and the future estimation of ECoV prevalence and incidence in various equine (sub) populations.

2. Materials and Methods

2.1. Equine Serum Panels

A total of 1138 equine serum samples were included in this study. The details of serum panels A–H (*n* = 1084) are shown in Table 1. They were retrieved from the serum bank at GD Animal health Deventer, the Netherlands. All of them were collected for the monitoring of other diseases independent to this study, and their ECoV exposure status was unknown. With the exception of panel H (collected from Iceland), all serum samples from panel A to G were collected from horses in the Netherlands. Additionally, panel I included 27 paired (acute- and convalescent-phase) serum samples that were collected during an ECoV outbreak in the USA (2014). All samples were stored at −20 °C until tested.

2.2. Cells and Virus

ECoV strain NC99 was propagated and titrated in human rectal adenocarcinoma (HRT-18G) cells. HRT-18G cells and human embryonic kidney 293 cells stably expressing the SV40 large T antigen (HEK-293T) were maintained in Dulbecco modified Eagle medium (DMEM, Lonza, Basel, Switzerland) containing glutamine and supplemented with 10% fetal bovine serum (FBS, Bodinco, Alkmaar, The Netherlands), penicillin (100 IU/mL), and streptomycin (100 µg/mL).

The ECoV NC99 and HRT-18G were obtained from Dr. Udeni B.R. Balasuriya, School of Veterinary Medicine, Louisiana State University, USA [3,4].

2.3. Plasmids Design and Protein Expression

The sequence of the S1 subunit of the spike protein of the ECoV NC99 strain (residue 1–762 of the amino acid sequence) was derived from Genbank (Genbank No.: EF446615.1). Human codon-optimized sequences encoding the ECoV S1 subunit were synthesized and fused to the Fc domain of mouse IgG2a, which was subsequently cloned into the pCAGGS mammalian expression vector as described before [26]. For ECoV S1-Fc protein production, expression plasmid was transfected into HEK-293T cells using polyethyleneimine (Polysciences, Inc., Warrington, PA, USA) in a ratio of 1:10. After 6 h of incubation, the transfection medium was removed and replaced by 293 SFM II expression medium (Gibco®, Life Technologies Inc., Grand Island, NY, USA). At six days post transfection, cell culture supernatants were harvested and the soluble S1 was purified from the culture medium using Protein A Sepharose beads (GE Healthcare Bio-Sciences AB, Uppsala, Sweden). Subsequently, the proteins were eluted using 0.1M citric acid, pH 3.0, and immediately neutralized with 1 M Tris-HCl, pH 8.8. The purity and integrity of proteins were analyzed by sodium dodecyl sulphate polyacrylamide gel electrophoresis (SDS-PAGE) and stained with GelCodeBlue stain reagent (ThermoFisher Scientific Inc., Waltham, MA, USA). Purified proteins were quantified by Nanodrop spectrophotometry (ThermoFisher Scientific Inc., Waltham, MA, USA) and by sodium dodecyl sulphate polyacrylamide gel electrophoresis (SDS-PAGE) with bovine serum albumin (BSA) as standard, then stored at −80 °C until further usage.

2.4. Virus Neutralization (VN) Assay

Equine sera (n = 231) were randomly selected from different serum panels (A–D) and tested for neutralizing antibody titers in an ECoV VN assay. Heat-inactivated equine sera (56 °C for 30 min) were serially diluted 2-fold in DMEM supplemented with 2% fetal bovine serum and mixed with an equal volume of ECoV NC99 strain (100 50% tissue culture infective doses ($TCID_{50}$)/well) in 96-well cell culture plates (Corning Inc., Kennebunk, ME, USA). Virus–serum mixtures were incubated at 37 °C for 60 min. Then 100 µL of the virus–serum mixture was added in duplicate to HRT-18G cells monolayers in 96-well cell culture plates. At six days post infection, a clear cytopathic effect (CPE) was observed and the virus neutralization titers (VNT) were determined. The VNT of sera were expressed as the reciprocals of the highest serum dilution that resulted in 90% neutralization of CPE. A titer of ≥8 was considered to be positive.

2.5. ECoV S1 ELISA Development

Two different formats were developed employing ECoV S1 protein, a so-called wet format ELISA (wELISA) and a dry format ELISA (dELISA).

2.5.1. ECoV S1 Wet Format ELISA (wELISA)

High-binding microtiter plates (Greiner Bio-one BV, Alphen aan den Rijn, The Netherlands) were coated with ECoV S1 protein (100 µL per well) in phosphate buffered saline (PBS, pH 7.4) overnight at 4 °C. The optimal protein amount and dilution of secondary antibody conjugate were determined by checkerboard titration. The protein concentration in use was 0.25 µg/mL. After three washes with PBS containing 0.05% Tween-20 (PBST), the plates were blocked with PBST containing 5% milk powder (Protifar, Nutricia, Zoetermeer, The Netherlands) for 2 h at 37 °C. Following blocking, plates were incubated with serum samples diluted 1:200 in PBST containing 5% milk powder for 1 h at 37 °C. After a washing step, 100 µL/well 1:20,000 diluted horseradish peroxidase (HRP)-conjugated goat anti-horse IgG (H&L) (Abnova, Taiwan, China) was added to detect bound antibodies and plates were incubated for 1 h at 37 °C. Subsequently, the plates were washed, and the peroxidase reaction was then visualized via incubating plated with TMB Super Slow One Component HRP Microwell Substrate (BioFX®, Surmodics IVD, Inc., Eden Prairie, MN, USA) for 10 min at room temperature. The reaction was stopped by adding 12.5% sulfuric acid (H_2SO_4 (VWR International BV, Amsterdam, The Netherlands)) and optical densities (OD) were immediately measured at 450 nm using an ELISA microplate reader (BioTek Instruments, Inc., Winooski, VT, USA). All serum samples were tested in duplicate.

2.5.2. ECoV S1 Dry Format ELISA (dELISA)

High-binding microtiter plates (Greiner Bio-one BV, Alphen aan den Rijn, The Netherlands) were coated with ECoV S1 protein (100 μL per well) in ammoniumcarbonate solution (9.8 g/L (VWR International BV, Amsterdam, The Netherlands)) overnight at 4 °C. Then 100 μL/well blocking solution (9.8 g/L ammoniumcarbonate + 4 g/L caseine (VWR International BV, Amsterdam, The Netherlands) + 20 g/L sucrose (Merck and Co., Inc., Kenilworth, NJ, USA)) was added and plates were incubated for one hour at room temperature. Subsequently, the contents of the plates were discarded, and plates were dried for four hours at 37 °C, vacuum sealed and stored at 4–8 °C. The optimal protein amount and dilution of secondary antibody conjugate were determined by checkerboard titration. The protein concentration in use was 0.13 μg/mL. Plates were incubated with serum samples 100 μL per well and diluted 1:200 in PBS + 0.05% Tween-20 + 2.5% dry milk (Bio-Rad Laboratories, Inc., Hercules, CA, USA) for 1 h at 37 °C. After a washing step (five times with PBST 300 μL/well on a Biotek automatic washing station), 100 μL per well 1:60,000 diluted horseradish peroxidase (HRP)-conjugated goat anti-horse IgG (H&L) (Abnova, Taiwan, China) was added to detect bound antibodies and incubated for 1 h at 37 °C. Subsequently, the plates were washed again using the same washing procedure and the peroxidase reaction was then visualized by incubating plates with TMB (IDEXX Laboratories, Westbrook, NJ, USA) for 15 min at room temperature. The reaction was stopped by adding 50 μL/well sulfuric acid (H_2SO_4 0.5 M (VWR International BV, Amsterdam, The Netherlands)) and optical densities (OD) were immediately measured at 450 nm using an ELISA microplate reader (BioTek Instruments, Inc., Winooski, VT, USA). All serum samples were tested in duplicate. S/P values were calculated with the formula: S/P = (OD Sample-OD Negative control)/(OD Positive control-OD Negative control).

2.6. Statistical Analysis

The correlation between OD values scored with two ELISA formats was measured by the Pearson correlation coefficient using Graph Pad Prism, version 7. The discriminating power of the two different ELISA formats was analyzed by performing receiver operator characteristic (ROC) analysis with 231 sera, which 94 were negative (VNT < 8) and 137 were positive (VNT ≥ 8). The cut-off value, diagnostic specificity and sensitivity were determined by ROC analysis using Sigmaplot. A minimum specificity of 99% was chosen for the selection of cut-off values. Additionally, the reproducibility of assays was evaluated by testing three samples with different OD values. Inter-assay coefficients of variation (CV) and intra-assay CV were determined testing each sample in triplicate on three different plates in three different runs and within the same plate, respectively.

3. Results

3.1. Determination of Neutralizing Antibodies

To identify equine sera containing ECoV-neutralizing antibodies, we screened a subset of 231 equine sera, composed of randomly selected serum samples from panels A–C, and all samples from panel D were screened in the VN assay. Of the 231 sera, 94 sera were tested as negative (titers < 8) and 137 positive samples (titers ranging from 8 to 4096).

Additionally, paired samples from 27 horses ($n = 54$, panel I) were tested in the VN assay. Twenty out of 27 sera collected from the first time point exhibit titers ranging from 12 to 2048. The convalescent serum samples were collected 21–28 days following the first round of sample collections, and all of them showed neutralization responses with titers ranging from 16 to 4096. Within these horses, seven of them showed seroconversion and 14 showed a significant (4-fold or greater: 2log2) increase in titer in the VN assay. To confirm the presence of ECoV specific IgG in Icelandic horses, 24 horse sera with positive ECoV S1 ELISA results (in panel H, S/P value > 0.5) were tested in ECoV neutralization assays. All of them had neutralizing antibodies with titers varying between 32 and 768.

3.2. Development of ECoV S1 ELISA

Besides the conventional wet ELISA format (wELISA) for general laboratory usage, a dry standardized ELISA format (dELISA) was also developed and validated to facilitate implementation as routine diagnostic method in different laboratories and possibly wider application as an ELISA kit. Both ELISA formats were developed for the detection of ECoV-specific antibodies in horse serum samples. The diagnostic performance of both ELISAs was evaluated using a subset of 231 horse sera with known VN results as described above. The Pearson correlation coefficient was calculated to assess the correlation between the OD values obtained with the two ELISA formats (Figure 1). Results indicate that OD values obtained with both ELISAs show a high degree of correlation, with correlation and regression coefficients close to 1 (R^2 = 0.939, regression coefficient = 0.9513, p < 0.0001). Thus, the performance of both ELISA formats is very similar.

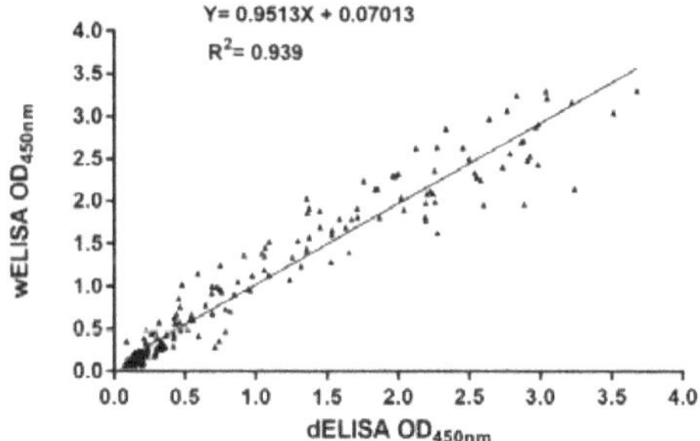

Figure 1. Correlation between optical density (OD) values obtained with wet format ELISA (wELISA) and dry format ELISA (dELISA).

Subsequently, the discriminating power of the wELISA and dELISA was evaluated via receiver operator characteristic (ROC) analysis. The ROC curves were plotted based on the previous classification of 231 sera into negative and positive by VN assays (Figure 2A,B). Then the optimal cut-off values, diagnostic specificity and sensitivity of both ELISA formats were determined by the established ROC curves. The ELISA results of the VNT-positive and negative samples are shown in Figure 2C,D. The diagnostic accuracy of both ELISA formats was considered to be high as the same area under the curve (AUC) values were observed (AUC = 0.985), with a relative sensitivity and specificity approximately 95% according to the Youden plot of wELISA and dELISA. Therefore, the test characteristics of both ELISA formats were assigned the same weight. In this study, a minimum specificity of 99% was chosen for the threshold of cut-off values for both ELISAs. Accordingly, the optimal cut-off for wELISA was an OD value of 0.35—for which, the sensitivity was 87% and the specificity was 99%. For dELISA, the test results were expressed as S/P values. A cut-off at an S/P value of 0.13 yielded a sensitivity of 85% and specificity of 99%, respectively.

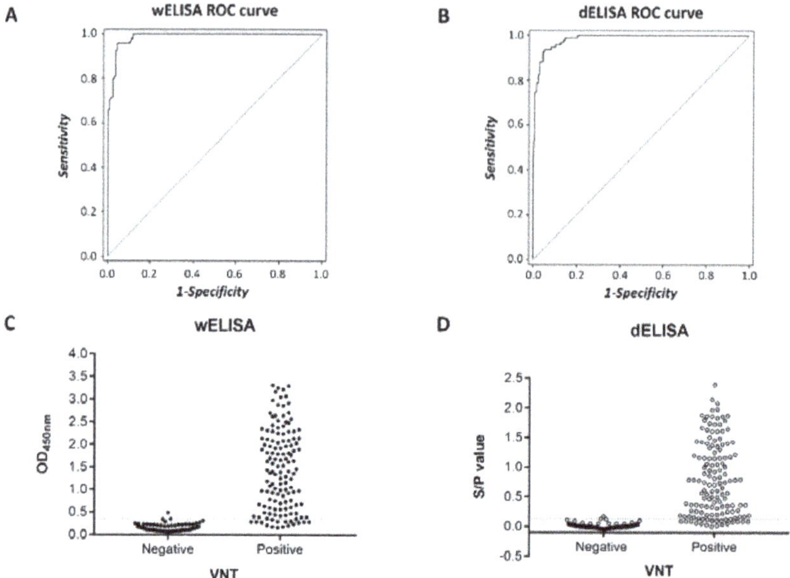

Figure 2. Receiver operating characteristic (ROC) analyses of equine coronavirus (ECoV) S1 ELISAs. ROC curves for wELISA (**A**) and dELISA (**B**) were plotted with positive (*n* = 137) and negative (*n* = 94) sera confirmed via VN assays. The area under the curve (AUC) is 0.985 for both ELISA formats. Distributions of wELISA (**C**) and dELISA (**D**) with confirmed sera are shown above. Calculated cut-off points are indicated by the vertical dashed lines. VN assays, virus neutralization assays; VNT, virus neutralization titer.

Furthermore, the inter- and intra-coefficient of variation (CV) of the three ECoV positive sera tested with both ELISA formats were lower than 12%. More specifically, the intra-assay CV of wELISA and dELISA ranged from 3.04% to 4.87% and from 5.4% to 7.7%, respectively, while the inter-assay CV of wELISA and dELISA varied from 4.9% to 10.26% and from 8.9% to 11.2%, respectively. Overall, these results indicate that the performances of both ELISA formats were very much equivalent and that the results of both ELISAs were strongly correlated to VN results.

3.3. Detection of Antibodies against ECoV in Horses during an Acute Outbreak

To determine the diagnostic performance of the ECoV S1 ELISA, 27 paired serum samples (panel I) collected from an acute ECoV outbreak were investigated by wELISA. The horses presented similar clinical signs as described in [27], and virus shedding was confirmed by qPCR analysis [7]. At the acute stage, 11 out of 27 horses were qPCR positive, while at the convalescent stage, this number had decreased to six. Serum samples were further validated by VN assay (Figure 3B; Table S1). Seven out of 27 horses showed seroconversion, while another 14 horses showed a significant (4-fold or greater) increase in VNT. Performing the wELISA (see Figure 3A; Table S1), the same seven out of 27 horses showed seroconversion; acute phase sera were negative (OD value < 0.35) whereas the convalescent phase sera all had OD values greater than 1.00 (1.14–2.90). Thus, seroconversion rates calculated from wELISA and VNT showed a 100% correlation (Table S1). For the horses that showed a 4-fold or greater increase in VNT (*n* = 14), nine of the acute phase sera had positive OD values between 0.35 and 0.70 (2x background) and also a higher than 2 (*n* = 2) to 4 (*n* = 7) fold increase in the OD value in the convalescent serum. Five of the VNT positive paired serum samples had OD values of >0.70 (twice the background OD value) in the acute phase serum. Two of these samples with an OD value of 1.12 and 1.41 respectively in wELISA also showed a greater than 2-fold increase in OD value. The three VNT positive samples with less than 2-fold increase in OD values already had high OD values in the acute phase serum as well as high VNT (mean OD value = 2.51, mean VNT = 8.30). For the six horses that did not show a significant rise in VNT, five serum samples collected at the acute stage already had high antibody levels as shown by ELISA and neutralization assay (mean OD value > 2.6, mean VNT > 9, Table S1). Further, the Pearson correlation coefficient was calculated to assess the overall correlation between the OD values obtained with wELISA and VNT (log2 titers) from acute and convalescent-phase sera of the 27 horses (Figure S1). Results indicate that OD values and VNT show a good degree of correlation (R^2 = 0.83, p < 0.0001). These data support the use of the wELISA as a diagnostic tool in case of suspected ECoV outbreaks.

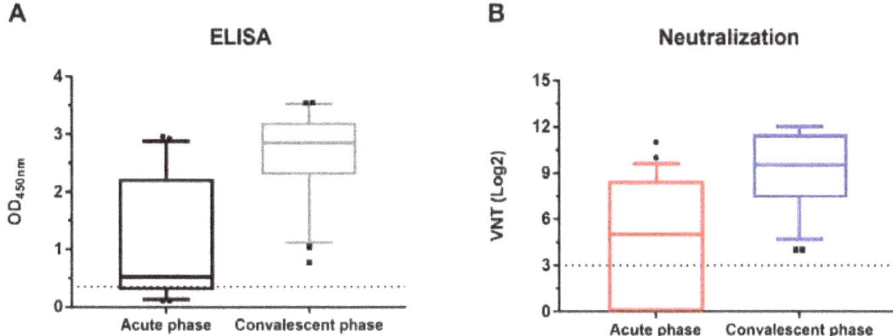

Figure 3. Antibodies response against ECoV from 27 horses during an acute outbreak. Boxplots show the ELISA reactivities (**A**) and VNT (**B**) of 27 horses from acute and convalescent-phase sera. Each cut-off is indicated by the dotted dashed line; VNT, virus neutralization titer.

3.4. ECoV Seroprevalence in Horses with Unknown ECoV Exposure

We further set out to determine the seroprevalence in horses with unknown ECoV exposure using the dELISA format. A total of 1084 serum samples (Table 1, panel A–H) were analyzed. With the exception of panel D, all sera were from adult horses (older than 36 months). Seroprevalence varied from 25.9% (panel D) to 82.8% (panel C) among these eight serum panels. The lowest number of positive samples was found in panel D which contained young horses (6-30 months old, average age: 8.38 months (95% CI 6.975–9.785)). In the other four serum panels (panel A, B, E and F) from Dutch horses, the historical serum samples (panel G) and samples from Iceland (panel H) higher seroprevalences were found (59.2–82.8%).

Table 1. Prevalence of ECoV S1-reactive antibodies in equine sera used in this study.

Panel	Samples Source/Project Names	Collection Year	Country	Numbers of Samples	Numbers of ECoV-S1 Positive Samples	Seroprevalence (%)
A	West Nile virus (WNV) surveillance	2016	The Netherlands	167	128	76.60
B	Equine infectious anemia (EIA) surveillance	2016	The Netherlands	112	80	71.40
C	Export horses	2016	The Netherlands	99	82	82.80
D	Influenza surveillance	2015	The Netherlands	81	21	25.90
E	WNV surveillance	2015	The Netherlands	176	145	82.40
F	EIA surveillance	2015	The Netherlands	184	109	59.20
G	Equine herpesvirus 1 and 4 diagnostic serum panel	1990	The Netherlands	165	93	56.40
H	Horse sera from Iceland	2018	Iceland	100	62	62.00

4. Discussion

Since the beginning of the 21st century, ECoV infections have been reported in horses, causing fever and enteric diseases [4]. More recently, infections in adult horses were reported with clinical signs of fever, anorexia, lethargy and, less commonly, specific signs of diarrhea and colic [7,8]. Nevertheless, information regarding the circulation of ECoV in the equine population, especially in Europe, is still limited [6,28]. Serological studies are useful tools to investigate ECoV prevalence in horse populations. In the present study, our aim was to develop a simple and reliable method for antibody detection against ECoV that can be used for diagnostics and sero-epidemiological studies.

As compared to virus neutralization assays, the ELISA method has the advantage of being reproducible, potentially high-throughput and much less laborious. In our study, we set up an ECoV S1-based ELISA method in two complementary formats. The conventional wELISA format is for general laboratory usage with simplified, easy to perform coating procedures. On the other hand, coated plates of the dELISA format could be stored for a longer time period, making it ideal for transportation and kit development. We showed that both formats performed equally well, and their results correlated nicely. When comparing with the VN assay by ROC analysis, our ELISA method with both formats was shown to have high accuracy. In our current study we applied wELISA for the analysis of the paired outbreak samples, while the dELISA was further validated and used for the high-throughput screening of larger amount of serum samples.

We utilize ECoV S1 as the viral antigen for antibody detection in this study. The S1 chimeric protein was expressed in mammalian cells, and hence both the protein conformation and modification (e.g., glycosylation) are mimicking the S proteins on the surface of virus particles [29]. As the most divergent and immunodominant component of coronaviruses, S1 has been widely used in the development of methods for specific coronavirus serological studies [19,20,26,30]. Our findings validate that ECoV S1 is a highly suitable antigen for the detection of antibodies against ECoV showing very good agreement between the ELISA and VN assays. Recently, similar conclusions were also drawn for the role of MERS S1 in MERS serology [31].

With our wELISA method, we were able to analyze paired samples that were collected during an ECoV outbreak. In the virus neutralization assay seroconversion or a 4-fold or greater increase in ECoV antibody titers could be detected in sera of 21 out of 27 horses within weeks of the initial observation of clinical disease and detection of viral RNA in feces. Of these 21 positive horses 18 showed seroconversion or a 2-fold or higher increase in OD values in the wELISA. The three remaining VNT positive samples had high OD values already in the acute phase serum. Of the six ECoV negative paired samples five had high VN antibody titers and OD values already at the acute phase. This might be due to late sampling of these horses or previous exposure to ECoV (Table S1). This study confirms that the ECoV S1 ELISA is a useful diagnostic test for the demonstration of a potential ECoV outbreak and should be considered as a useful adjunct to investigation of fecal samples by qPCR.

We also determined the seroprevalence of serum samples collected from horses with unknown ECoV exposure via our dELISA. Results showed that the overall seroprevalence in the different cohorts tested is 25.9%–82.8%. These percentages are in agreement with the study performed by Hemida et al. [32], in which they detected coronavirus infections in horses in Saudi Arabia and Oman and they found that 74% of them had detectable neutralizing antibodies to ECoV. A lower percentage (9.6%) of positive animals was found in another ECoV seroprevalence study conducted in the USA [33]. Several factors might contribute to these differences in results. There is only limited information regarding ECoV prevalence in Europe including the Netherlands [6,28], and it is possible that the overall ECoV distribution differs between continents. Moreover, our study employs eukaryotically expressed ECoV S1 protein as coating antigens, while in the US study chimeric S2 protein expressed in Escherichia coli was used. The expression in mammalian cells guarantees a more native configuration of the protein, in particular of glycosylated antigens such as the coronavirus spike protein. Reports had shown that both coronavirus S1 and S2 subunit elicit antibody responses, but the level of immune responses triggered by them may differ [34,35]. Furthermore, the criteria for determining the cut-off value are different for the two studies. In our study we defined positive and negative samples on the basis of a VN assay, whereas the US study used negative qRT-PCR and absence of clinical signs as criteria to define horses as ECoV negative. In this way, seropositive horses may have contributed to higher cut-off values and potentially a lower sensitivity of the assay.

In our study, we noticed differences in seroprevalence between young and adult horses. In the group of young horses (panel D, Table 1), the lowest seroprevalence was found. Young horses may initially be protected against ECoV infection by maternal antibodies and may become gradually more susceptible as maternal antibodies wane. The risk of becoming infected increases with age. This hypothesis is further supported by the age distribution of PCR-confirmed ECoV infection cases: foals (age 0–6 months) have the lowest infection rates, and the infection rate increases with age [10].

We also observed a significant percentage of seropositive horse serum samples collected back in 1990 (panel G, Table 1). ECoV-like viruses were detected in the 70s and 80s by electron microscopy in feces of horses with enteric disease, but virus isolation and characterization was not reported [36–39]. The history of ECoV presence, especially in Europe, is possibly much longer than currently understood [6]. Intriguingly, we noticed that Icelandic horses also are seropositive against ECoV (panel H, Table 1). Twenty-four serum samples showed high ECoV ELISA reactivity (S/P value > 0.5) and also had neutralizing antibodies with VNT varying between 32 and 768. The horse population of Iceland has been geographically isolated for more than 1000 years and is free from most common equine contagious diseases such as equine influenza, equine herpesvirus 1, strangles and equine viral arteritis [40]. To date, no prior studies of ECoV prevalence in horses from Iceland had been performed. This is the first evidence of the existence of ECoV infection in Iceland.

In conclusion, we developed a high-throughput, reliable and specific ELISA method to study humoral immune responses in horses against ECoV. With this method, we are able to perform the serodiagnosis of ECoV infection and assess the seroprevalence within horse populations in the future.

Supplementary Materials: The following are available online at http://www.mdpi.com/1999-4915/11/12/1109/s1. Table S1: Detection of antibodies to ECoV in equine serum samples during an ECoV outbreak by wELISA in winter 2014; Figure S1. Correlation between the OD values obtained with wELISA and virus neutralization titers (VNT) of 27 horses from acute and convalescent-phase sera.

Author Contributions: Conceptualization, K.v.M. and H.E.; Formal analysis, S.Z.; Funding acquisition, F.v.K.; Investigation, S.Z., C.S. and N.S.; Methodology, S.Z., K.v.M. and H.E.; Project administration, K.v.M. and H.E.; Resources, S.B. and N.P.; Supervision, F.v.K., B.-J.B., K.v.M. and H.E.; Visualization, S.Z.; Writing—original draft, S.Z.; Writing—review and editing, S.B., N.P., F.v.K., B.-J.B., K.v.M. and H.E.

Funding: S.Z. was supported by a grant from the Chinese Scholarship Council (File No. 201606910061).

Acknowledgments: We are grateful to Udeni B.R. Balasuriya (Louisiana Animal Disease Diagnostic Laboratory and Department of Pathobiological Sciences, School of Veterinary Medicine, Louisiana State University, Baton Rouge, Louisiana, USA) for providing strain NC99 and HRT-18G cells and to Sigríður Björnsdóttir (Icelandic Food And Veterinary Authority, Selfoss, Iceland) and Vilhjálmur Svansson (Institute for Experimental Pathology, University of Iceland, Reykjavik, Iceland) for providing sera from Icelandic horses. We also thank Heleen Zweerus for her practical assistance.

Conflicts of Interest: The authors declare no conflict of interest. The funders had no role in the design of the study; in the collection, analyses, or interpretation of data; in the writing of the manuscript, or in the decision to publish the results. The funders had no role in the design of the study; in the collection, analyses, or interpretation of data; in the writing of the manuscript, or in the decision to publish the results.

References

1. Cui, J.; Li, F.; Shi, Z.L. Origin and evolution of pathogenic coronaviruses. *Nat. Rev. Microbiol.* **2019**, *17*, 181–192. [CrossRef]

2. Su, S.; Wong, G.; Shi, W.; Liu, J.; Lai, A.C.K.; Zhou, J.; Liu, W.; Bi, Y.; Gao, G.F. Epidemiology, Genetic Recombination, and Pathogenesis of Coronaviruses. *Trends Microbiol.* **2016**, *24*, 490–502. [CrossRef] [PubMed]

3. Zhang, J.; Guy, J.S.; Snijder, E.J.; Denniston, D.A.; Timoney, P.J.; Balasuriya, U.B.R. Genomic characterization of equine coronavirus. *Virology* **2007**, *369*, 92–104. [CrossRef] [PubMed]

4. Guy, J.S.; Breslin, J.J.; Breuhaus, B.; Vivrette, S.; Smith, L.G. Characterization of a coronavirus isolated from a diarrheic foal. *J. Clin. Microbiol.* **2000**, *38*, 4523–4526.

5.   Oue, Y.; Ishihara, R.; Edamatsu, H.; Morita, Y.; Yoshida, M.; Yoshima, M.; Hatama, S.; Murakami, K.; Kanno, T. Isolation of an equine coronavirus from adult horses with pyrogenic and enteric disease and its antigenic and genomic characterization in comparison with the NC99 strain. *Vet. Microbiol.* **2011**, *150*, 41–48. [CrossRef]

6.   Miszczak, F.; Tesson, V.; Kin, N.; Dina, J.; Balasuriya, U.B.R.; Pronost, S.; Vabret, A. First detection of equine coronavirus (ECoV) in Europe. *Vet. Microbiol.* **2014**, *171*, 206–209. [CrossRef]

7.   Pusterla, N.; Mapes, S.; Wademan, C.; White, A.; Ball, R.; Sapp, K.; Burns, P.; Ormond, C.; Butterworth, K.; Bartol, J.; et al. Emerging outbreaks associated with equine coronavirus in adult horses. *Vet. Microbiol.* **2013**, *162*, 228–231. [CrossRef]

8.   Oue, Y.; Morita, Y.; Kondo, T.; Nemoto, M. Epidemic of equine coronavirus at obihiro racecourse, Hokkaido, Japan in 2012. *J. Vet. Med. Sci.* **2013**, *75*, 1261–1265. [CrossRef]

9.   Nemoto, M.; Schofield, W.; Cullinane, A. The First Detection of Equine Coronavirus in Adult Horses and Foals in Ireland. *Viruses* **2019**, *11*, 946. [CrossRef]

10.   Pusterla, N.; Vin, R.; Leutenegger, C.M.; Mittel, L.D.; Divers, T.J. Enteric coronavirus infection in adult horses. *Vet. J.* **2018**, *231*, 13–18. [CrossRef]

11.   Pusterla, N.; James, K.; Mapes, S.; Bain, F. Frequency of molecular detection of equine coronavirus in faeces and nasal secretions in 277 horses with acute onset of fever. *Vet. Rec.* **2019**, *184*, 385. [CrossRef] [PubMed]

12.   Fielding, C.L.; Higgins, J.K.; Higgins, J.C.; Mcintosh, S.; Scott, E.; Giannitti, F.; Mete, A.; Pusterla, N. Disease Associated with Equine Coronavirus Infection and High Case Fatality Rate. *J. Vet. Intern. Med.* **2015**, *29*, 307–310. [CrossRef] [PubMed]

13.   Berryhill, E.H.; Magdesia, K.G.; Aleman, M.; Pusterla, N. Clinical presentation, diagnostic findings, and outcome of adult horses with equine coronavirus infection at a veterinary teaching hospital: 33 cases (2012–2018). *Vet. J.* **2019**, *248*, 95–100. [CrossRef] [PubMed]

14.   Sanz, M.G.; Kwon, S.Y.; Pusterla, N.; Gold, J.R.; Bain, F.; Evermann, J. Evaluation of equine coronavirus fecal shedding among hospitalized horses. *J. Vet. Intern. Med.* **2019**, *33*, 918–922. [CrossRef]

15.   Nemoto, M.; Oue, Y.; Morita, Y.; Kanno, T.; Kinoshita, Y.; Niwa, H.; Ueno, T.; Katayama, Y.; Bannai, H.; Tsujimura, K.; et al. Experimental inoculation of equine coronavirus into Japanese draft horses. *Arch. Virol.* **2014**, *159*, 3329–3334. [CrossRef]

16.   Perera, R.A.; Wang, P.; Gomaa, M.R.; El-Shesheny, R.; Kandeil, A.; Bagato, O.; Siu, L.Y.; Shehata, M.M.; Kayed, A.S.; Moatasim, Y.; et al. Seroepidemiology for MERS coronavirus using microneutralisation and pseudoparticle virus neutralisation assays reveal a high prevalence of antibody in dromedary camels in Egypt, June 2013. *Eurosurveillance* **2013**, *18*, 20574. [CrossRef]

17.   Payne, D.C.; Iblan, I.; Rha, B.; Alqasrawi, S.; Haddadin, A.; Al Nsour, M.; Alsanouri, T.; Ali, S.S.; Harcourt, J.; Miao, C.; et al. Persistence of antibodies against middle east respiratory syndrome coronavirus. *Emerg. Infect. Dis.* **2016**, *22*, 1824–1826. [CrossRef]

18.   Tråvén, M.; Näslund, K.; Linde, N.; Linde, B.; Silván, A.; Fossum, C.; Hedlund, K.O.; Larsson, B. Experimental reproduction of winter dysentery in lactating cows using BCV—Comparison with BCV infection in milk-fed calves. *Vet. Microbiol.* **2001**, *81*, 127–151. [CrossRef]

19.   Dortmans, J.C.F.M.; Li, W.; van der Wolf, P.J.; Buter, G.J.; Franssen, P.J.M.; van Schaik, G.; Houben, M.; Bosch, B.J. Porcine epidemic diarrhea virus (PEDV) introduction into a naive Dutch pig population in 2014. *Vet. Microbiol.* **2018**, *221*, 13–18. [CrossRef]

20.   Reusken, C.B.E.M.; Haagmans, B.L.; Müller, M.A.; Gutierrez, C.; Godeke, G.J.; Meyer, B.; Muth, D.; Raj, V.S.; De Vries, L.S.; Corman, V.M.; et al. Middle East respiratory syndrome coronavirus neutralising serum antibodies in dromedary camels: A comparative serological study. *Lancet Infect. Dis.* **2013**, *13*, 859–866. [CrossRef]

21.   Leung, G.M.; Chung, P.H.; Tsang, T.; Lim, W.; Chan, S.K.K.; Chau, P.; Donnelly, C.A.; Ghani, A.C.; Fraser, C.; Riley, S.; et al. SARS-CoV antibody prevalence in all Hong Kong patient contacts. *Emerg. Infect. Dis.* **2004**, *10*, 1653–1656. [CrossRef] [PubMed]

22.   Walls, A.C.; Tortorici, M.A.; Bosch, B.-J.; Frenz, B.; Rottier, P.J.M.; DiMaio, F.; Rey, F.A.; Veesler, D. Cryo-electron microscopy structure of a coronavirus spike glycoprotein trimer. *Nature* **2016**, *531*, 114–117. [CrossRef] [PubMed]

23.   Hulswit, R.J.G.; de Haan, C.A.M.; Bosch, B.-J. Coronavirus Spike Protein and Tropism Changes. In *Advances in virus research*; Elsevier: Amsterdam, The Netherlands, 2016; Volume 96, pp. 29–57.

24. Belouzard, S.; Millet, J.K.; Licitra, B.N.; Whittaker, G.R. Mechanisms of Coronavirus Cell Entry Mediated by the Viral Spike Protein. *Viruses* **2012**, *4*, 1011–1033. [CrossRef]
25. Meyer, B.; Drosten, C.; Müller, M.A. Serological assays for emerging coronaviruses: Challenges and pitfalls. *Virus Res.* **2014**, *194*, 175–183. [CrossRef]
26. Zhao, S.; Li, W.; Schuurman, N.; van Kuppeveld, F.; Bosch, B.-J.; Egberink, H. Serological Screening for Coronavirus Infections in Cats. *Viruses* **2019**, *11*, 743. [CrossRef]
27. Kooijman, L.J.; Mapes, S.M.; Pusterla, N. Development of an equine coronavirus-specific enzyme-linked immunosorbent assay to determine serologic responses in naturally infected horses. *J. Vet. Diagn. Invest.* **2016**, *28*, 414–418. [CrossRef]
28. Bryan, J.; Marr, C.M.; Mackenzie, C.J.; Mair, T.S.; Fletcher, A.; Cash, R.; Phillips, M.; Pusterla, N.; Mapes, S.; Foote, A.K. Detection of equine coronavirus in horses in the United Kingdom. *Vet. Rec.* **2019**, *184*, 123. [CrossRef]
29. Wurm, F.M. Production of recombinant protein therapeutics in cultivated mammalian cells. *Nat. Biotechnol.* **2004**, *22*, 1393–1398. [CrossRef]
30. Reusken, C.; Mou, H.; Godeke, G.; van der Hoek, L.; Meyer, B.; Müller, M.; Haagmans, B.; de Sousa, R.; Schuurman, N.; Dittmer, U.; et al. Specific serology for emerging human coronaviruses by protein microarray. *Eurosurveillance* **2013**, *18*, 20441. [CrossRef]
31. Okba, N.M.A.; Raj, V.S.; Widjaja, I.; GeurtsvanKessel, C.H.; de Bruin, E.; Chandler, F.D.; Park, W.B.; Kim, N.-J.; Farag, E.A.B.A.; Al-Hajri, M.; et al. Sensitive and Specific Detection of Low-Level Antibody Responses in Mild Middle East Respiratory Syndrome Coronavirus Infections. *Emerg. Infect. Dis.* **2019**, *25*, 1868–1877. [CrossRef]
32. Hemida, M.G.; Chu, D.K.W.; Perera, R.A.P.M.; Ko, R.L.W.; So, R.T.Y.; Ng, B.C.Y.; Chan, S.M.S.; Chu, S.; Alnaeem, A.A.; Alhammadi, M.A.; et al. Coronavirus infections in horses in Saudi Arabia and Oman. *Transbound. Emerg. Dis.* **2017**, *64*, 2093–2103. [CrossRef] [PubMed]
33. Kooijman, L.J.; James, K.; Mapes, S.M.; Theelen, M.J.P.; Pusterla, N. Seroprevalence and risk factors for infection with equine coronavirus in healthy horses in the USA. *Vet. J.* **2017**, *220*, 91–94. [CrossRef] [PubMed]
34. He, Y.; Zhou, Y.; Wu, H.; Luo, B.; Chen, J.; Li, W.; Jiang, S. Identification of immunodominant sites on the spike protein of severe acute respiratory syndrome (SARS) coronavirus: Implication for developing SARS diagnostics and vaccines. *J. Immunol.* **2004**, *173*, 4050–4057. [CrossRef] [PubMed]
35. He, Y.; Li, J.; Heck, S.; Lustigman, S.; Jiang, S. Antigenic and immunogenic characterization of recombinant baculovirus-expressed severe acute respiratory syndrome coronavirus spike protein: Implication for vaccine design. *J. Virol.* **2006**, *80*, 5757–5767. [CrossRef]
36. Bass, E.P.; Sharpee, R.L. Coronavirus and gastroenteritis in foals. *Lancet (London, England)* **1975**, *2*, 822. [CrossRef]
37. Durham, P.J.K.; Stevenson, B.J.; Farquharson, B.C. Rotavirus and coronavirus associated diarrhoea in domestic animals. *N. Z. Vet. J.* **1979**, *27*, 30–32. [CrossRef]
38. Huang, J.C.; Wright, S.L.; Shipley, W.D. Isolation of coronavirus-like agent from horses suffering from acute equine diarrhoea syndrome. *Vet. Rec.* **1983**, *113*, 262–263. [CrossRef]
39. Mair, T.S.; Taylor, F.G.; Harbour, D.A.; Pearson, G.R. Concurrent cryptosporidium and coronavirus infections in an Arabian foal with combined immunodeficiency syndrome. *Vet. Rec.* **1990**, *126*, 127–130.
40. Björnsdóttir, S.; Harris, S.R.; Svansson, V.; Gunnarsson, E.; Sigurðardóttir, Ó.G.; Gammeljord, K.; Steward, K.F.; Newton, J.R.; Robinson, C.; Charbonneau, A.R.L.; et al. Genomic Dissection of an Icelandic Epidemic of Respiratory Disease in Horses and Associated Zoonotic Cases. *MBio* **2017**, *8*. [CrossRef]

Communication

Equine Rhinitis A Virus Infection in Thoroughbred Racehorses—A Putative Role in Poor Performance?

Helena Back [1], John Weld [2], Cathal Walsh [3] and Ann Cullinane [4,*]

[1] Department of Virology, Immunology and Parasitology, National Veterinary Institute, SE-751-89 Uppsala, Sweden; Helena.Back@mpa.se
[2] Riverdown, Barrettstown, Newbridge, Co., Kildare W12HD83, Ireland; johnweld1@eircom.net
[3] Department of Mathematics and Statistics, University of Limerick, Castletroy, Limerick V94 T9PX, Ireland; Cathal.Walsh@ul.ie
[4] Virology Unit, The Irish Equine Centre, Johnstown, Naas, Co. Kildare W91RH93, Ireland
* Correspondence: ACullinane@irishequinecentre.ie; Tel.: +353-45-866266; Fax: +353-45-866273

Received: 9 September 2019; Accepted: 15 October 2019; Published: 18 October 2019

Abstract: The aim of this study was to identify respiratory viruses circulating amongst elite racehorses in a training yard by serological testing of serial samples and to determine their impact on health status and ability to race. A six-month longitudinal study was conducted in 30 Thoroughbred racehorses (21 two-year-olds, five three-year-olds and four four-year-olds) during the Flat racing season. Sera were tested for the presence of antibodies against equine herpesvirus 1 and 4 (EHV-1 and EHV-4) and equine rhinitis viruses A and B (ERAV and ERBV) by complement fixation (CF) and equine arteritis virus (EAV) by ELISA. Antibodies against equine influenza (EI) were measured by haemagglutination inhibition (HI). Only ERAV was circulating in the yard throughout the six-month study period. Seroconversion to ERAV frequently correlated with clinical respiratory disease and was significantly associated with subsequent failure to race ($p = 0.0009$). Over 55% of the two-year-olds in the study seroconverted to ERAV in May and June. In contrast, only one seroconversion to ERAV was observed in the older horses. They remained free of any signs of respiratory disease and raced successfully throughout the study period. The importance of ERAV as a contributory factor in the interruption of training programmes for young horses may be underestimated.

Keywords: equine rhinitis virus A; Thoroughbred racehorses; loss of performance

---

1. Introduction

Equine respiratory infection together with lameness are the most common reasons for loss of training days and inability to race [1]. The main viruses associated with respiratory disease and loss of performance in racehorses are equine influenza virus (EI), equine herpesvirus 1 and 4 (EHV-1 and EHV-4) and equine rhinitis viruses A and B (ERAV and ERBV).

EI is a highly contagious virus and outbreaks of influenza may necessitate the cancellation of race meetings and other equestrian events [2]. However, in 1981, mandatory vaccination against EI was introduced for Thoroughbred racehorsesin Ireland and since then, no race meetings have been cancelled due to EI. However, outbreaks continue to occur in vaccinated horses, and young racehorses are particularly susceptible [3–5]. The return to athletic normality can be prolonged by damage to the mucociliary clearance mechanism and secondary bacterial infections [6].

EHV-1 and -4 circulate in horse populations worldwide and are associated with respiratory disease in horses, immunosuppression, reduced performance, abortion and occasionally, neurological disease [7–9]. Although there is no mandatory vaccination programme for EHV-1 and -4 in Ireland, some trainers vaccinate their horses in an effort to reduce the virus challenge in their yards.

ERAV and ERBV have also been identified worldwide in both healthy horses and horses with clinical respiratory signs [10–16]. A seroprevalence of 57% and 71% for ERAV and ERBV was recorded for Thoroughbred yearlings in Kentucky [17] and seroconversion to ERAV is common among young horses on entry to training yards [18–20]. Furthermore, ERBV was isolated in 30% of the horses with acute respiratory diseases in a Canadian survey [10].

There is a need for serological surveys to identify the viruses associated with economic loss in the racing industry. Longitudinal surveillance studies contribute to our understanding of the epidemiology of equine respiratory diseases in different countries and in different populations of horses. The aim of this serological study was to identify respiratory viruses circulating amongst elite racehorses of different ages in a training yard in Ireland and to determine their impact on health status and ability to race.

2. Materials and Methods

A six-month longitudinal study was conducted in 30 Thoroughbred racehorses during the Flat season. All the horses originated from the same stud farm and were in training with the same trainer. There were three different age groups of horses; the main group, i.e., 21 (70%) horses were two years of age, five (17%) of the horses were three years old and four (13%) were four years old. The two-year-old horses were in training for their first season. All the horses had been vaccinated against EI and EHV1 prior to entering the training yard.

At the request of the owner, whole blood samples were collected by the veterinary surgeon at monthly intervals from April to September. These were for routine serological screening and archiving as the "acute" samples for the testing of paired sera ("acute" and "convalescent") in the event of an outbreak of respiratory disease. Additional blood samples were collected from horses with respiratory disease in mid-May.

Similarly to previous longitudinal studies of Thoroughbred racehorses in training [21–23], sera were tested for presence of antibodies against ERAV, ERBV, EHV-1 and EHV-4 using the Complement fixation (CF) test (Table S1). The test was performed as described by Thomson et al. [24], using guinea pig complement and sensitized sheep erythrocytes. Serial twofold dilutions of sera from 1:5 to 1:640 were tested. Sera were tested for antibodies against EI H7N7 and H3N8 viruses using the Haemagglutination Inhibition (HI) test. The sera were pretreated with potassium periodate, inactivated for 30 min at 56 °C (±1 °C) and tested as described previously [25]. Seroconversion was defined as a 4-fold or higher rise in CF or HI antibody titers. Sera were tested for the presence of antibodies against equine arteritis virus (EAV) by indirect ELISA (ID Screen® Equine Viral Arteritis Indirect -Grabels, France).

A Chi-squared test was used to test the association of seroconversion with subsequent failure to race. The data was summarised in 2 × 2 contingency tables and an analysis was conducted in the R statistical software package version 3.5.1. The statistical significance was set at $\alpha = 0.05$.

3. Results

The numbers of samples taken during the study period were distributed over the age groups as follows: 154 (71%) were from two-year-old horses, 35 (16%) were from the three-year-old horses and 28 (13%) of the samples were collected from four-year-old horses.

The two-year-old horses entered the training yard where the older horses resided in the beginning of April. At the time of first sampling, all the horses tested by HI ($n = 26$) had antibody titres against EI that were consistent with vaccination. All the horses had undetectable or low antibody titres (≤20) against EHV1 and 4. During the study period, five horses seroconverted to EI and 12 to EHV-1 and/or EHV-4 in response to vaccination. The remaining horses were not vaccinated against these viruses during the study period and no other seroconversions to EI, EHV-1 or EHV-4 were detected. All the horses were seronegative for EAV throughout the study period. Thus, there was no serological evidence of natural exposure to these viruses during the study period.

With the exception of two of the two-year-old horses (antibody titres of 80 and 40) all the horses had undetectable or low antibody titres (≤20) against ERAV at the time of first sampling in April. This was also true for ERBV, with the exception of one two-year-old with an antibody titre of 40. The first seroconversion to ERAV was observed at the second sampling occasion in the beginning of May. One seronegative two-year-old colt had mounted a significant antibody response to ERAV (0 to 160) within four weeks of arriving in the training yard. He did not exhibit clinical respiratory signs but was described by the trainer as very slow at work. Within the following two weeks, clinical signs, including inappetence, dullness, nasal discharge, limb oedema, enlarged submandibular lymph nodes and occasional coughing, were observed in seven of the in-contact two-year-old horses. They were returned to the stud farm of origin to recuperate where blood samples were collected for serological testing. Four of the horses seroconverted to ERAV during the first two weeks in May, two were seropositive with stable titres, and one seroconverted at a later time point.

The ERAV serological results are summarized in Figure 1. Four additional seroconversions in two-year-old horses were identified by the end of May. Three of these horses had acute respiratory disease and were moved back to the stud farm. The fourth horse was subclinically infected and remained in training but was slow at work. By the end of June, four new seroconversions to ERAV were detected but only one horse had clinical respiratory signs and was returned home. One subclinically infected horse had previously seroconverted in early May.

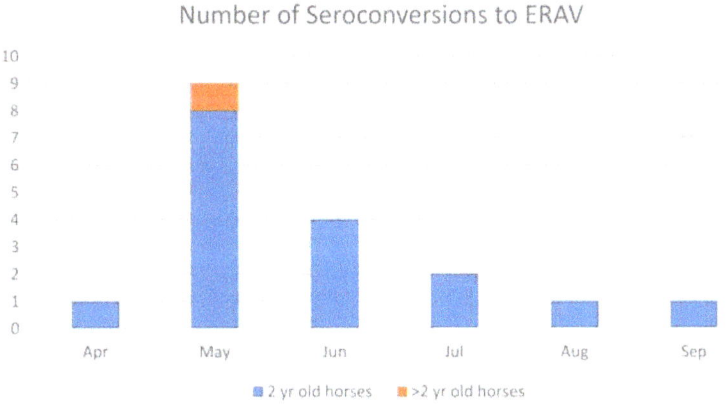

Figure 1. The number of horses that seroconverted to equine rhinitis A virus (ERAV) each month during the study period (April to September).

The number of seroconversions to ERAV decreased with time, with two in July, one of which was a horse that had previously seroconverted in the first week of May. One horse seroconverted in August and one in September, which had seroconverted previously in May. In contrast to the initial exposure to virus, these seroconversions in autumn were not associated with clinical respiratory signs or loss of performance.

No seroconversions to ERAV or ERBV and no clinical respiratory signs were observed in the three-year-old horses during the study period. One four-year-old horse seroconverted to ERAV at the end of May, but no clinical signs were observed and he won a race five days before the seropositive blood sample was collected.

In total, 18 seroconversions to ERAV were detected during the study period and 17 of them were in two-year-old horses. The majority, i.e. 13 (72%) of the seroconversions occurred in May and June. Only one of the two-year-old horses raced in May prior to being sent back to the stud farm to recover from respiratory disease. The two-year-old horses did not start racing after the respiratory episode until July, and, as can be seen in Figure 2, the percentage of two year-old-horses participating in race meetings was strikingly low when compared to the other two age groups. In total, during the study period, the five three-year-old horses raced 20 times and were placed first, second or third 12 times. The four older horses raced 12 times with six places. Only six of the 21 two-year-old horses raced. However, from 18 starts, they were placed 11 times. There was a significant association between seroconversion to ERAV and subsequent failure to race ($p = 0.009$).

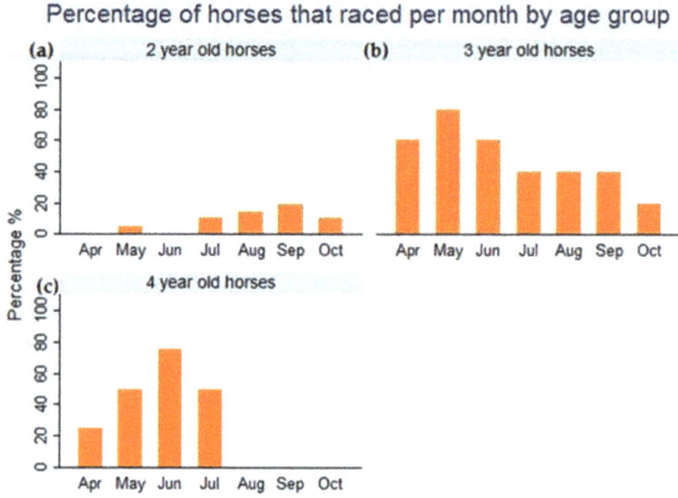

Figure 2. The percentage of horses that raced per month during the study period. The percentage of (**a**) two-year-old horses (**b**) three-year-old horses and (**c**) four-year-old horses that raced is shown.

The time from seroconversion to ERAV until the titers decreased to insignificant levels ranged from three weeks to five months, with a median value of two months and a mean value of 2.3 months. An association was observed between the antibody level and the rate of decline to original level.

Only one seroconversion to ERBV was observed in this study, a two-year-old horse seroconverted in June without any associated clinical respiratory signs.

4. Discussion

In this study, exposure to ERAV, as determined by serological testing, was associated with respiratory disease, loss of training days and failure to race in young racehorses. Monthly screening indicated that the majority of two-year-old horses were exposed to ERAV during the three months after entering the yard. The clinical signs that resulted in an interruption to training were primarily observed after the horses had been in the yard for several weeks and were being prepared for their first race. The susceptibility of young horses to rhinitis virus infection when moved to a new environment and starting to comingle with other horses has been reported previously [26,27]. In a seven-year serological study of racehorses in Japan, 69% of horses that seroconverted to ERAV were two years of age [16] and Black et al. [18] reported that 43% of horses seroconverted to ERAV within 7 months of entering a training stable in Australia. A previous study in a training yard in Ireland indicated that ERAV infection was largely confined to two-year-old horses and was most prevalent in late winter and spring [19]. However, we believe that this is the first investigation of the association between ERAV infection with interrupted training and failure to race. Over 55% of the two-year-olds in the study seroconverted to ERAV in May and June and none of them raced at that time. In fact, only 29% of the two-year-olds raced in their first season.

The results of this study suggest that the importance of ERAV in the interruption of training programmes for young horses may be underestimated. Acute febrile respiratory disease following ERAV infection has been reported previously [11,15,28,29]. In an experimental study, ERAV-inoculated ponies developed respiratory tract disease characterized by pyrexia, nasal discharge, adventitious lung sounds, and enlarged mandibular lymph nodes, which corresponded with an increase in antibody titres against the virus [29]. The clinical signs observed during this study were associated with the seroconversions and there was no evidence of other viral infections. ERAV may have been the primary cause of the respiratory disease and subsequent training loss, but it is more likely that the virus was a contributory factor in a multifactorial disorder. The serology tests used in this study are sensitive and specific, but it is possible that other less common viruses, such as adenovirus and coronavirus, or even as-yet unidentified equine viruses, may have played a role. Previous studies have demonstrated the complexity of poor performance syndrome in Thoroughbred racehorses and that bacteria are more common and may be aetiologically more important than viruses [21–23]. Unfortunately, the possible role of bacterial infection as a cofactor was not investigated in this study and the statistically significant association of ERAV with failure to race during the study period does not conclusively demonstrate causation. However, it is essential to determine which infectious agents are prevalent in populations that suffer interruption to training as a first step in assessing their true impact.

Lack of previous exposure to ERAV and stress associated with the change of environment and intensive training may have contributed to disease susceptibility. In contrast to the two-year-old horses, no seroconversions to ERAV were observed in the three-year-old horses—they remained free of any signs of respiratory disease and raced successfully throughout the study period. The seroconversion to ERAV in the four-year-old horse and the reinfection of two of the younger horses were not associated with clinical signs or reduced performance. Experimental studies have indicated that ERAV infection stimulated a protective response and that reinfection is asymptomatic [29].

The serology results indicate that ERAV was circulating in the yard throughout the six month study period. The dynamics of ERAV infection in horse populations are poorly understood but virus persists in horses even in the presence of high levels of antibodies [26,28]. ERAV can be isolated from the blood for a few days post infection and from the nasopharynx and faeces for up to a month [28]. Studies in the USA, Ireland and Australia detected ERAV in post-race urine samples at frequencies of 17%, 29% and 23%, respectively, which led to a suggestion that the persistent presence of ERAV in urine may contribute to its maintenance in training yards [17,30,31]. ERBV was not detected in urine [17,30]. During this study period, only one horse seroconverted to ERBV. This was consistent with an earlier study in Ireland in which ERAV was found to be more commonly detected than ERBV [30]. Co-circulation of ERAV and ERBV was also reported previously [30].

The low seroprevalence of ERAV and ERBV detected in the older horses at the beginning of this study may be due to the use of the CF test rather than the virus neutralization test (VNT) favored by some investigators [13,18,32]. The CF test is useful for the diagnosis of acute infections, as positive CF titres are often an indication of immunoglobulin M (IgM) antibodies or very high levels of IgG. Neutralizing antibodies which are mainly IgG persist for many years and the VNT is the test of choice for seroprevalence studies [33]. Burrows (1969) [34] reported that 59% of mares had neutralizing antibody against ERAV, in contrast to 10% of foals and yearlings, and suggested that most infection occurs during the period of training and racing. The relationship between neutralizing antibodies and clinical protection has been established Diaz-Mendez et al. [29]. It is likely that the older horses in this study that were seronegative by the CF test were exposed to ERAV in their first year of training and had neutralizing antibodies at the start of the study that were sufficient to protect them from infection and subsequent seroconversion, as measured by the CF test. However, this hypothesis remains unproven and it is also possible that although they were in the same yard as the two-year-old horses, they may not have been exposed to the virus. In this study, the CF antibodies declined to the original levels within months, confirming that they are not persistent and thus, a useful indicator of recent exposure to virus. As no relationship between CF antibodies and protection has been proposed, an investigation of an outbreak of ERAV including the measurement of neutralizing antibodies would give useful insight into protection.

In summary, this study demonstrates that routine serological monitoring of young racehorses by CF test revealed that ERAV is potentially an important contributor to training loss in young racehorses and merits further investigation with larger study populations and a more comprehensive testing regime. Unfortunately, only five nasopharyngeal samples were collected during this study, and all were negative for EHV1, EHV4, EI, EAV, ERAV and ERAB by real time PCR (data not shown). None of the nasal swabs were collected at the optimal time for detection of ERAV, which experimental infections suggest is from one to 12 days post-infection [31]. Further investigations in which samples are methodically collected for virus detection, serology and bacteriological investigation would be beneficial in elucidating the role of ERAV and other pathogens in the poor performance in young racehorses. However, all such field studies are dependent on the cooperation of the owner and the availability of horses.

ERAV is not contained in any of the vaccines that are currently available. As no clinical signs were attributable to reinfection, it is likely that immunization would be effective. The results of this study suggest that strategic vaccination of young horses against this virus could reduce economic loss due to respiratory disease and interruption of training schedules.

Supplementary Materials: The following are available at http://www.mdpi.com/1999-4915/11/10/963/s1.

Author Contributions: Conceptualization, H.B., J.W. and A.C.; methodology, A.C.; formal analysis, H.B., A.C. and C.W.; investigation, H.B., J.W. and A.C.; resources, H.B., J.W. and A.C.; writing—original draft preparation, H.B. and A.C.; writing—review and editing, A.C.; supervision, A.C.; project administration, A.C.; funding acquisition, H.B. and A.C.

Funding: The acquisition of data was funded by the owner and the Irish Equine Centre. Helena Beck's study visit to the Irish Equine Centre was funded by the National Veterinary Institute, Uppsala, Sweden. Ann Cullinane's contribution was supported by the Department of Agriculture, Food and the Marine under the Equine Technical Support Scheme 2019 Ref 35/IEC/2019.

Acknowledgments: The authors wish to thank the staff in the serology laboratory of the Irish Equine Centre, the owner and trainer without whom this study would not have been possible and Marie Garvey for assistance with manuscript preparation.

Conflicts of Interest: The authors declare no conflict of interest.

References

1. Wilsher, S.; Allen, W.R.; Wood, J.L. Factors associated with failure of thoroughbred horses to train and race. *Equine Vet. J.* **2006**, *38*, 113–118. [CrossRef] [PubMed]
2. Cullinane, A.; Newton, J.R. Equine influenza-A global perspective. *Vet. Microbiol.* **2013**, *167*, 205–214. [CrossRef] [PubMed]
3. Newton, J.R.; Daly, J.M.; Spencer, L.; Mumford, J.A. Description of the outbreak of equine influenza (H3N8) in the United Kingdom in 2003, during which recently vaccinated horses in Newmarket developed respiratory disease. *Vet. Rec.* **2006**, *158*, 185–192. [CrossRef] [PubMed]
4. Gildea, S.; Arkins, S.; Cullinane, A. Management and environmental factors involved in equine influenza outbreaks in Ireland 2007-2010. *Equine Vet. J.* **2011**, *43*, 608–617. [CrossRef] [PubMed]
5. Gildea, S.; Fitzpatrick, D.A.; Cullinane, A. Epidemiological and virological investigations of equine influenza outbreaks in Ireland (2010-2012). *Influenza Other Respir. Viruses* **2013**, *7* (Suppl. 4), 61–72. [CrossRef]
6. Willoughby, R.; Ecker, G.; McKee, S.; Riddolls, L.; Vernaillen, C.; Dubovi, E.; Lein, D.; Mahony, J.B.; Chernesky, M.; Nagy, E.; et al. The effects of equine rhinovirus, influenza virus and herpesvirus infection on tracheal clearance rate in horses. *Can. J. Vet. Res.* **1992**, *56*, 115–121.
7. Allen, G.P.; Kydd, J.; Slater, J.D.; Smith, K.C. Equid herpesvirus 1 and equid herpesvirus 4 infections. In *Infectious Diseases on Livestock*, 2nd ed.; Coetzer, J.A.W., Tustin, R.C., Eds.; Oxford Press: Cape Town, South Africa, 2004; pp. 829–859.

8.   Pusterla, N.; Kass, P.H.; Mapes, S.; Johnson, C.; Barnett, D.C.; Vaala, W.; Gutierrez, C.; McDaniel, R.; Whitehead, B.; Manning, J. Surveillance programme for important equine infectious respiratory pathogens in the USA. *Vet. Rec.* **2011**, *169*, 12. [CrossRef]

9.   Burgess, B.A.; Tokateloff, N.; Manning, S.; Lohmann, K.; Lunn, D.P.; Hussey, S.B.; Morley, P.S. Nasal shedding of equine herpesvirus-1 from horses in an outbreak of equine herpes myeloencephalopathy in Western Canada. *J. Vet. Intern. Med.* **2012**, *26*, 384–392. [CrossRef]

10.   Carman, S.; Rosendal, S.; Huber, L.; Gyles, C.; McKee, S.; Willoughby, R.A.; Dubovi, E.; Thorsen, J.; Lein, D. Infectious agents in acute respiratory disease in horses in Ontario. *J. Vet. Diagn. Invest.* **1997**, *9*, 17–23. [CrossRef]

11.   Li, F.; Drummer, H.E.; Ficorilli, N.; Studdert, M.J.; Crabb, B.S. Identification of noncytopathic equine rhinovirus 1 as a cause of acute febrile respiratory disease in horses. *J. Clin. Microbiol.* **1997**, *35*, 937–943.

12.   Plummer, G. An equine respiratory virus with enterovirus properties. *Nature* **1962**, *195*, 519–520. [CrossRef] [PubMed]

13.   Studdert, M.; Gleeson, L. Isolation and characterisation of an equine rhinovirus. *Zentralbl Veterinarmed B* **1978**, *25*, 225–237. [CrossRef] [PubMed]

14.   Diaz-Mendez, A.; Viel, L.; Hewson, J.; Doig, P.; Carman, S.; Chambers, T.; Tiwari, A.; Dewey, C. Surveillance of equine respiratory viruses in Ontario. *Can. J. Vet. Res.* **2010**, *74*, 271–278. [PubMed]

15.   Ditchfield, J.; Macpherson, L. Properties and classification of 2 new rhinoviruses recovered from horses in Toronto. *Cornell. Vet.* **1965**, *55*, 181–189.

16.   Sugiura, T.; Matsumura, T.; Imagawa, H.; Fukunaga, Y. A seven year serological study of viral agents causing respiratory infection with pyrexia among racehorses in Japan. In *Equine Infectious Diseases V*; Powell, D.G., Ed.; University of Kentucky Publications: Lexington, NY, USA, 1988; pp. 258–261.

17.   McCollum, W.; Timoney, P. Studies on the seroprevalence and frequency of equine rhinovirus-I and-II infection in normal horse urine. In *Equine Infectious Diseases VI*; Plowright, W., Rossdale, P.D., Wade, J.F., Eds.; R&W Publications: Newmarket, UK, 1992; pp. 83–87.

18.   Black, W.D.; Wilcox, R.S.; Stevenson, R.A.; Hartley, C.A.; Ficorilli, N.P.; Gilkerson, J.R.; Studdert, M.J. Prevalence of serum neutralising antibody to equine rhinitis A virus (ERAV), equine rhinitis B virus 1 (ERBV1) and ERBV2. *Vet. Microbiol.* **2007**, *119*, 65–71. [CrossRef]

19.   Klaey, M.; Sanchez-Higgins, M.; Leadon, D.P.; Cullinane, A.; Straub, R.; Gerber, H. Field case study of equine rhinovirus 1 infection: clinical signs and clinicopathology. *Equine Vet. J.* **1998**, *30*, 267–269. [CrossRef]

20.   Powell, D.G.; Burrows, R.; Spooner, P.; Mumford, J.; Thomson, G. A study of infectious respiratory disease among horses in Great Britain, 1971-1976. In *Equine Infectious Diseases IV*; Bryans, J.T., Gerber, H., Eds.; Veterinary Publications: Princeton, NJ, USA, 1978; pp. 451–459.

21.   Burrell, M.H.; Wood, J.L.; Whitwell, K.E.; Chanter, N.; Mackintosh, M.E.; Mumford, J.A. Respiratory disease in thoroughbred horses in training: the relationships between disease and viruses, bacteria and environment. *Vet. Rec.* **1996**, *139*, 308–313. [CrossRef]

22.   Cardwell, J.; Smith, K.; Wood, J.; Newton, J. A longitudinal study of respiratory infections in British National Hunt racehorses. *Vet. Rec.* **2013**, *172*, 637. [CrossRef]

23.   Wood, J.L.; Newton, J.R.; Chanter, N.; Mumford, J.A. Association between respiratory disease and bacterial and viral infections in British racehorses. *J. Clin. Microbiol.* **2005**, *43*, 120–126. [CrossRef]

24.   Thomson, G.R.; Mumford, J.A.; Campbell, J.; Griffiths, L.; Clapham, P. Serological detection of equid herpesvirus 1 infections of the respiratory tract. *Equine Vet. J.* **1976**, *8*, 58–65. [CrossRef]

25.   Cullinane, A.; Weld, J.; Osborne, M.; Nelly, M.; McBride, C.; Walsh, C. Field studies on equine influenza vaccination regimes in thoroughbred foals and yearlings. *Vet. J.* **2001**, *161*, 174–185. [CrossRef] [PubMed]

26.   Hofer, B.; Steck, F.; Gerber, H.; Löhrer, J.; Nicolet, J.; Paccaud, M. An Investigation of the Etiology of Viral Respiratory Disease in a Remount Depot. In *Equine Infectious Diseases III*; Bryans, J.T., Gerber, H., Eds.; Karger: Basel, Switzerland, 1973; pp. 527–545.

27.   Burrows, R. Equine rhinovirus and adenovirus infections. In Proceedings of the 24th Annual Convention of the American Association of Equine Practitioners, St. Louis, MO, USA, 1978; pp. 299–306.

28.   Plummer, G. Studies on an equine respiratory virus. *Vet. Rec.* **1962**, *74*, 967–970.

29.   Diaz-Mendez, A.; Hewson, J.; Shewen, P.; Nagy, E.; Viel, L. Characteristics of respiratory tract disease in horses inoculated with equine rhinitis A virus. *Am. J. Vet. Res.* **2014**, *75*, 169–178. [CrossRef] [PubMed]

30. Quinlivan, M.; Maxwell, G.; Lyons, P.; Arkins, S.; Cullinane, A. Real-time RT-PCR for the detection and quantitative analysis of equine rhinitis viruses. *Equine Vet. J.* **2010**, *42*, 98–104. [CrossRef] [PubMed]
31. Lynch, S.E.; Gilkerson, J.R.; Symes, S.J.; Huang, J.A.; Hartley, C.A. Persistence and chronic urinary shedding of the aphthovirus equine rhinitis A virus. *Comp. Immunol. Microbiol. Infect. Dis.* **2013**, *36*, 95–103. [CrossRef] [PubMed]
32. De Boer, G.F.; Osterhaus, A.D.; van Oirschot, J.T.; Wemmenhove, R. Prevalence of antibodies to equine viruses in the Netherlands. *Tijdschr. Diergeneeskd.* **1979**, *104*, 65–74. [CrossRef] [PubMed]
33. Burrows, R.; Goodridge, D. Observations of picornavirus, adenovirus, and equine herpesvirus infections in the Pirbright pony herd. In *Equine Infectious Diseases IV*; Bryans, J.T., Gerber, H., Eds.; Veterinary Publications: Princeton, NJ, USA, 1978; pp. 155–164.
34. Burrows, R. Equine rhinoviruses. In Proceedings of the Equine Infectious Diseases II, Paris, France, 16–18 June 1969; Bryans, J.T., Gerber, H., Eds.; Karger: Basel, Switzerland, 1970; pp. 154–164.

Article

Equine Mx1 Restricts Influenza A Virus Replication by Targeting at Distinct Site of its Nucleoprotein

Urooj Fatima [1], Zhenyu Zhang [1], Haili Zhang [1], Xue-Feng Wang [1], Ling Xu [1], Xiaoyu Chu [1], Shuang Ji [1,2] and Xiaojun Wang [1,*]

[1] State Key Laboratory of Veterinary Biotechnology, Harbin Veterinary Research Institute, Chinese Academy of Agricultural Sciences, Harbin 150069, China; uroojfatima.hvri@yahoo.com (U.F.); zhangzhenyu@caas.cn (Z.Z.); zhanghaili@caas.cn (H.Z.); wangxuefeng@caas.cn (X.-F.W.); 18266423976@163.com (L.X.); chuxiaoyuer@foxmail.com (X.C.); jishuang9096@163.com (S.J.)
[2] Advanced Institute for Medical Sciences, Dalian Medical University, Dalian 116044, China
* Correspondence: wangxiaojun@caas.cn; Tel.: +86-0451-5105-1749

Received: 25 September 2019; Accepted: 29 November 2019; Published: 2 December 2019

Abstract: Interferon-mediated host factors myxovirus (Mx) proteins are key features in regulating influenza A virus (IAV) infections. Viral polymerases are essential for viral replication. The Mx1 protein has been known to interact with viral nucleoprotein (NP) and PB2, resulting in the influence of polymerase activity and providing interspecies restriction. The equine influenza virus has evolved as an independent lineage to influenza viruses from other species. We estimated the differences in antiviral activities between human MxA (huMxA) and equine Mx1 (eqMx1) against a broad range of IAV strains. We found that huMxA has antiviral potential against IAV strains from non-human species, whereas eqMx1 could only inhibit the polymerase activity of non-equine species. Here, we demonstrated that NP is the main target of eqMx1. Subsequently, we found adaptive mutations in the NP of strains A/equine/Jilin/1/1989 (H3N8$_{JL89}$) and A/chicken/Zhejiang/DTID-ZJU01/2013 (H7N9$_{ZJ13}$) that confer eqMx1 resistance and sensitivity respectively. A substantial reduction in Mx1 resistance was observed for the two mutations G34S and H52N in H3N8$_{JL89}$ NP. Thus, eqMx1 is an important dynamic force in IAV nucleoprotein evolution. We, therefore, suggest that the amino acids responsible for Mx1 resistance should be regarded as a robust indicator for the pandemic potential of lately evolving IAVs.

Keywords: MxA; equine Mx1; influenza A viruses; polymerase activity; interspecies transmission; nucleoprotein; equine influenza

---

1. Introduction

Influenza A viruses (IAVs), commonly termed flu viruses, belong to the *Orthmyxoviridae* family and are broadly associated with acute febrile respiratory disease in many animals. Wild aquatic birds are believed to be the reservoir for these viruses [1,2]. However, although IAVs were initially limited to wild waterfowl, these viruses crossed the species barrier and have been spreading to other populations, including mammalian species, and have established independent lineages [1,3,4].

IAV possess a segmented and negative sense single-stranded RNA genome. The eight genomic segments of IAV encode at least 10 viral proteins. IAV derives its envelope from the host cell plasma membrane and contains three transmembrane proteins: Two surface glycoproteins termed as hemagglutinin (HA) and neuraminidase (NA), along with matrix protein 2 (M2). Two non-structural proteins NS1 and NS2, collectively termed as nuclear export protein (NEP) are associated with matrix protein 1 (M1) and viral ribonucleoprotein complexes (vRNPs). The eight vRNPs comprise of eight negative-strand RNA segments associated with the nucleoprotein (NP) and three RNA-dependent RNA polymerase (RdRp) subunits (PA, PB1, and PB2) [5]. Viral vRNPs are considered as minimal functional units required for early transcription (viral mRNA synthesis) and viral replication (vRNA synthesis) [6].

Myxovirus resistance proteins, including MxA and MxB in humans (named Mx1 and Mx2 in other animals), belong to dynamin-like GTPases family and are considered to be cell-autonomous host restriction factors of the innate immune system against many viral pathogens. Both the type-I and type-III interferons stimulate MxA and MxB expression during innate immune signaling [7,8]. The MxA protein is an indicator gene that is induced following an interferon action and halts a broad range of viral pathogens including DNA and RNA viruses (mainly *Orthomyxoviruses* such as influenza [9,10], measles [11], La Crosse [12], and *Hantaan* viruses [13]). HuMxA is a potent interspecies barrier for influenza viruses from other species [8,10,14].

MxA is a dynamin-like large GTPase comprised of an N-terminal globular GTPase domain, a bundle signaling element (BSE), and a C-terminal helical stalk. MxA can form stable tetramers and oligomers, which assemble in a criss-cross manner via the stalk [15,16]. Although the exact mechanism of Mx1-mediated immunity to influenza viruses is still unclear, a proposed possible mechanism suggests that, upon entry of viral infectious particles, MxA recognizes the incoming vRNPs and starts to self-assemble into rings, resulting in a higher-order oligomeric complex that blocks the vRNP function [16,17]. Viral NP is known to be a target of human and mouse MxA whereas PB2, which is associated with NP in the viral nucleocapsid, may serve as an additional target [6,18,19].

MxA represents a considerable barrier against the zoonotic introduction of avian influenza viruses into the human population [7]. Interestingly, several investigations reported the antiviral properties of MxA from a human and mouse [10,20,21], porcine [22,23], and bovine [14,24–26] origin. The interspecies transmission would have occurred when avian-origin IAV acquired certain mutations in the NP which overcome the MxA restriction [20,22,27]. A few of these mutations were already found in circulating IAV strains before they were even reported to cross the species barrier [20,22], but in many cases, the adaptational mutations occurred subsequent to infection and transmission. Indeed, recent studies indicated that IAVs which successfully established stable lineages in humans acquired adaptive mutations in the NP [20,28]. It was reported that a mutated H7N7 IAV carrying human signature NP mutations was more virulent in transgenic mice than the parental virus [29].

The equine influenza virus underwent an independent evolution pattern that can be compared with viruses from other species [30]. Equine IAVs can be classified into two main subtypes: H7N7 (a prototype virus A/equine/H7N7/Prague/56 that was isolated in 1956) [31] and H3N8 (a prototype virus A/equine/H3N8/Miami/63 that was isolated in 1963). Equine H7N7 viruses have not been isolated since 1990 and it is believed that they are no longer circulating in the equine population. Very little genetic exchange between the equine H3N8 virus subtype and viruses from other host species has been reported [32], which was thought to signify that horses were an endpoint until the H3N8 virus was acknowledged as being of canine and equine origin. Whether the equine Mx1 is functional and plays any role in interspecies restriction remains largely known. In this study, we investigated the interspecies restriction of eqMx1 to IAVs from different species. We found that different IAV strains were Mx1 or MxA sensitive to varying degrees. We further identified that 1) viral NP interacts with eqMx1 and determines the sensitivity to Mx1, and 2) single amino acid substitution at site 52 of the NP has a key role in determining this interaction and binding to Mx1. Our results suggest that the IAVs in the equine population have gained important mutations in the NP which enable a higher chance of eqMx1 resistance.

2. Materials and Methods

2.1. Cells and Viruses

Human embryonic kidney 293T (HEK293T) and Madin–Darby canine kidney (MDCK) cells were maintained in Dulbecco's Modified Eagle's Medium (DMEM) (Sigma Aldrich, St. Louis, MO, USA) supplemented with 10% foetal bovine serum (FBS, Wisent, Canada) and 1% antibiotics (100 units/mL of penicillin and 100 μg/mL of streptomycin; ThermoFisher, Waltham, MA, USA). The equine monocyte-derived macrophages (eMDMs) were prepared from equine peripheral blood mononuclear cells (PBMCs) as described previously [33]. Briefly, three healthy equids (horses) were used to collect the blood. The buffy coat was separated from blood samples by centrifugation at 1000 rpm for 10 min. Later, the buffy coat was used to isolate the PBMCs by centrifugation using a HybriMax Histopaque cushion (Sigma Aldrich, St. Louis, MO, USA) (d = 1.077 g/cm^3). The collected PBMCs were washed three times with PBS, resuspended, and maintained in RPMI 1640 (HyClone, Logan, Utah, USA) supplemented with 30% horse serum and 30% fetal bovine serum (FBS) (HyClone, Logan, Utah, USA). Aliquots of these isolated PBMCs were later seeded into tissue culture flasks at a density of 5×10^6 cells/cm^2 and were further incubated in a humidified chamber at 37 °C with 5% CO_2. At 24 hr post-incubation, the non-adherent cells which were found floating in the medium were discarded and the adherent monocytes were again incubated for the next 3 days in order to allow their differentiation into eMDMs. The cell culture atmosphere was maintained at 37 °C with 5% CO_2.

Wild type Influenza A/equine/Jilin/1/89 (H3N8$_{JL89}$) and recombinant viruses were generated using a plasmid-based reverse genetics system as previously described [34,35]. The generated wild-type virus and JL89-H52N-NP mutant virus were further propagated in 9-day-old embryonated chicken eggs at 35 °C for 72 hr, and allantoic fluid was harvested and centrifuged, and a hemagglutination test was performed to confirm the presence of the virus. Viruses from positive-testing samples were filtered through a 0.2 μm syringe filter (Millex, Merck, Ireland), and the viral titers were determined using the TCID$_{50}$ method of Reed and Muench [36]. Viruses were subsequently stored in small aliquots at −80 °C until further use.

2.2. Plasmids and Antibodies

In order to clone the cDNA derived from the total RNA of eMDMs, we used reverse-transcription PCR (RT-PCR) to amplify eqMx1. The eMDMs were treated with equine IFN-α1 (100 ng/μL) (Kingfisher Biotech, Minnesota, USA) for 24 hr. The primer sets were constructed following the genome sequences of *Equus caballus* myxovirus (Mx) dynamin-like GTPase 1 (Mx1), transcript variant X1, gene (GenBank accession no. XM_005606071.3), and the sequence of primers can be found in supplementary data (Table S1). Then, the fragments were successfully cloned in pcDNA3.1-HA vector (pcDNA3.1 (+) vector (Invitrogen, ThermoFisher, Waltham, MA, USA) that possessed 2X HA tags situated on the C-terminus and also p3X-Flag-CMV vector (Sigma Aldrich, St. Louis, MO, USA) that contained 3X Flag tags located at the N-terminus. The huMxA gene was acquired from Summus Co. (China) and was also successfully cloned using pcDNA3.1 (+).

The regents used in this study were kindly provided by the following personals: H1N1 human influenza virus A/WSN/1933 (WSN) by Dr. Kawoaka, and plasmids of H7N9 A/chicken/Zhejiang/DTID-ZJU01/2013 (H7N9$_{ZJ13}$) and H1N1 human IAV A/Sichuan/01/2009 (H1N1$_{SC09}$) by Dr. Hualan Chen. The equine influenza viruses H3N8$_{XJ07}$ (A/equine/Xinjiang/1/2007) and H3N8$_{JL89}$ were acquired from the previously preserved viral stock in our lab.

Site-directed mutants of NP sequences were generated using overlapping PCR and identified by DNA sequencing. Mutants of pcAGGS-H7N9-NP and pEF-JL89-NP were constructed according to the online In-Fusion® HD Cloning Kit (Clontech, Felicia, CA, USA) user manual (http://www.clontech.com/CN/Products/Cloning_and_Competent_Cells/Clonin_Kits/xxclt_searchResults.jsp). Briefly, the fragments of the pCAGGS/pcDNA3.1 vector and each target gene were amplified and were then fused using the In-Fusion® HD Enzyme kit (Clontech, Felicia, CA, USA). To create the N-H7N9xJL89-C plasmid, pcAGGS-H7N9-NP was used as the template to amplify the pCAGGS vector along with the N-terminal (1–200) of H7N9. This sequence was then fused with the JL89-NP C-terminal (201–480) fragment, where pEF-JL89-NP was used to amplify the inserted C-terminal (201–480) fragments. To obtain N-JL89xH7N9-C plasmids, pEF-JL89-NP was used as the template to amplify the pEF-JL89 vector backbone sequence along with the N-terminal (1–200) of JL89 and the inserted C-terminal (201–480) fragment of H7N9 was amplified from the pCAGGS-H7N9-NP sequence. The amplified vector and insert fragments were then fused together as described earlier. Single point mutations were also created by overlapping PCR in the same way. The constructed plasmids were transformed into DH5-alpha or Stable II competent cells according to the manufacturer's protocol (ThermoFisher, Waltham, MA, USA). Once the transformation was successful, the plasmids were extracted using PureLink HiPure plasmids DNA purification kits (Invitrogen, ThermoFisher, Waltham, MA, USA). All generated plasmids were confirmed by PCR and sequencing. The sequence of all the primers is available in supplementary data (Tables S2 and S3).

2.3. Polymerase Reconstitution Assay/Minireplicon Assay

In order to determine polymerase activity, we transfected the minigenome reporter (enclosing a firefly luciferase gene linked with non-coding regions of the HA gene of influenza containing human polI as the promoter and a mice terminator sequence) with NP and viral polymerase expression plasmids. We generated the chimeric mutants of the NP gene by overlapping PCR and verified it by the sequencing technique. To predict the viral polymerase activity due to mutations in the NP, a previously described protocol was followed [37]. Briefly, HEK293T cells were seeded in 24-well plates and co-transfected with expression plasmids of all viral polymerases i.e., PB1 (40 ng), PB2 (40 ng), PA (20 ng), and NP (80 ng), together with 40 ng of minigenome reporter (FF-Luc) and 10 ng of *Renilla* luciferase expression plasmids (pRL-TK, as an internal control) using polyjet transfection reagent (Invitrogen, ThermoFisher, Waltham, MA, USA) according to the manufacturers' protocol. Cells were maintained at 37 °C. To perform the cell lysis, we employed 200 μL of 1X reporter lysis buffer (Promega, Madison, USA) after 24 hr of transfection. To measure the luciferase activities of *Renilla* and Firefly, we employed the commercially available Dual-luciferase kit (Promega, Madison, USA) using Centro XS LB 960 luminometer (Berthold Technologies, Oak Ridge, Tennessee, USA). The levels of polymerase protein expression were detected using western blotting, with specific mouse monoclonal antibodies for NP and anti-HA tag antibody (Sigma Aldrich, St. Louis, MO, USA) for eqMx1-HA and huMxA-HA proteins. In the present study, all experimentation was performed independently at least thrice. The calculated results demonstrate mean ± SEM within one experiment.

2.4. Co-Immunoprecipitation Assay

Cells were incubated in a T-75 cell culture flask (ThermoFisher, Waltham, MA, USA) until a confluency of 70% was attained. HEK293T cells were then co-transfected with p3XFLAG-tagged eqMx1 (4 μg) together with PB1, PB2 (2 μg each), PA, FF-Luc (1 μg each), and NP expression plasmids either from (1) H7N9, and its mutants H7N9-S34G and N52H; or (2) H3N8$_{JL89}$ and its mutants JL89-G34S and H52N; or (3) an empty control vector (4 μg each) by using PolyJet™ In Vitro DNA transfection reagent (SignaGen, Rockville, MD, USA) according to the manufacturer's instructions. At24 hr post-transfection, co-immunoprecipitation was performed as previously described [38]. Briefly, transfected cells were lysed using an ice-cold lysis buffer (50 mM of Tris-HCL [pH 8], 150 mM of NaCl, 5 Mm of EDTA, and 1% NP-40), protease inhibitors (1:100), and freshly prepared 2 M N-ethylmaleimide (NEM) at 1:80 (Sigma Aldrich, St. Louis, MO, USA) and centrifuged at 4 °C for 10 min at 13,000× *g*. Post centrifugation, using Anti-FLAG M2 magnetic beads (Sigma Aldrich, St. Louis, MO, USA, M8823) or Anti-NP magnetic beads (made by incubation of MCE protein A/G magnetic beads along with the NP antibody from our lab) the lysates were co-incubated for 2 hr at 4 °C. Using a magnetic separator, the resins were harvested and washed thrice with cold PBS. In order to elute the resin-bound materials, we added a 3X Flag peptide and incubated the samples at 4 °C for 30 min, and then the eluted samples were boiled and SDS-PAGE was performed. The samples were shifted on the nitrocellulose membrane. 5% skim milk in Tris-buffered saline (TBS) was used as a blocking buffer for 2 hr and then the membranes were incubated with specific primary antibodies-TBST (TBS with 0.05% Tween20) at room temperature for 2 hr. Washing was performed thrice and the membranes were then incubated in TBST, containing secondary antibody (Sigma Aldrich, St. Louis, MO, USA, 1:10,000) at room temperature for 1 hr.

Subsequently, membranes were washed three times in PBST and scanned using a LI-COR Odyssey Imaging System (LI-COR, Lincoln, NE, USA).

2.5. Virus Growth Kinetics

The cells were seeded in 6-well plates and the next day, cells were transfected with pcDNA-3.1 eqMx1. 24 hr post-transfection, an infection experiment was performed as previously described [39]. Briefly, after 24 hr, the culture medium was removed and the cells were gently washed twice with preheated phosphate-buffered saline (PBS) and then infected for 1 hr at 37 °C with wild-type (H3N8$_{JL89}$) and/or mutant viruses (JL89-H52N-NP) at a multiplicity of infection (MOI) of 0.001. Viral dilutions were made in DMEM, supplemented with 1 µg/mL of TPCK added trypsin (Sigma Aldrich, St. Louis, MO, USA) and 0.25% BSA (Roche, Basel, Switzerland). Plates were gently shaken every 15 min so that the virus remained equally distributed within the cells. Later, excess virus was removed and cells were gently washed three times with preheated PBS and fresh infection media was added. Finally, the supernatants were collected at 0, 12, 24, 36, and 48 hr post-infection and debris was removed by centrifugation at 1000× g for 5 min. The amount of virus in each supernatant was determined either by plaque assay or by determining tissue culture infectivity dose and is expressed here either as PFU per mL or TCID$_{50}$/mL respectively.

2.6. Measurement of Gene Expression Using RT-qPCR

Total RNA from the collected supernatants was extracted using the RNeasy plus Minikit (Qiagen, Venlo, Netherland) and were subjected to a one-step real-time quantitative PCR (RT-qPCR) analysis using the AgPath-ID™ One-Step RT-PCR reagents (ThermoFisher, Waltham, MA, USA) according to the manufacturer's protocol. FAM-labeled probes (EIV-Tq-P) 5′-TCAGGCCCCCTCAAAGCCGA-TAMRA and specific primers targeting the M-gene of IAV, 5′-AGATGAGYCTTCTAACCGAGGTCG-3′ (EIV-Tq forward) and 5′-TGCAAANACATCYTCAAGTCTCTG-3′ (EIV-Tq-reverse) were used for the amplification of RNA, and relative mRNA expression levels were determined using double-standard curve methods.

2.7. Immunofluorescence Assay (IFA)

Briefly, MDCK cells were infected with wild-type (H3N8$_{JL89}$) and/or mutant viruses (JL89-H52N-NP) at an MOI of 0.001 and collected supernatants were used to infect MDCK cells in 96-well plates at various dilutions for 48 hr. Cells were fixed with chilled absolute ethanol for 30 min. The fixed cells were incubated with the anti-NP antibody (from our lab) for 2 hr at 37 °C in a humidified chamber, followed by three washes with phosphate-buffered saline (PBS), and then incubated with fluorescein isothiocyanate- (FITC) labeled goat anti-mouse IgG (Sigma Aldrich, St. Louis, MO, USA) for 1 hr 37 °C. Following three washes with PBS, the cells were examined under an inverted fluorescence microscope (Nikon TE200, Tokyo, Japan).

2.8. Evolutionary Analysis

In order to determine the evolutionary relationships between NPs from IAV strains from different hosts, a total of 153 representative nucleotide sequences of viral nucleoproteins from IAVs with different host specificities (including avian, porcine, canine, equine, and human) were retrieved from the GenBank and aligned manually using Bioedit alignment tools. The evolutionary history was inferred by the maximum likelihood method based on the JTT matrix-based model.

2.9. Statistical Analysis

In order to perform the statistical analyses, we used GraphPad Prism, version 5 (https://www.graphpad.com/scientific-software/prism/) (San Diego, California, USA). Employing the statistical approaches of one-way ANOVA and Dunnett's post-test, the statistical differences were analyzed. All experiments were thrice conducted independently. The error bars present standard deviation (SD) or standard error of the mean (SEM) in each experimental group that is documented in the figure legends. The abbreviations are designated as NS not significant, $p > 0.05$, * $0.01 \leq p < 0.05$, ** $0.001 \leq p < 0.01$, *** $0.0001 \leq p < 0.001$, and **** $p < 0.0001$ respectively.

3. Results

3.1. Identification of Cross-Species Antiviral Potential of Human and Equine Mx1

MxA (or Mx1) as one of the major interferon-stimulated genes (ISGs) is responsible for counteracting many invading viruses, especially those belonging to *Orthomyxoviridae*. To identify the eqMx1's function in the restriction of IAVs, we first cloned the equine Mx1 from eMDMs. By using bioinformatics prediction software (https://www.uniprot.org, SWISS-MODEL, and Phyre2), eqMx1 was predicted to be structurally similar to huMxA [15] (Figure 1A). The eMDMs were treated with equine IFN-α1 (100 ng/μL) (Kingfisher Biotech, Minnesota, USA) for 24 hr to enhance the basic expressing level of eqMx1, and a 1983-bp fragment was amplified and cloned into a pcDNA3.1-HA vector [pcDNA3.1 (+)] (Invitrogen, ThermoFisher, Waltham, MA, USA) with 2X HA tags at the C-terminal. The expressions of eqMx1 and huMxA were verified by western blotting (Figure 1B). Polymerase assay was then used to evaluate Mx1's effect on the replications of IAV strains from different host species, including human IAVs H1N1 (A/Sichuan/01/2009 and A/WSN/1933), avian IAVs H7N9 (A/chicken/Zhejiang/DTID-ZJU01/2013) and H5N1 (A/chicken/Scotland/1959), and equine IAVs H3N8 (A/equine/Xinjiang/1/2007 and A/equine/Jilin/1/1989). Mx1 or MxA resistance is defined here as the comparative activity of the viral polymerase in the presence of Mx1 divided by the activity calculated in the absence of Mx1. The results suggest that huMxA has antiviral potential against the IAV strains from non-human species i.e., avian and equine strains, whereas eqMx1 could only restrict the polymerase activity of non-equine species i.e., avian and human. The antiviral activity of eqMx1 to avian IAVs (H7N9$_{ZJ13}$ and H5N1$_{(R)}$) was slightly stronger than that of huMxA (Figure 1C,D). The expression of all the plasmids was determined with western blotting (Figure 1E). The results suggest that huMxA has antiviral potential against the IAV strains of avian and equine but not a human origin, whereas eqMx1 could inhibit the polymerase activity of avian and human but not equine stains.

3.2. The Viral Nucleoprotein as a Possible Target of Mx1 Action

Viral NP is a major target of MxA and the adaptive mutations of NP enable the virus to escape from MxA restriction [20,21,41]. Analysis of a total of 153 representative nucleotide sequences of viral NPs from IAVs with different host specificities (including avian, swine, canine, equine, and human) showed that the NP proteins generate diverse lineages in a host-species-specific manner (Figure 2). Considering the diversity of Mx1 proteins and the different activities on viruses from different hosts (Figure 1), we hypothesized that the NP from equine viruses may have evolved a specific signature of adaptation to counter eqMx1, and this "signature" of the equine virus may not be able to overcome the restriction of human MxA.

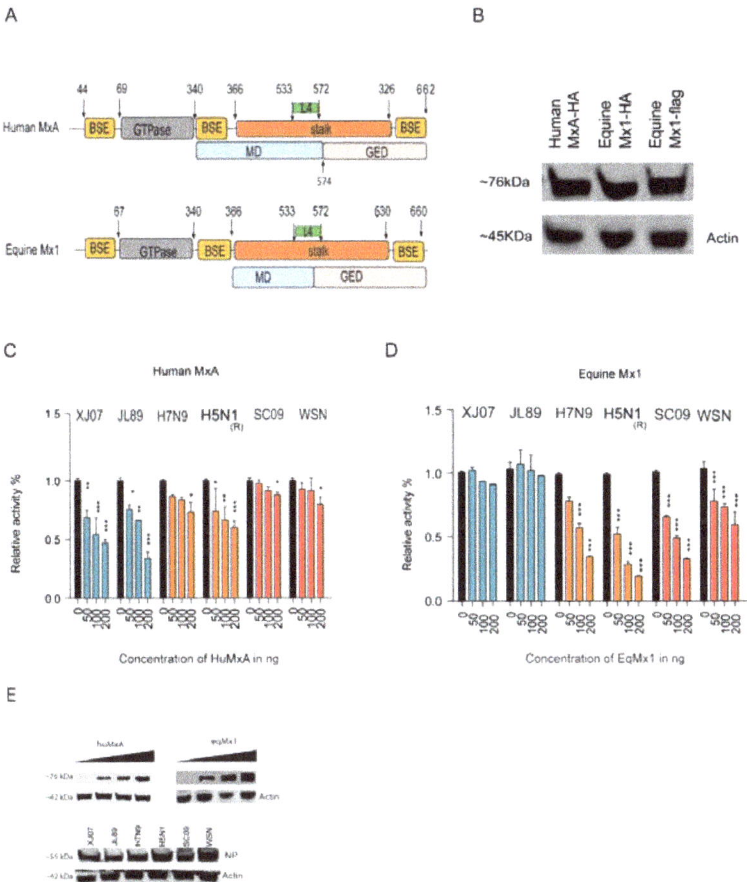

Figure 1. Antiviral activities of human MxA (huMxA) and equine Mx1 (eqMx1). (**A**) Structure-based domain representation of human MxA and equine Mx1. The GTPase is shown in grey, the middle domain (MD) is blue, and the GTPase effector domain (GED) is shown in pink. The GTPase domain is connected to the MD and GED through bundle signaling element (BSE) (yellow) and the stalk is colored orange. The arrows in the schematic denote the first and last visible residues in the structure. (**B**) Expression of plasmids by western blotting. The expression of huMxA-pcDNA3.1-HA, eqMx1-pcDNA3.1-HA, and p3X-FLAG-eqMx1 plasmids. The figures represent the bands of expressed proteins as visualized using western blotting, with antibodies used against the hemagglutinin (HA) tag and flag tag. (**C,D**) Relative luciferase activities of different influenza A virus (IAV) strains. Human embryonic kidney 293T (HEK293T) cells were co-transfected with expression plasmids of PB1 (40 ng), PB2 (40 ng), PA (20 ng), and nucleoprotein (NP) (80 ng) from different IAV strains, together with 40 ng of minigenome reporter (FF-luc) and 10 ng of *Renilla* luciferase expression plasmids (pRL-TK, as an internal control) in the presence of HA-tagged pcDNA-3.1 eq-mx1 (**C**) or pcDNA-3.1 huMxA (**D**) and an empty control vector at increasing concentrations of 0, 50, 100, and 200 ng each. After 24 hr of transfection, cells were lysed using a 1X reporter lysis buffer. Firefly and *Renilla* luciferase activities were measured. The resulting relative activity in the presence of either H7N9$_{ZJ13}$ or H3N8$_{JL89}$ was set to 100%. (**E**) The western blot analysis was performed to determine the expression levels of eqMx1 or huMxA and NP expression plasmids. (Statistical differences between samples are indicated, according to a one-way ANOVA followed by a Dunnett's test; NS = not significant, * $0.01 \leq p < 0.05$, ** $0.001 \leq p < 0.01$, *** $p < 0.001$. Error bars represent the SEM within one representative experiment).

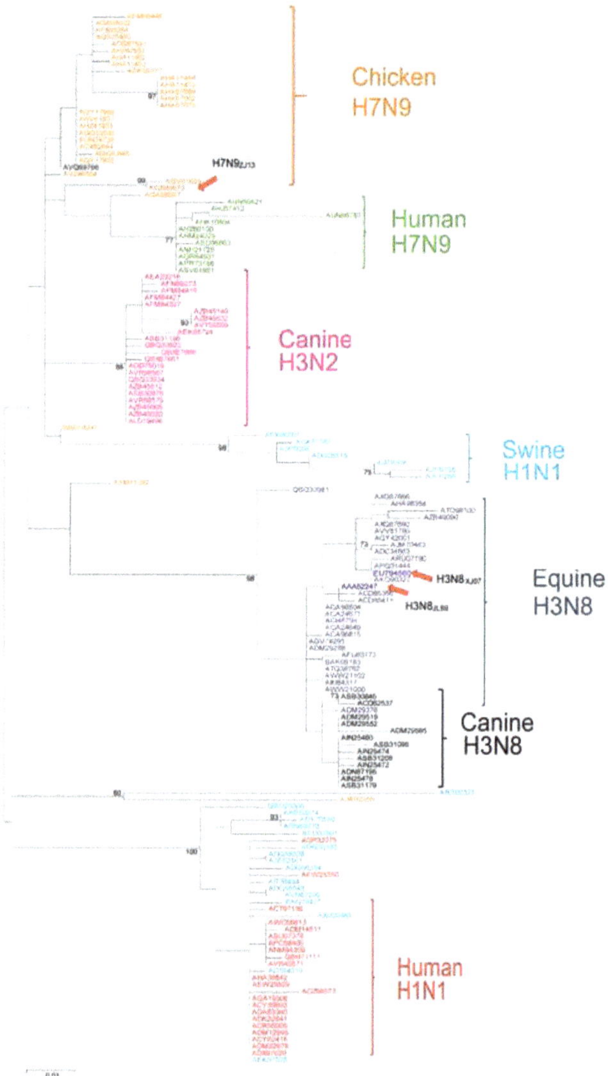

Figure 2. Phylogenetic analysis demonstrates the relationship of influenza A virus NP from distinct hosts. Molecular phylogenetic analysis of evolutionary relationships among different viral strains was inferred using the maximum likelihood strategy. The analysis was conducted using the JTT matrix-based model. The tree with the highest log likelihood (2519.6632) is shown. The percentage of trees in which the associated taxa clustered together is shown next to the branches. The initial tree(s) for the heuristic search was obtained automatically by applying neighborhood Join and BioNJ algorithms to a matrix of pairwise distances estimated using the JTT model and then selecting the topology with the superior log-likelihood value. A discrete Gamma distribution was used to model the evolutionary rate differences among sites [five categories (+G, parameter = 0.4056)]. The tree is drawn to scale, with branch lengths representing the number of substitutions per site. The analysis involved 153 sequences. All positions containing gaps and missing data were eliminated. The final dataset comprised a total of 304 positions. Evolutionary analyses were conducted in MEGA7 software [40].

To confirm this idea, we used a mini-replicon system-based approach to identify the target sequence of NP for eqMx1. We swapped expression plasmids between the different strains $H7N9_{ZJ13}$ and $H3N8_{JL89}$ to determine whether an Mx1-sensitive mini-replicon system could be converted into a more resistant system and vice versa (Figure 3A). The replacement PB1, PB2, or PA from $H3N8_{JL89}$ into $H7N9_{ZJ13}$, did not change the polymerase activity of the re-assorted complex to a large extent. However, the replacement by $H3N8_{JL89}$ NP resulted in a complete reversal of the polymerase activity of $H7N9_{ZJ13}$ equivalent to the actual $H3N8_{JL89}$ polymerase system (Figure 3B). Thus, our approach identified the IAV NP as a possible target of eqMx1.

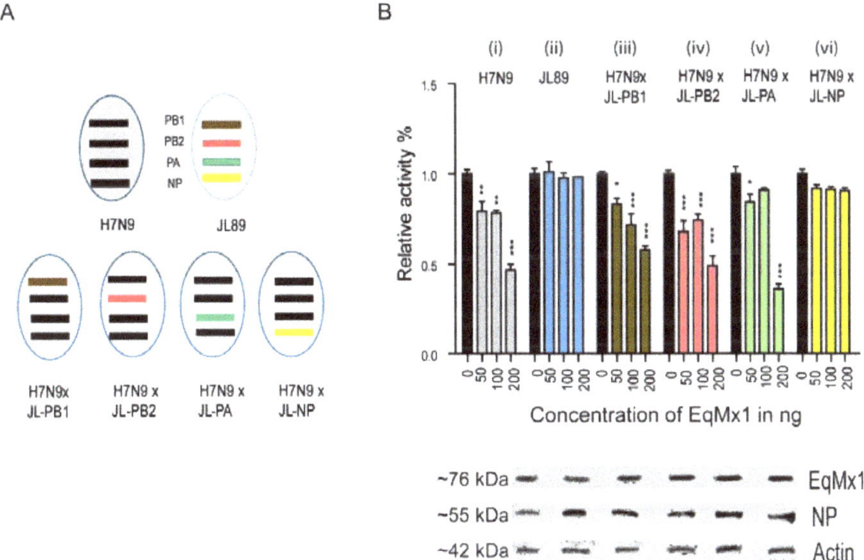

Figure 3. The viral NP is responsible for the sensitivity to eqMx1. (**A**) Schematic representation of the assortment of viral polymerases. The four polymerases (PB1, PB2, PA, and NP) from chicken $H7N9_{ZJ13}$ were swapped one by one with the equivalent polymerases from A/equine/Jilin/1/1989 ($H3N8_{JL89}$), and each group of assorted plasmids was co-transfected into HEK293T. (**B**) Relative Luciferase activities of combination sets of IAV against eqMx1. Cells were co-transfected with expression plasmids of the polymerase PB1 (40 ng), PB2 (40 ng), PA (20 ng), and NP (80 ng) in six different groups, (i) all polymerases of $H7N9_{ZJ13}$, (ii) all polymerases of $H3N8_{JL89}$, (iii) PB2, PA, NP of $H7N9_{ZJ13}$ and PB1 of $H3N8_{JL89}$, (iv) PB1, PA, NP of $H7N9_{ZJ13}$ and PB2 of $H3N8_{JL89}$, (v) PB1, PB2, NP of $H7N9_{ZJ13}$ and PA of $H3N8_{JL89}$, (vi) PB1, PB2, PA of $H7N9_{ZJ13}$ and NP of $H3N8_{JL89}$, together with 40 ng of minigenome reporter (FF-luc) and 10 ng of *Renilla* luciferase expression plasmids (pRL TK, as an internal control) in the presence of HA-tagged pcDNA-3.1 eqMx1 or an empty control vector at an increasing concentration 0, 50, 100, and 200 ng each. After 24 hr of transfection, cells were lysed using a 1X reporter lysis buffer. Firefly and *Renilla* luciferase activities were measured. The resulting relative activity in the presence of either $H7N9_{ZJ13}$ or $H3N8_{JL89}$ was set to 100%. The western blot analysis shown in (**B**) was performed to determine the expression levels of eqMx1 and NP. (Statistical differences between samples are indicated, according to a one-way ANOVA followed by a Dunnett's test; NS = not significant, * $0.01 \leq p < 0.05$, ** $0.001 \leq p < 0.01$, *** $p < 0.001$. Error bars represent the SEM within one representative experiment).

3.3. Identification of Residues in NP of the H3N8$_{JL89}$ Influenza A Virus Responsible for Resistance to eqMx1

Next, we evaluated the key residues of NP from the H3N8$_{JL89}$ IAV that confer eqMx1 resistance in the context of the H7N9$_{ZJ13}$ polymerase. The sequence analysis comparing the NPs of H7N9$_{ZJ13}$ and H3N8$_{JL89}$ of avian origin showed 16 amino acids (AAs) variations between the two strains (Figure 4A). Based on the sequence alignment obtained in Figure 4A, two different chimeras were constructed by fusing two NP proteins from H3N8$_{JL89}$ and H7N9$_{ZJ13}$ (Figure 4B) and tested their polymerase activity against eqMx1. An artificial chimera N-JL89xH7N9-C comprised of the N-terminal 200 AAs of the H3N8$_{JL89}$ NP and the C-terminal domain (AAs position 201 to 480) of the H7N9$_{ZJ13}$ NP, behaved like the full-length H3N8$_{JL89}$ NP, indicating that the eight different AAs at the C-terminal of the NP protein did not contribute to the Mx1 resistance phenotype (Figure 4C). However, a chimeric construct N-H7N9xJL89-C with the replacement of the N-terminal of H3N8$_{JL89}$-NP by 200 AAs from H7N9$_{ZJ13}$ made it more sensitive towards inhibition by eqMx1 in contrast to the actual H3N8$_{JL89}$-NP. Conversely, the opposite mutant N-JL89xH7N9-C that had an insertion of 200 AAs from H3N8$_{JL89}$ into the N-terminal of H7N9$_{ZJ13}$-NP lost its activity as compared to the wild type H7N9$_{ZJ13}$-NP, indicating that the N-terminal played a key role in the determination of resistance or sensitivity towards eqMx1 (Figure 4C).

Out of eight substitutions of NP, two positions V85A and T197I were not selected for mutations as when the sequence of two strains of H3N8, H3N8$_{JL89}$, and H3N8$_{XJ07}$, were compared, the H3N8$_{XJ07}$ (GenBank accession no. EU794560.1) contained the same amino acids at these positions as that of H7N9$_{ZJ13}$, suggesting that these sites were not responsible for eqMx1 restriction (Figure 5A). The other six residues were selected for single point mutations in both H3N8$_{JL89}$ NP and H7N9$_{ZJ13}$ NP. In H3N8$_{JL89}$ NP, the mutants were G34S, G50S, H52N, K77R, M105V, and V186I. We found mutations G34S (albeit not statistically significant) and H52N altered the restriction of eqMx1, whereas the mutants G50S, K77R, M105V, and V186I remained completely unaffected by eqMx1. The effect of these point mutations was also confirmed by creating reverse mutations in the H7N9$_{ZJ13}$ NP. As expected, the reverse effect was observed for the mutation at site 52 of H7N9$_{ZJ13}$ NP (N52H) (Figure 5B). However, interestingly, mutating S34G in the H7N9$_{ZJ13}$ NP had no distinguishable alteration in the activity towards eqMx1, rather a markedly less resistant phenotype was observed and the mutant behaved like a wild type H7N9$_{ZJ13}$ NP. Similarly, mutations at S50G, R77K, V105M, and I186V of H7N9$_{ZJ13}$ NP showed no effect and mutation I186V showed more inhibition towards eqMx1. These results indicate that position 52 of equine IAV is important for NP adaptation to eqMx1.

In order to confirm the importance of site 52, common to both H3N8$_{JL89}$ NP and H7N9$_{ZJ13}$ NP, the specific mutations were introduced into H5N1 NP and tested for its influence on eqMx1 resistance in the H5N1 polymerase reconstitution assay. H5N1 was chosen because it is another avian IAV that is blocked by eqMx1. H5N1 contains Tyrosine (Y) at position 52 (Figure 5C). As expected, point mutations in H3N8$_{JL89}$ NP (H52Y) and H5N1 NP (Y52H) showed a reverse effect against eqMx1, confirming the importance of position 52. In order to determine the effect of the two identified sites 34 and 52 in H3N8$_{JL89}$ in resistance against huMxA, the polymerase activity of three H3N8$_{JL89}$ NP mutants G34S, H52N, and H52Y, as well as two H7N9$_{ZJ13}$ NP mutants S34G and N52H, were measured. The mutants G34S, H52N, and H52Y showed less pronounced polymerase activity than the wild type H3N8$_{JL89}$ NP but still could not escape the antiviral activity of huMxA (Figure 5D). Similarly, reverse mutations in H7N9$_{ZJ13}$ NP, S34G, and N52H did not alter the sensitivity to huMxA (Figure 5E). A decent expression level of all the NP mutant plasmids was detected using western blotting. Together, these results suggested that the amino acids at position 52 of NPs from different IAVs conferred Mx1 resistance.

A

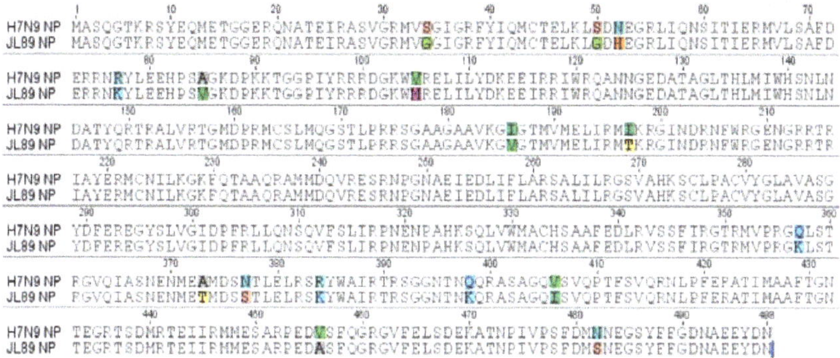

B

C

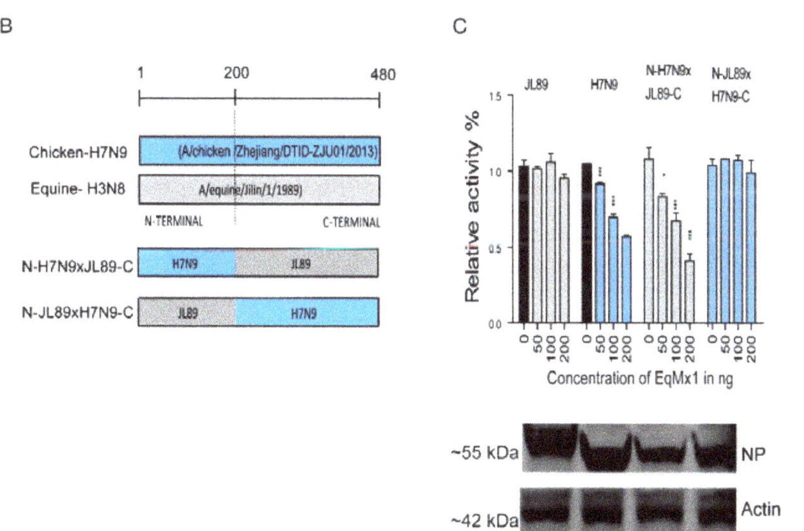

Figure 4. The N terminal of the NP determines the resistance to the restriction of eqMx1. (**A**) The sequence alignment of NP from H7N9$_{ZJ13}$ and H3N8$_{JL89}$ showed the differences in AAs at 16 different positions. (**B**) The schematic presentation of mutant NP sequences generated using overlapping PCR. The NP sequences of H3N8$_{JL89}$ and H7N9$_{ZJ13}$ were divided into two segments at the N and C terminals (AAs position N = 1–200; C = 201–480). The figure shows the nucleotide positions where the site-directed mutants of NP sequences were generated. (**C**) The relative luciferase activity of chimeric clones of NP against eqMx1. Constructed site-directed mutants of the NP sequences were tested for their polymerase activities in the presence of an HA-tagged pcDNA-3.1 eqMx1 or pcDNA-3.1-HA empty control vector, transfection complexes of viral polymerases and reporter plasmids were made and the test was performed as described earlier. The expression levels of eqMx1 and NP proteins were assessed using western blotting. (Statistical differences between samples are indicated, according to a one-way ANOVA followed by a Dunnett's test; NS = not significant, * $0.01 \leq p < 0.05$, *** $p < 0.001$. Error bars represent the SEM within one representative experiment).

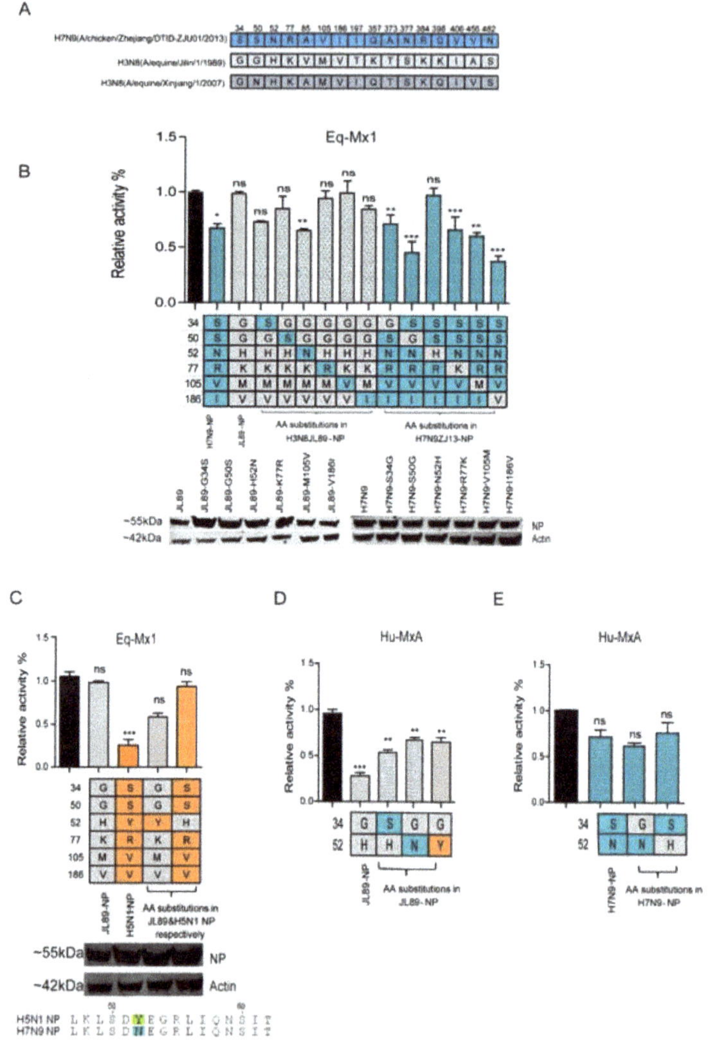

Figure 5. Single amino acid mutation of the NP confers resistance to eqMx1. (**A**) The different AAs in the NP of H7N9$_{ZJ13}$, H3N8$_{JL89}$, and H3N8$_{XJ07}$. (**B–E**) HEK 293T cells were transfected with a firefly minigenome reporter, *Renilla* expression control, and respectively indicated point mutants of the NP from either H3N8$_{JL89}$ polymerase (**B**) or H7N9$_{ZJ13}$ polymerase (**B**) or H5N1 polymerase (**C**) in the presence of eqMx1 (0, 200 ng). Polymerase activity of mutants with point mutations at sites 34 and 52 of the NP from both H3N8$_{JL89}$ (**D**) and H7N9$_{ZJ13}$ (**E**) was measured against huMxA (0, 200 ng). Luciferase activity was measured at 24 hr post-transfection. The polymerase activity observed in the presence of the HA-tagged eqMx1 was normalized to an empty control vector (black bar). The resulting relative activity in the presence of either H7N9$_{ZJ13}$ (blue) or H3N8$_{JL89}$ (grey) was set to 100%. The sequence analysis of the aforementioned IAV strains is shown in C. (**A–E**) data are firefly Luciferase gene activity normalized to that of *Renilla*. (Statistical differences between cells are indicated, following a one-way ANOVA and subsequent Dunnett's test; NS = not significant, * $0.01 \leq p < 0.05$, ** $0.01 \leq p < 0.01$, *** $p < 0.001$. Error bars represent the SEM of the replicates within one representative experiment). The expression levels of all NP point mutant proteins were assessed by western blotting.

3.4. The Amino Acid Position 52 of NP Affects its Interaction with eqMx1

To further confirm the interaction between NP and eqMx1 and the importance of site 52 of the NP in determining this interaction, we conducted a co-immunoprecipitation (Co-IP) experiment using different NP proteins and eqMx1 protein. Flag-tagged eqMx1 was co-expressed in HEK293T cells, individually or in combination with expression plasmids of PB1, PB2, PA, viral reporter RNA (FF-luc), and NP from either H3N8$_{JL89}$ and mutants (H3N8$_{JL89}$-G34S-NP, H3N8$_{JL89}$-H52N-NP) or H7N9$_{ZJ13}$ and its mutants (H7N9$_{ZJ13}$-S34G-NP, H7N9$_{ZJ13}$-N52H-NP), to mimic the viral vRNPs. Twenty four hours post-transfection, total cell lysates were harvested from transfected cells, and western blotting was performed after FLAG-IP using Anti-FLAG M2 magnetic beads (Sigma Aldrich, St. Louis, MO, USA, M8823). Results showed that when the eqMx1 was co-expressed with either H3N8$_{JL89}$-NP or H3N8$_{JL89}$-G34S-NP, no interaction was identified. Expectedly, the mutant H3N8$_{JL89}$-H52N-NP showed a strong interaction with eqMx1 (Figure 6A). In contrast, Flag-tagged eqMx1 was co-immunoprecipitated with NP from H7N9$_{ZJ13}$ and its mutant –S34G when they were co-expressed. This interaction was lost when site 52 mutated in H7N9$_{ZJ13}$ to–N52H (Figure 6B). This finding gave evidence to the hypothesis that position 52 could be the most important site for interaction between eqMx1 and NP of IAV.

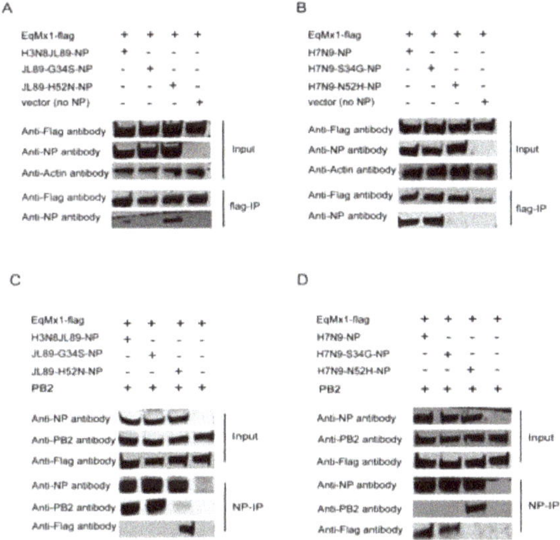

Figure 6. Co-immunoprecipitation of eqMx1 and NP/PB2. We carried out this experiment by doing a pull-down of the NP followed by the analysis of co-immunoprecipitated PB2 using HEK293T cells that were co-transfected with p3XFLAG- eqMx1 (4 µg) together with PB1, PB2 (2 µg each), PA, FF-Luc (1 µg each), and NP expression plasmids from H7N9$_{ZJ13}$ or its mutants H7N9 $_{ZJ13}$-S34G, N52H; or H3N8$_{JL89}$ or its mutants H3N8$_{JL89}$-G34S, H52N; or an empty control vector (4 µg each). Total lysates were made 24 hr after transfection, HEK293T cells lysates were assayed by FLAG-IP (**A,B**) or NP-IP (**C,D**) and blotted with the indicated antibodies. The PB2 antibody was acquired from NOVUS Biologicals, Centennial, Colorado, USA (NBP2-42879). For western blot analysis, actin was used as a loading control. All experiments were repeated three times and representative figures are shown.

To confirm this result, the reverse Co-IP experiment with Anti-NP magnetic beads (MCE protein A/G magnetic beads and NP antibody from our lab) was done (Figure 6C,D) and the results confirmed the results obtained in Figure 6A,B. It has been previously reported that interaction between PB2 and NP is crucial for the establishment of an enzymatically-active influenza RdRp in RNP complexes [42], and that enzymatically-active Mx1 blocks viral replication by blocking the PB2-NP interface, signifying its direct effect on the vRNP complex. Therefore, we analyzed the PB2-NP interaction in the IP complexes in the presence of eqMx1. No PB2 was found co-immunoprecipitated with H3N8$_{JL89}$-H52N-NP, H7N9$_{ZJ13}$-NP, and H7N9$_{ZJ13}$-S34G. On the other hand, in the complex where there was no interaction between eqMx1 and NP (H3N8$_{JL89}$,-G34S and H7N9$_{ZJ13}$,-N52H) PB2 was co-immunoprecipitated with the said variants of NP (Figure 6C,D; lanes 4 to 6). This result supported the model in which eqMx1 interacts with the influenza RNP and affects its assembly by disturbing the PB2-NP interaction [6]. Together these results reveal a fundamental function of site 52 in altering IAV polymerase activity in the presence of eqMx1, as well as its interaction with eqMx1.

3.5. A Single Mutation at Position 52 in H3N8$_{JL89}$ NP Allows the Virus to escape from eqMx1 Restriction

To further evaluate the importance of 52H of the NP protein in the resistance of eqMx1, we rescued viruses H3N8$_{JL89}$ and H3N8$_{JL89}$ with an H52N substitution in the NP using a reverse genetic system as previously described [34]. Viral titers were calculated by means of the Reed and Muench methodology [36]. MDCK cells expressing HA-tagged pcDNA3.1-eqMx1 or HA-tagged pcDNA3.1 empty control vector, were infected at an MOI of 0.001 for a duration of 1 hr with these harvested viruses and the supernatants were collected at 0, 12, 24, 36, and 48 hr post-infection, and RNA was extracted. Real-time quantitative PCR (qPCR) was performed to determine relative mRNA expression levels. We found that the wild type H3N8$_{JL89}$ had almost the same number of copies in the presence or absence of eqMx1 during replication, whereas the number of copies of mutant virus (JL89-H52N) were drastically reduced in the cells expressing eqMx1, indicating that this mutant virus strain could be blocked by eqMx1, in contrast to the wild type H3N8$_{JL89}$ virus (Figure 7A). Additionally, viral titers were measured using TCID$_{50}$ /mL and the results were in accordance with the qPCR analysis (Figure 7B). From these data we can conclude that a single position in the NP of the H3N8$_{JL89}$ virus (position 52) is essential for viral replication and to allow the virus to escape the restriction effect of eqMx1.

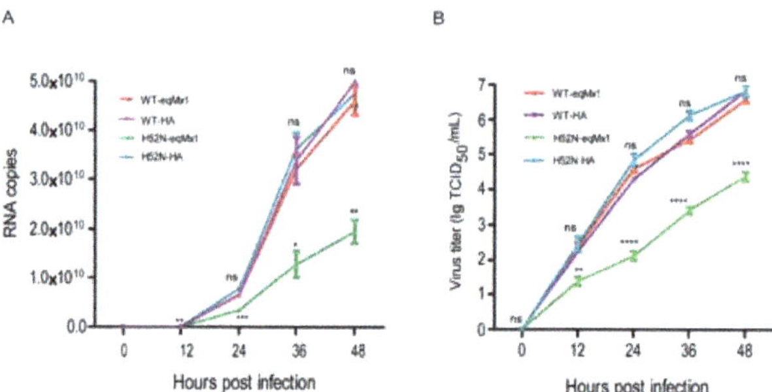

Figure 7. The replication abilities of viruses with different NPs underexpression of eqMx1. (**A**) MDCK cells expressing HA-tagged eqMx1 were infected with wild type (H3N8$_{JL89}$) or mutant viruses (H3N8$_{JL89}$-H52N-NP) at an MOI of 0.001 and the supernatants were collected at 0, 12, 24, 36, and 48 hr post-infection. Total RNA from the collected supernatants was extracted using the RNeasy plus mini kit (Qiagen) and subjected to one-step real-time quantitative PCR (qPCR) analysis using the AgPath-ID™ One-Step RT-PCR reagents according to manufacturer's protocol. Relative mRNA expression levels were determined using double-standard curve methods. All the experiments were performed three times and with three replicates with means ± SE shown. (**B**) MDCK cells expressing eqMx1 were infected with wild type (H3N8$_{JL89}$) or mutant viruses (H3N8$_{JL89}$-H52N-NP) at an MOI of 0.001 and the supernatants were collected at 0, 12, 24, 36, and 48 hr post-infection. These supernatants were subsequently used to infect MDCK cells at different dilutions (10^{-1} to 10^{-11}) with at least four repeats. 48 hr post-infection, immunofluorescence assays (IFA) were performed using specific antibodies against viral NP (from our lab) and FITC-labelled secondary antibody. Finally the viral titers were calculated using Reed and Muench methodology and results are shown as TCID$_{50}$/mL. The results of a single experiment performed with four repeats are shown and results were subsequently confirmed in three separate experiments. Error bars indicate standard deviations (SD).

4. Discussion

The interferon-induced huMxA protein signifies a key interspecies barrier for a large number of zoonotic viruses, including IAVs [8]. Equine Mx1, encoded by a gene located on chromosome 26, is a homolog of MxA from humans or other hosts and has been poorly studied [43,44]. A recent study on the comparison of antiviral activity of Mx1 from different host species includes the corresponding gene from equines and suggests that its antiviral activity against the influenza virus was similar to Mx1 from water buffalo [14]. Here we firstly compared the antiviral potentials of Mx1 proteins from human and equine hosts and identified the potential eqMx1 to inhibit IAV from different species in a species-specific manner. We further identified that NP is responsible for interaction with eqMx1 and confers resistance to the restriction of eqMx1. A single amino acid mutation H52N of NP alters this resistance and interaction to eqMx1 and changed the viral replication ability when eqMx1 was present.

Equine H3N8$_{JL89}$ is predicted to be evolved from an avian origin and equine IAV epizootics has a relation to IAV epidemics in humans and canines [45]. The equine IAV H3N8 strain has been seen to affect dog populations, indicating its efficiency in switching stable hosts and to evolve under antigenic drift [46]. The H3N8 virus has also been isolated from a swine population in China, but this has not been confirmed as a stable host switch, and sequence and phylogenetic analyses of eight gene segments showed that the two swine isolates were of equine origin and most closely related to European equine H3N8 influenza viruses from the early 1990s [47]. All this evidence is of rare cases and the equine IAVs keep an independent evolutionary lineage compared with IAVs from other species.

NP is known as a key module of the vRNP complex and is indispensable for virus replication. NP is the most conserved protein within IAV viral proteins and was identified as a determinant of viral sensitivity towards Mx1 proteins in previously conducted studies, for example, site mutations at 48Q, 98K, and 99K in A/swine/Belzig/2/2001 are sufficient to provide resistance to huMxA [20–22]. By using polymerase reconstitution assays, we found two positions in H3N8$_{JL89}$ NP, G34, and H52, which could be a possible target for H3N8$_{JL89}$ IAV. In our study, we showed that in the H7N9$_{ZJ13}$ NP the mutation N52H enabled it to escape the eqMx1 restriction. Interestingly, the S34G substitution in the H7N9$_{ZJ13}$ did not reverse the activity against eqMx1. Substituting tyrosine (Y) at position 52 with histidine (H) in H5N1 altered the polymerase activity of H5N1 drastically and eqMx1 almost lost its restriction affect towards H5N1-Y52H. A similar substitution in H3N8$_{JL89}$ NP (H3N8$_{JL89}$-H52Y) produced a reverse effect and eqMx1 inhibited the polymerase activity, confirming the importance of this site. We concluded that site 52 is more universal for eqMx1 as it was found to be of equal importance in both IAV strains, whereas site 34 was found to alter the polymerase activity only of H3N8$_{JL89}$ against eqMx1. The sequence alignment of unique sequences of H7N9 and H3N8 predicted that position 52 was highly conserved in both strains, whereas site 34 was more prone to mutations in both strains, reiterating the importance of site 52. None of the mutations at sites 34 and 52 in either H3N8$_{JL89}$ or H7N9$_{ZJ13}$ had a significant effect against huMxA but showed a slight alternation of the restriction instead. These sites may be important in escaping the huMxA antiviral activity when in combination with other as yet unidentified sites.

The NP and PB2 have been found to have a major role in blocking the antiviral effect of murine Mx1 [18–20]. As previously reported, the interaction between PB2 and PB1 is not influenced by the presence of the Mx1 protein. Mx1 inhibits the interaction between PB2 and NP, probably leaving the ternary RdRp complex intact, suggesting that Mx1 inhibits the interaction between the RdRp and the NP protein in vRNP complexes [6,18]. Our results helped to explain how eqMx1 could block viral replication by interacting with viral NPs, thus blocking NP-PB2 interaction and explaining how a single mutation could affect the binding of two proteins.

The interaction of Mx1 with PB2 and NP could be direct or indirect. Other cellular proteins might be involved in this interaction and could act as a bridge between Mx1 and the vRNP complex. Only a few proteins have been reported to interact with the Mx1 protein, and these interacting proteins are potential candidates as bridging factors [48]. Multiple cellular proteins interact with PB2 and/or NP and might fulfill this function. The aforementioned experiments clearly demonstrate an interaction between Mx1 and PB2 or NP when all components of the minireplicon system are present, but Verhelst, Judith, et al. have reported that this interaction between Mx1 and PB2 or NP could occur in the presence or absence of other viral proteins. The authors also reported that interactions between Mx1 and PB2 or NP on the other cannot be detected in the absence of NEM [6]. We also observed a similar result during the IP experiment (data not shown). NEM might preserve the conformation of viral proteins or a cellular factor that mediates the interaction between Mx1 and PB2 or NP. It is also possible that the interactions are too transient or too weak, and inhibiting the GTPase activity of Mx1 by using NEM could stabilize these interactions.

The HEK293T cell line was selected as the first choice for the transfection of plasmids throughout the study except for virus infection experiments because the HEK293T cells are well used in previous studies [6,38,49,50]. Since influenza viruses are the pathogens that cause major respiratory tract infections, the lung epithelial cells might be used for in vitro analysis of host-virus interaction at the cellular level. We tried to transfect human lung epithelial A549 cells but found the transfection efficiency was very low (only about 10–20%). Additionally, when compared with the endogenous MxA in both cell lines using MX1 rabbit polyclonal antibody (Protientech, Rosemont, Illinois, USA), A549 cells did express notable levels of endogenous human MxA that might interfere with our results (see Figure S1). A validation of interaction between eqMx1 and viral NP, and the site-specific antagonism of different viral strains, in lung epithelial cells of equine origin might be a future perspective in order to clarify the mode of action of equine Mx1 in vitro.

To summarize, like its other homologs, eqMx1 could be an efficient barrier against the transmission of IAVs into the equine population. The anti-influenza activities of eqMx1 were species-specific and the NP was the major target of eqMx1. The key sites on $H3N8_{JL89}$ NP determined the sensitivity and resistance to eqMx1. Further extensive studies are required to demonstrate the exact molecular mechanism involved in the structure-specific interaction and evolution, which may deliver new aims for antiviral intervention.

Supplementary Materials: The following are available online at http://www.mdpi.com/1999-4915/11/12/1114/s1, Table S1: Equine Mx1 primers, Table S2: Chimeric clones primers, Table S3: Point Mutation primers, Figure S1 western blot image of huMx1 endogenous expression level in A549 and 293T cells.

Author Contributions: Conceptualization, X.W., and U.F.; methodology, U.F. conducted the experiments, S.J., constructed the Mx1 plasmids and L.X. and X.C. raised NP antibodies; software, U.F., H.Z., Z.Z., and X.-F.W.; validation, H.Z., and Z.Z.; writing—original draft preparation, U.F.; writing—review and editing, X.W.

Funding: This work was supported by the grant from the Natural Science Foundation of China to Hualan Chen for the project entitled "Animal infection disease (No. 31521005)" and Zhenyu Zhang for the project "The study on mechanism of influenza A virus NS1 protein binding with host protein LRPPRC affect the virus replication (No. 31702296)".

Acknowledgments: We thank Hu Zhe for providing the plasmids and advice. We are grateful to Arslan Mehboob and Wang Meiyue for their discussions.

Conflicts of Interest: The authors declare no conflict of interest.

References

1. Taubenberger, J.K.; Kash, J.C. Influenza virus evolution, host adaptation, and pandemic formation. *Cell Host Microbe* **2010**, *7*, 440–451. [CrossRef] [PubMed]
2. Horimoto, T.; Kawaoka, Y. Pandemic threat posed by avian influenza A viruses. *Clin. Microbiol. Rev.* **2001**, *14*, 129–149. [CrossRef]
3. Short, K.R.; Richard, M.; Verhagen, J.H.; van Riel, D.; Schrauwen, E.J.; van den Brand, J.M.; Mänz, B.; Bodewes, R.; Herfst, S. One health, multiple challenges: The inter-species transmission of influenza A virus. *One Health* **2015**, *1*, 1–13. [CrossRef]
4. Webster, R.G.; Bean, W.J.; Gorman, O.T.; Chambers, T.M.; Kawaoka, Y. Evolution and ecology of influenza A viruses. *Microbiol. Mol. Biol. Rev.* **1992**, *56*, 152–179.

5.   Mostafa, A.; Abdelwhab, E.M.; Mettenleiter, T.C.; Pleschka, S. Zoonotic potential of influenza A viruses: A comprehensive overview. *Viruses* **2018**, *10*, 497. [CrossRef]

6.   Verhelst, J.; Parthoens, E.; Schepens, B.; Fiers, W.; Saelens, X. Interferon-inducible protein Mx1 inhibits influenza virus by interfering with functional viral ribonucleoprotein complex assembly. *J. Virol.* **2012**, *86*, 13445–13455. [CrossRef]

7.   Haller, O.; Staeheli, P.; Schwemmle, M.; Kochs, G. Mx GTPases: Dynamin-like antiviral machines of innate immunity. *Trends Microbiol.* **2015**, *23*, 154–163. [CrossRef]

8.   Verhelst, J.; Hulpiau, P.; Saelens, X. Mx proteins: Antiviral gatekeepers that restrain the uninvited. *Microbiol. Mol. Biol. Rev.* **2013**, *77*, 551–566. [CrossRef]

9.   Xiao, H.; Killip, M.J.; Staeheli, P.; Randall, R.E.; Jackson, D. The human interferon-induced MxA protein inhibits early stages of influenza A virus infection by retaining the incoming viral genome in the cytoplasm. *J. Virol.* **2013**, *87*, 13053–13058. [CrossRef]

10. Pavlovic, J.; Haller, O.; Staeheli, P. Human and mouse Mx proteins inhibit different steps of the influenza virus multiplication cycle. *J. Virol.* **1992**, *66*, 2564–2569.

11. Schneider-Schaulies, S.; Schneider-Schaulies, J.; Schuster, A.; Bayer, M.; Pavlovic, J.; Ter Meulen, V. Cell type-specific MxA-mediated inhibition of measles virus transcription in human brain cells. *J. Virol.* **1994**, *68*, 6910–6917.

12. Kochs, G.; Janzen, C.; Hohenberg, H.; Haller, O. Antivirally active MxA protein sequesters La Crosse virus nucleocapsid protein into perinuclear complexes. *Proc. Natl. Acad. Sci. USA* **2002**, *99*, 3153–3158. [CrossRef]

13. Frese, M.; Kochs, G.; Feldmann, H.; Hertkorn, C.; Haller, O. Inhibition of bunyaviruses, phleboviruses, and hantaviruses by human MxA protein. *J. Virol.* **1996**, *70*, 915–923.

14. Dam Van, P.; Desmecht, D.; Garigliany, M.-M.; Bui Tran Anh, D.; Van Laere, A.-S. Anti-Influenza A Virus Activities of Type I/III Interferons-Induced Mx1 GTPases from Different Mammalian Species. *J. Interferon Cytokine Res.* **2019**, *39*, 274–282. [CrossRef]

15. Gao, S.; von der Malsburg, A.; Dick, A.; Faelber, K.; Schröder, G.F.; Haller, O.; Kochs, G.; Daumke, O. Structure of myxovirus resistance protein a reveals intra-and intermolecular domain interactions required for the antiviral function. *Immunity* **2011**, *35*, 514–525. [CrossRef]

16. Gao, S.; von der Malsburg, A.; Paeschke, S.; Behlke, J.; Haller, O.; Kochs, G.; Daumke, O. Structural basis of oligomerization in the stalk region of dynamin-like MxA. *Nature* **2010**, *465*, 502. [CrossRef]

17. Daumke, O.; Gao, S.; von der Malsburg, A.; Haller, O.; Kochs, G. Structure of the MxA stalk elucidates the assembly of ring-like units of an antiviral module. *Small GTPases* **2010**, *1*, 502–506. [CrossRef]

18. Stranden, A.M.; Staeheli, P.; Pavlovic, J. Function of the mouse Mx1 protein is inhibited by overexpression of the PB2 protein of influenza virus. *Virology* **1993**, *197*, 642–651. [CrossRef]

19. Huang, T.; Pavlovic, J.; Staeheli, P.; Krystal, M. Overexpression of the influenza virus polymerase can titrate out inhibition by the murine Mx1 protein. *J. Virol.* **1992**, *66*, 4154–4160.

20. Mänz, B.; Dornfeld, D.; Götz, V.; Zell, R.; Zimmermann, P.; Haller, O.; Kochs, G.; Schwemmle, M. Pandemic influenza A viruses escape from restriction by human MxA through adaptive mutations in the nucleoprotein. *PLoS Pathog.* **2013**, *9*, e1003279. [CrossRef]

21. Dittmann, J.; Stertz, S.; Grimm, D.; Steel, J.; García-Sastre, A.; Haller, O.; Kochs, G. Influenza A virus strains differ in sensitivity to the antiviral action of Mx-GTPase. *J. Virol.* **2008**, *82*, 3624–3631. [CrossRef] [PubMed]

22. Dornfeld, D.; Petric, P.P.; Hassan, E.; Zell, R.; Schwemmle, M. Eurasian Avian-Like Swine Influenza A Viruses Escape Human MxA Restriction through Distinct Mutations in Their Nucleoprotein. *J. Virol.* **2019**, *93*, e00997-18. [CrossRef] [PubMed]

23. Asano, A.; Ko, J.H.; Morozumi, T.; Hamashima, N.; Watanabe, T. Polymorphisms and the antiviral property of porcine Mx1 protein. *J. Vet. Med. Sci.* **2002**, *64*, 1085–1089. [CrossRef]

24. Babiker, H.; Nakatsu, Y.; Yamada, K.; Yoneda, A.; Takada, A.; Ueda, J.; Hata, H.; Watanabe, T. Bovine and water buffalo Mx2 genes: Polymorphism and antiviral activity. *Immunogenetics* **2007**, *59*, 59–67. [CrossRef]

25. Baise, E.; Pire, G.; Leroy, M.; Gérardin, J.; Goris, N.; Clercq, K.D.; Kerkhofs, P.; Desmecht, D. Conditional expression of type I interferon-induced bovine Mx1 GTPase in a stable transgenic vero cell line interferes with replication of vesicular stomatitis virus. *J. Interferon Cytokine Res.* **2004**, *24*, 513–521. [CrossRef]

26. Yamada, K.; Nakatsu, Y.; Onogi, A.; Ueda, J.; Watanabe, T. Specific intracellular localization and antiviral property of genetic and splicing variants in bovine Mx1. *Viral Immunol.* **2009**, *22*, 389–395. [CrossRef]
27. Riegger, D.; Hai, R.; Dornfeld, D.; Mänz, B.; Leyva-Grado, V.; Sánchez-Aparicio, M.T.; Albrecht, R.A.; Palese, P.; Haller, O.; Schwemmle, M. The nucleoprotein of newly emerged H7N9 influenza A virus harbors a unique motif conferring resistance to antiviral human MxA. *J. Virol.* **2015**, *89*, 2241–2252. [CrossRef]
28. Götz, V.; Magar, L.; Dornfeld, D.; Giese, S.; Pohlmann, A.; Höper, D.; Kong, B.-W.; Jans, D.A.; Beer, M.; Haller, O. Influenza A viruses escape from MxA restriction at the expense of efficient nuclear vRNP import. *Sci. Rep.* **2016**, *6*, 23138. [CrossRef]
29. Deeg, C.M.; Hassan, E.; Mutz, P.; Rheinemann, L.; Götz, V.; Magar, L.; Schilling, M.; Kallfass, C.; Nürnberger, C.; Soubies, S. In vivo evasion of MxA by avian influenza viruses requires human signature in the viral nucleoprotein. *J. Exp. Med.* **2017**, *214*, 1239–1248. [CrossRef]
30. Mino, S.; Mojsiejczuk, L.; Guo, W.; Zhang, H.; Qi, T.; Du, C.; Zhang, X.; Wang, J.; Campos, R.; Wang, X. Equine Influenza Virus in Asia: Phylogeographic Pattern and Molecular Features Reveal Circulation of an Autochthonous Lineage. *J. Virol.* **2019**, *93*. [CrossRef]
31. Sovinova, O.; Tumova, B.; Pouska, F.; Nemec, J. Isolation of a virus causing respiratory disease in horses. *Acta Virol.* **1958**, *2*, 52. [PubMed]
32. Gorman, O.; Bean, W.; Kawaoka, Y.; Donatelli, I.; Guo, Y.; Webster, R. Evolution of influenza A virus nucleoprotein genes: Implications for the origins of H1N1 human and classical swine viruses. *J. Virol.* **1991**, *65*, 3704–3714. [PubMed]
33. Lin, Y.-Z.; Cao, X.-Z.; Li, L.; Li, L.; Jiang, C.-G.; Wang, X.-F.; Ma, J.; Zhou, J.-H. The pathogenic and vaccine strains of equine infectious anemia virus differentially induce cytokine and chemokine expression and apoptosis in macrophages. *Virus Res.* **2011**, *160*, 274–282. [CrossRef]
34. Zimmermann, P.; Mänz, B.; Haller, O.; Schwemmle, M.; Kochs, G. The viral nucleoprotein determines Mx sensitivity of influenza A viruses. *J. Virol.* **2011**, *85*, 8133–8140. [CrossRef] [PubMed]
35. Pleschka, S.; Jaskunas, R.; Engelhardt, O.G.; Zürcher, T.; Palese, P.; Garcia-Sastre, A. A plasmid-based reverse genetics system for influenza A virus. *J. Virol.* **1996**, *70*, 4188–4192.
36. Reed, L.J.; Muench, H. A simple method of estimating fifty per cent endpoints12. *Am. J. Epidemiol.* **1938**, *27*, 493–497. [CrossRef]
37. Zhang, H.; Zhang, Z.; Wang, Y.; Wang, M.; Wang, X.; Zhang, X.; Ji, S.; Du, C.; Chen, H.; Wang, X. Fundamental contribution and host range determination of ANP32 protein family in influenza A virus polymerase activity. *bioRxiv* **2019**, 529412.
38. Verhelst, J.; De Vlieger, D.; Saelens, X. Co-immunoprecipitation of the mouse Mx1 protein with the influenza a virus nucleoprotein. *JoVE (J. Vis. Exp.)* **2015**, e52871. [CrossRef]
39. Wang, M.; Zhang, Z.; Wang, X. Strain-specific antagonism of the human H1N1 influenza A virus against equine tetherin. *Viruses* **2018**, *10*, 264. [CrossRef]
40. Kumar, S.; Tamura, K.; Nei, M. MEGA3: Integrated software for molecular evolutionary genetics analysis and sequence alignment. *Brief. Bioinform.* **2004**, *5*, 150–163. [CrossRef]
41. Ashenberg, O.; Padmakumar, J.; Doud, M.B.; Bloom, J.D. Deep mutational scanning identifies sites in influenza nucleoprotein that affect viral inhibition by MxA. *PLoS Pathog.* **2017**, *13*, e1006288. [CrossRef]
42. Biswas, S.K.; Boutz, P.L.; Nayak, D.P. Influenza virus nucleoprotein interacts with influenza virus polymerase proteins. *J. Virol.* **1998**, *72*, 5493–5501.
43. Lear, T.; Breen, M.; Ponce de Leon, F.; Coogle, L.; Ferguson, E.; Chambers, T.; Bailey, E. Cloning and chromosomal localization of MX1 and ETS2 to chromosome 26 of the horse (Equus caballus). *Chromosome Res.* **1998**, *6*, 333–334. [CrossRef]
44. Heinz, H.; Marquardt, J.; Schuberth, H.-J.; Adolf, G.; Leibold, W. Proteins induced by recombinant equine interferon-β1 within equine peripheral blood mononuclear cells and polymorphonuclear neutrophilic granulocytes. *Vet. Immunol. Immunopathol.* **1994**, *42*, 221–235. [CrossRef]
45. Taubenberger, J.; Morens, D. Pandemic influenza–including a risk assessment of H5N1. *Rev. Sci. Tech. (Int. Off. Epizoot.)* **2009**, *28*, 187. [CrossRef]
46. Crawford, P.; Dubovi, E.J.; Castleman, W.L.; Stephenson, I.; Gibbs, E.; Chen, L.; Smith, C.; Hill, R.C.; Ferro, P.; Pompey, J. Transmission of equine influenza virus to dogs. *Science* **2005**, *310*, 482–485. [CrossRef]

47. Tu, J.; Zhou, H.; Jiang, T.; Li, C.; Zhang, A.; Guo, X.; Zou, W.; Chen, H.; Jin, M. Isolation and molecular characterization of equine H3N8 influenza viruses from pigs in China. *Arch. Virol.* **2009**, *154*, 887–890. [CrossRef]
48. Engelhardt, O.G.; Ullrich, E.; Kochs, G.; Haller, O. Interferon-induced antiviral Mx1 GTPase is associated with components of the SUMO-1 system and promyelocytic leukemia protein nuclear bodies. *Exp. Cell Res.* **2001**, *271*, 286–295. [CrossRef]
49. Patzina, C.; Haller, O.; Kochs, G. Structural requirements for the antiviral activity of the human MxA protein against Thogoto and influenza A virus. *J. Biol. Chem.* **2014**, *289*, 6020–6027. [CrossRef]
50. Giese, S.; Ciminski, K.; Bolte, H.; Moreira, É.A.; Lakdawala, S.; Hu, Z.; David, Q.; Kolesnikova, L.; Götz, V.; Zhao, Y. Role of influenza A virus NP acetylation on viral growth and replication. *Nat. Commun.* **2017**, *8*, 1259. [CrossRef]

MDPI

St. Alban-Anlage 66

4052 Basel

Switzerland

Tel. +41 61 683 77 34

Fax +41 61 302 89 18

www.mdpi.com

Viruses Editorial Office

E-mail: viruses@mdpi.com

www.mdpi.com/journal/viruses